Elementary Statistics

Elementary Statistics

Everett Davies

NY Research Press
New York

Published by NY Research Press
118-35 Queens Blvd., Suite 400,
Forest Hills, NY 11375, USA
www.nyresearchpress.com

Elementary Statistics
Everett Davies

International Standard Book Number: 978-1-63238-705-9 (Hardback)

Cataloging-in-Publication Data

Elementary statistics / Everett Davies.
p. cm.
Includes bibliographical references and index.
ISBN 978-1-63238-705-9
1. Statistics. 2. Mathematical statistics. I. Davies, Everett.
QA276.12 .E44 2019
519.5--dc23

Printed in China.

Contents

Preface

Statistics is a branch of mathematics, which deals with the collection, analysis, presentation and organization of data. It can involve the design of surveys and experiments for the collection of data. Statistical methods, such as descriptive statistics and inferential statistics are of importance in data analysis. Descriptive statistics summarizes sample data using indexes like mean or standard deviation. Inferential statistics draws conclusions from data that are prone to random variations. Some statistical tests and procedures include the analysis of variance, chi-squared test, correlation, factor analysis, regression analysis, etc. Statistics is applied in the fields of machine learning, data mining, statistical computing, etc. This textbook is a valuable compilation of topics, ranging from the basic to the most complex theories and principles in the field of statistics. It studies, analyzes and uphold the pillars of this discipline and its utmost significance in modern times. For someone with an interest and eye for detail, this book covers the most significant topics in the field of statistics.

Given below is the chapter wise description of the book:

Chapter 1- Statistics is a field of mathematics. It is concerned with the organization, presentation, analysis and collection of data. The aim of this chapter is to provide an introduction to statistics through a study of applied and computational statistics.

Chapter 2- Descriptive statistics is the analysis and summarization of data using indexes. The measures of central tendency, variability or dispersion are used for describing a data set. The topics elucidated in this chapter cover some of the important aspects of descriptive statistics, such as mean, median, mode, mean log deviation, etc.

Chapter 3- Statistical inference is a data analysis method that is used for the deduction of properties of an underlying probability distribution. This chapter closely examines the vital aspects of statistical inference, such as frequentist inference, Bayesian inference, fiducial inference, decision theory, etc.

Chapter 4- Probability refers to the likelihood for the occurrence of an event. It is a number that has value between 0 and 1. The concepts and axioms of probability are of significant use in statistics. An elaborate study of the varied principles of probability has been provided in this chapter, which includes topics that cover the diverse axioms of probability, random variable, conditional probability, probability distributions, etc.

Chapter 5- The collection of data in statistics occurs through observational and experimental studies or via sampling. There can be many classifications of statistical data. This chapter discusses the varied types of statistical data, such as quantitative and qualitative data, primary and secondary data, etc. as well as the collection methods of survey method, census method, etc.

Chapter 6- A statistical hypothesis is a hypothesis, which can be tested by observing a process modeled using a set of random variables. Statistical hypothesis testing is a form of statistical inference. All the important aspects of statistical hypothesis testing, such as paired difference test, null hypothesis, P-value, etc. have been included in this chapter.

Chapter 7- Regression analysis is a framework of statistical processes for the estimation of the relationships between variables in statistical modeling. This is an important chapter, which will analyze the principles and types of regression analysis, such as linear regression, nonlinear regression, nonparametric regression, semiparametric regression analysis, etc.

Indeed, my job was extremely crucial and challenging as I had to ensure that every chapter is informative and structured in a student-friendly manner. I am thankful for the support provided by my family and colleagues during the completion of this book.

Everett Davies

1

Introduction to Statistics

Statistics is a field of mathematics. It is concerned with the organization, presentation, analysis and collection of data. The aim of this chapter is to provide an introduction to statistics through a study of applied and computational statistics.

Statistics is the science concerned with developing and studying methods for collecting, analyzing, interpreting and presenting empirical data. Statistics is a highly interdisciplinary field; research in statistics finds applicability in virtually all scientific fields and research questions in the various scientific fields motivate the development of new statistical methods and theory. In developing methods and studying the theory that underlies the methods statisticians draw on a variety of mathematical and computational tools.

Two fundamental ideas in the field of statistics are uncertainty and variation. There are many situations that we encounter in science (or more generally in life) in which the outcome is uncertain. In some cases the uncertainty is because the outcome in question is not determined yet (e.g., we may not know whether it will rain tomorrow) while in other cases the uncertainty is because although the outcome has been determined already we are not aware of it (e.g., we may not know whether we passed a particular exam).

Probability is a mathematical language used to discuss uncertain events and probability plays a key role in statistics. Any measurement or data collection effort is subject to a number of sources of variation. By this we mean that if the same measurement were repeated, then the answer would likely change. Statisticians attempt to understand and control (where possible) the sources of variation in any situation.

Applied Statistics

Statistics is an aggregate of facts affected to a marked extent by multiplicity of causes numerically expressed, enumerated or estimated according to a reasonable standard of accuracy, collected in a systematic manner for a predetermined purpose and placed in relation to each other. Statistics is the science, which deals with the method of collecting, classifying and interpreting numerical data collected.

Levels of Measurement

In statistics, there are four types of level of measurement or measurement scales used in applied statistics methods:

- Nominal

- Ordinal
- Interval
- Ratio

They have various degrees of usefulness in applied statistics methods.

1. Nominal Measurements: It contain no significant rank sequence amongst values.
2. Ordinal Measurements: It contain inaccurate differences between successive values, but have a significant order to those values.
3. Interval Measurements: It have significant distances between measurements described, but have no consequential zero value defined.
4. Ratio Measurements: It have together a zero value described and the distances between different measurements classified. The maximum flexibility in applied statistical methods that will be used for examining the data. Since variables conforming only to the nominal or ordinal measurements will not be logically measured.

Statistics Terms

Some statistics terms are given below:

Null Hypothesis

Interpretation of applied statistical methods will frequently occupy the development of a null hypothesis. In that, the statement is that, whatever is future as a reason has no effect on the variable measured. The most excellent illustration for a trainee is the predicament meet by a jury trial. The reflection comes because of suspicion of the guilt.

Errors

Working from a null hypothesis, the following two basic forms of error is standard.

Type I Errors

This type of error occurs where the null hypothesis is incorrectly discarded giving a false positive.

Type II Errors

This type of error occurs where the null hypothesis fails to be discarded and a real difference between populations is missed.

Limitations of Statistics

1. Statistics can be used to study numerical facts. It cannot be used for measuring qualitative data like intelligence or beauty.

2. Statistics cannot study individual items. In Statistics we can consider only a group of values. It is not possible to say the average mark of a student in a particular subject.
3. It is only a method of solving a problem.
4. Only experts can understand statistics.
5. It is true only to an average. It need not be accurate in all cases.

Applied Statistics has its applications in both business and economics.

In Business

Business statistics is the science of using data to make good decisions when faced with uncertainty. Several departments typically make use of business statistics, such as finance, marketing and customer service. Statistical techniques commonly found in businesses include using descriptive statistics, probability, binomial and normal distributions, hypotheses tests, confidence intervals, linear regression and correlation.

Managers frequently use business statistics when forecasting or making decisions that involve a level of uncertainty. It takes a certain amount of skill and training in order to interpret and apply business statistics. Business majors within the U.S. are usually required to complete a course in business statistics before graduation. Understanding and applying business statistics can give managers and businesses a competitive advantage.

Businesses that use statistics have a competitive advantage when making decisions backed by data. For instance, if a manager believes that an item will be obsolete in the future and wants to slash its price in half to sell out, he may find after performing a linear regression that the product will sell out at full price within two months. If he slashes the prices now, the company would be reducing its profit margin for no reason. In addition, managers typically compute the mean, or average, when they should also consider the mode or median. Outliers could be present, making the mean and the decisions based on that number inaccurate.

The most common departments that use business statistics include finance, marketing and customer service. The finance department benefits from using business statistics in order to perform financial statement analysis, forecasts and correlations. The marketing department can use it for forecasting sales and demand, as well as looking for trends and finding the probability of a certain event occurring. Customer service can increase their performance by using business statistics for determining which retention offers or techniques are working best, as well as reviewing the performance of personnel.

Techniques used for business statistics vary depending on the organization. Business schools teach their students descriptive statistics, probability, binomial and normal distributions, hypotheses tests, confidence intervals, linear regression and correlation. Descriptive statistics includes calculating and knowing when is the most appropriate time to use the mean, median and mode. Probability and distributions are used for determining how likely an event will occur. Hypotheses tests, confidence intervals, linear regression and correlations are usually used when making predictions or testing if a theory is true.

In Economics

Economics is concerned with the production and distribution of wealth as well as complex institutional setup connected with the consumption, saving and investment of income.

Statistical data and statistical methods provide valuable assistance for the study and solution of economic problems and also help in designing economic policies. The science of Economics is becoming statistical in its method. It is almost impossible to find a problem in Economics, which does not require an extensive use of statistical data. The use of statistical methods increases with the increase in the usage of economic theory. The laws in Economics like law of demand, law of supply etc. are established with the help of statistical methods.

In Retail

Retail statistics are different metrics a company will use to measure its strength. While retail companies can use a mix of standard and specialized statistics, the industry often focuses on a few universal metrics. The most common retail statistics include sales revenue growth, gross profit percentage, sales returns and allowances, same store sales, and employee turnover. Each of these can provide internal and external users with insight of the inner workings in the company. The statistics also provide companies with the ability to benchmark against other companies.

Sales growth represents the increase in total revenues sales from one period to the next. Retail companies can compare the current month against the previous month or against the same month in the previous year. The most basic formula is current sales less previous period sales divided by current sales. For example, a retail firm with $100,000 US Dollars (USD) in the current month and $85,000 USD in the same month last year has sales growth of 15 percent. Many retail companies will keep a trend analysis with several months' worth of sales growth retail statistics readily available, which allows them to determine when they can expect increases or decreases in sales.

Companies will often calculate a gross profit percentage for all products or individual lines. Owners and managers will do this by subtracting the costs of the product from the sales price and dividing the figure by the sales price. The retail statistics method is also similar for calculating the gross profit percentage for the entire month. Total sales less cost of goods sold divided by sales will compute the gross profit percentage for the current period.

Returns and allowances represent goods returned by customers to the retail store. When customers return goods, the figure goes against the company's sales figure for the month. If the company is unable to sell the good at full price to another customer, then the product is essentially worthless and will often result in lost money. Retail statistics using this metric are fairly basic; the company will divide the total returns and allowance for the current period by sales. This provides a percentage of all sales the company expects to be returned by customers.

Employee turnover is another statistic more important to managers than external users. High turnover represents a working environment deemed unfavorable by workers, and will often cost a company money as it will have to place job ads, interview employees, and conduct employment tests for multiple individuals. Training new workers is often much more expensive than retaining current employees. This statistic is important because retail companies do not often have sufficient gross profits to continually replace employees.

Applied Statistics for the Behavioral Sciences

Statistics courses typically introduce parametric methods, which have more assumptions about the data than nonparametric methods. Statistics finds application in many other branches like banking, insurance, planning, administration, broking, research etc.

Nominal Measurement of Data

This type of measurement includes the qualitative characteristics of data. In this case we do not consider the magnitude of the data. Statistical operations cannot be applied on this type of data.

Example:

Categorization of people based on the hair color is an example of Nominal data. Here the hair color of the person selected is checked. The different categories will be black, brown, blonde, and red.

Ordinal Measurement of Data

The ordinal measurement of data is based on the ranks of data. In this case, we order the data based on some quantity. It is not necessary that the difference between the two ranks is the same. In this case also we cannot apply the statistical operations.

Example:

Students can be ranked based on marks. But the difference between the marks need not be the same. So the data can be considered ordinal data.

Interval Measurement of Data

The interval measurement of data is the measurement based on both the ranking and meaningful intervals between scale points. It not only orders the measurement, but also clearly defines the distance between the orders. The main disadvantage of the interval measurement of data is that it does not have zero as the point of reference.

Example:

The difference between the intervals 10-20 is same as the difference between 50-60. In both cases, the difference is 10 and also we know that 50-60 is higher than 20-30. So, in this case, we know both the order and also the distance between the intervals.

Ratio Measurement of Data

When we include a meaningful zero in the interval measurement of data, it becomes the ratio measurement of data. Hence the ratio measurement of data is very similar to the interval measurement. The ratio measurement of data is the best in all levels for the measurement of data. Zero is considered as the point of reference in this type of measurement of data. It is not compulsory that zero is the value of one of the items in this case. In the ratio level of measurement, the distance between the orders on the scale will be equal, and the ranks are assigned to the items, according to their size or magnitudes.

Example:

If we say a baby of one year is 1.5 times heavier than a newly born baby. Here the baby's weight is clearly a ratio data. We know that a baby cannot have zero weight. But the ratio is meaningful.

Nominal Measurement - Based on attributes

Ordinal Measurement - Based on ranks

Interval Measurement - Based on order with no meaningful 0

Ratio Measurement - Based on order with a meaningful 0

Applied Nonparametric Statistics

Nonparametric statistics has probably become the leading methodology for researchers performing data analysis. Applied Nonparametric Statistics in reliability is focused on the use of modern statistical methods for the estimation of dependability measures of reliability systems that operate under different conditions. There are various stages involved in a statistical investigation. They can be the collection of data, organization of data, presentation of the data, analysis and interpretation of data.

Every step is equally important and needs to be carefully observed. Statistics was considered as a branch of economics earlier. But now, it has developed itself into a full size independent status with all types of quantitative analysis relating to any department of enquiry being included in it. This is because statistics helps in drawing conclusions from the facts affected by the multiplicity of causes in any department of enquiry.

Modern Applied Statistics

The field of statistics involves methods and theory as they are applied to numerical data or observations with the objective of making a rational decision in the face of uncertainty. Let's now understand the applications of statistics in the sector of physical science and state.

Statistics and Physical Science

Statistics plays a great role in the study of physical sciences especially astronomy, geology, physics, meteorology and chemistry. The use of statistical methods is not only very essential but to some extent is unavoidable. Conversion of heterogeneous data into homogenous, interpretation methods and drawing inferences are places where physical sciences depend to a great extent on the science of statistics.

Statistics and State or Country

Earlier statistics was said to be science of king (However they didn't term it as statistics). The kings relied heavily on statistics for knowing man power for military purposes. Statistics was used as a tool to study the functions for the welfare of the states to frame suitable policies and corrective actions. Statistics was used by the government in understanding the census, finances, and other recurrences. Even now the governments across the globe rely on statistical data for analyzing, understanding and taking important decisions towards development of the country.

Computational Statistics

The interface between computer science and statistics is known as computational statistics. It is the area of scientific computing which is related to the mathematical science of statistics. This area is rapidly growing. It is concerned with the development of efficient methods for obtaining numerical solutions to statistically formulated problems.

Studies related to computational statistics involve the use of a variety of computationally intensive statistical techniques and also investigate the role of computation as a tool of discovery. Furthermore, it covers topics like, estimation of functions, numerical optimization in statistical inference, which involves, expectation-maximization algorithm and Fisher scoring, bootstrap methods, random number generation, jackknife methods, Monte Carlo methods, randomization methods, tools for identification of structure in data, etc.

These days, computational statistics is very popular. This is because, without a computer one cannot perform any realistic statistical data analysis having a large data set. To perform statistical data analysis, the on-line statistical calculators such as Excel have a few problems. They are slow, they depend on the cyberspace connection, and, they are very limited and nowhere equal to commercial off-the-shelf statistical computer-packages, whereas, statistical computer-packages are very widely used and so are powerful tools for data analysis with excellent data management capabilities. That's why, the demand of computational statistics is increasing day-by-day.

References

- What-is-statistics: stat.uci.edu, Retrieved 30 June 2018
- What-are-business-statistics: wisegeek.com, Retrieved 10 July 2018
- Applied-statistics, statistics: math.tutorvista.com, Retrieved 29 April 2018
- What-are-the-different-types-of-retail-statistics: wisegeek.com, Retrieved 14 May 2018

2

Descriptive Statistics

Descriptive statistics is the analysis and summarization of data using indexes. The measures of central tendency, variability or dispersion are used for describing a data set. The topics elucidated in this chapter cover some of the important aspects of descriptive statistics, such as mean, median, mode, mean log deviation, etc.

Descriptive statistics are used to describe the basic features of the data in a study. They provide simple summaries about the sample and the measures. Together with simple graphics analysis, they form the basis of virtually every quantitative analysis of data.

Descriptive statistics are typically distinguished from inferential statistics. With descriptive statistics you are simply describing what is or what the data shows. With inferential statistics, you are trying to reach conclusions that extend beyond the immediate data alone. For instance, we use inferential statistics to try to infer from the sample data what the population might think. Or, we use inferential statistics to make judgments of the probability that an observed difference between groups is a dependable one or one that might have happened by chance in this study. Thus, we use inferential statistics to make inferences from our data to more general conditions; we use descriptive statistics simply to describe what's going on in our data.

Descriptive Statistics are used to present quantitative descriptions in a manageable form. In a research study we may have lots of measures. Or we may measure a large number of people on any measure. Descriptive statistics help us to simplify large amounts of data in a sensible way. Each descriptive statistic reduces lots of data into a simpler summary. For instance, consider a simple number used to summarize how well a batter is performing in baseball, the batting average. This single number is simply the number of hits divided by the number of times at bat (reported to three significant digits). A batter who is hitting is getting a hit one time in every three at bats. One batting is hitting one time in four. The single number describes a large number of discrete events. Or, consider the scourge of many students, the Grade Point Average (GPA). This single number describes the general performance of a student across a potentially wide range of course experiences.

Every time you try to describe a large set of observations with a single indicator you run the risk of distorting the original data or losing important detail. The batting average doesn't tell you whether the batter is hitting home runs or singles. It doesn't tell whether she's been in a slump or on a streak. The GPA doesn't tell you whether the student was in difficult courses or easy ones, or whether they were courses in their major field or in other disciplines. Even given these limitations, descriptive statistics provide a powerful summary that may enable comparisons across people or other units.

Univariate Analysis

Univariate analysis involves the examination across cases of one variable at a time. There are three major characteristics of a single variable that we tend to look at:

- The distribution
- The central tendency
- The dispersion

In most situations, we would describe all three of these characteristics for each of the variables in our study.

The Distribution: The distribution is a summary of the frequency of individual values or ranges of values for a variable. The simplest distribution would list every value of a variable and the number of persons who had each value. For instance, a typical way to describe the distribution of college students is by year in college, listing the number or percent of students at each of the four years. Or, we describe gender by listing the number or percent of males and females. In these cases, the variable has few enough values that we can list each one and summarize how many sample cases had the value. But what do we do for a variable like income or GPA? With these variables there can be a large number of possible values, with relatively few people having each one. In this case, we group the raw scores into categories according to ranges of values. For instance, we might look at GPA according to the letter grade ranges. Or, we might group income into four or five ranges of income values.

Table: Frequency distribution table.

Category	Percent
Under 35	9%
36-45	21
46-55	45
56-65	19
66+	6

One of the most common ways to describe a single variable is with a frequency distribution. Depending on the particular variable, all of the data values may be represented, or you may group the values into categories first (e.g., with age, price, or temperature variables, it would usually not be sensible to determine the frequencies for each value. Rather, the value are grouped into ranges and the frequencies determined). Frequency distributions can be depicted in two ways, as a table or as a graph. Above table shows an age frequency distribution with five categories of age ranges defined. The same frequency distribution can be depicted in a graph as shown in figure. This type of graph is often referred to as a histogram or bar chart.

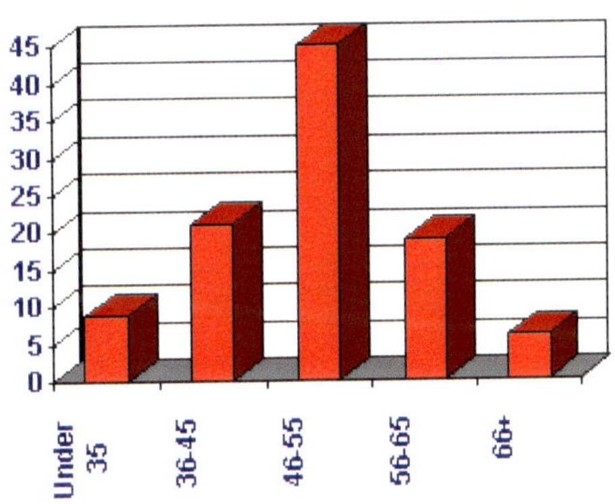

Figure: Frequency distribution bar chart

Distributions may also be displayed using percentages. For example, you could use percentages to describe the:

- Percentage of people in different income levels
- Percentage of people in different age ranges
- Percentage of people in different ranges of standardized test scores

Central Tendency: The central tendency of a distribution is an estimate of the "center" of a distribution of values. There are three major types of estimates of central tendency:

- Mean
- Median
- Mode

The Mean or average is probably the most commonly used method of describing central tendency. To compute the mean all you do is add up all the values and divide by the number of values. For example, the mean or average quiz score is determined by summing all the scores and dividing by the number of students taking the exam. For example, consider the test score values:

15, 20, 21, 20, 36, 15, 25, 15

The sum of these 8 values is 167, so the mean is 167/8 = 20.875.

The Median is the score found at the exact middle of the set of values. One way to compute the median is to list all scores in numerical order, and then locate the score in the center of the sample. For example, if there are 500 scores in the list, score #250 would be the median. If we order the 8 scores shown above, we would get:

15, 15, 15, 20, 20, 21, 25, 36

There are 8 scores and score #4 and #5 represent the halfway point. Since both of these scores are 20, the median is 20. If the two middle scores had different values, you would have to interpolate to determine the median.

The mode is the most frequently occurring value in the set of scores. To determine the mode, you might again order the scores as shown above, and then count each one. The most frequently occurring value is the mode. In our example, the value 15 occurs three times and is the model. In some distributions there is more than one modal value. For instance, in a bimodal distribution there are two values that occur most frequently.

Notice that for the same set of 8 scores we got three different values -- 20.875, 20, and 15 -- for the mean, median and mode respectively. If the distribution is truly normal (i.e., bell-shaped), the mean, median and mode are all equal to each other.

Dispersion: Dispersion refers to the spread of the values around the central tendency. There are two common measures of dispersion, the range and the standard deviation. The range is simply the highest value minus the lowest value. In our example distribution, the high value is 36 and the low is 15, so the range is 36 - 15 = 21.

The Standard Deviation is a more accurate and detailed estimate of dispersion because an outlier can greatly exaggerate the range (as was true in this example where the single outlier value of 36 stands apart from the rest of the values. The Standard Deviation shows the relation that set of scores has to the mean of the sample. Again let's take the set of scores:

15, 20, 21, 20, 36, 15, 25, 15

to compute the standard deviation, we first find the distance between each value and the mean. We know from above that the mean is 20.875. So, the differences from the mean are:

15 - 20.875 = -5.875

20 - 20.875 = -0.875

21 - 20.875 = +0.125

20 - 20.875 = -0.875

36 - 20.875 = 15.125

15 - 20.875 = -5.875

25 - 20.875 = +4.125

15 - 20.875 = -5.875

Notice that values that are below the mean have negative discrepancies and values above it have positive ones. Next, we square each discrepancy:

-5.875 * -5.875 = 34.515625

-0.875 * -0.875 = 0.765625

+0.125 * +0.125 = 0.015625

-0.875 * -0.875 = 0.765625

15.125 * 15.125 = 228.765625

-5.875 * -5.875 = 34.515625

+4.125 * +4.125 = 17.015625

-5.875 * -5.875 = 34.515625

Now, we take these "squares" and sum them to get the Sum of Squares (SS) value. Here, the sum is 350.875. Next, we divide this sum by the number of scores minus 1. Here, the result is 350.875 / 7 = 50.125. This value is known as the variance. To get the standard deviation, we take the square root of the variance (remember that we squared the deviations earlier). This would be SQRT (50.125) = 7.079901129253.

Although this computation may seem convoluted, it's actually quite simple. To see this, consider the formula for the standard deviation:

$$\sqrt{\frac{\Sigma(X-\bar{X})}{(\quad 1)}}$$

where

X = each score

$\bar{X}$ = the mean of average

N = the number or values

Σ = means we sum across the values

In the top part of the ratio, the numerator, we see that each score has the mean subtracted from it, the difference is squared, and the squares are summed. In the bottom part, we take the number of scores minus 1. The ratio is the variance and the square root is the standard deviation. In English, we can describe the standard deviation as:

The square root of the sum of the squared deviations from the mean divided by the number of scores minus one.

Although we can calculate these univariate statistics by hand, it gets quite tedious when you have more than a few values and variables. Every statistics program is capable of calculating them easily for you. For instance, put the eight scores into SPSS and got the following table as a result:

N	8
Mean	20.8750
Median	20.0000
Mode	15.00
Std. Deviation	7.0799
Variance	50.1250
Range	21.00

which confirms the calculations above.

The standard deviation allows us to reach some conclusions about specific scores in our distribution. Assuming that the distribution of scores is normal or bell-shaped (or close to it!), the following conclusions can be reached:

- Approximately 68% of the scores in the sample fall within one standard deviation of the mean
- Approximately 95% of the scores in the sample fall within two standard deviations of the mean
- Approximately 99% of the scores in the sample fall within three standard deviations of the mean

For instance, since the mean in our example is 20.875 and the standard deviation is 7.0799, we can from the above statement estimate that approximately 95% of the scores will fall in the range of 20.875-(2*7.0799) to 20.875+(2*7.0799) or between 6.7152 and 35.0348. This kind of information is a critical stepping stone to enabling us to compare the performance of an individual on one variable with their performance on another, even when the variables are measured on entirely different scales.

Grouped and Ungrouped Data

Ungrouped data which is also known as raw data is data that has not been placed in any group or category after collection. Data is categorized in numbers or characteristics therefore, the data which has not been put in any of the categories is ungrouped. For example, when conducting census and you want to analyze how many women above the age of 45 are in a particular area, you first need to know how many people reside in that area.

The number of individuals residing in that area is ungrouped data or raw information because nothing has been categorized. We can therefore conclude that ungrouped data is data used to show information on an individual member of a sample or population.

Some of the advantages of ungrouped data are as follows;

- Most people can easily interpret it.
- When the sample size is small, it is easy to calculate the mean, mode and median.
- It does not require technical expertise to analyze it.

Array: An arrangement of ungrouped data in ascending or descending order of magnitude is called an array.

Arranging the marks of 25 students in ascending order, we get the following array.

8, 10, 11, 12, 14, 16, 16, 16, 20, 24, 25, 25, 25, 29, 30, 33, 35, 36, 37, 40, 40, 42, 45, 45, 48.

To prepare a frequency distribution table for raw data using tally mark

We take each observation from the data, one at a time, and indicate the frequency (the number of times the observation has occurred in the data) by small line, called tally marks. For our convenience, we write tally marks in bunches of five, the fifth one crossing the fourth diagonally. In the table so formed, the sum of all the frequency is equal to the total number of observations in the given data.

Example: The sale of shoes of various sizes at a shop, on a frequency day is given below:

7 8 5 4 9 8 5 7 6 8 9 6 7 9

8 7 9 9 6 5 8 9 4 5 5 8 9 6

The above data is clearly raw data.

From this data, we may construct a frequency table, as given below:

Size	Tally marks	Frequency
4	||	2
5	~~||||~~	5
6	||||	4
7	||||	4
8	~~||||~~ |	6
9	~~||||~~ ||	7
total		28

Frequency Table

Grouped data is the type of data which is classified into groups after collection. The raw data is categorized into various groups and a table is created. The primary purpose of the table is to show the data points occurring in each group. For instance, when a test is done, the results are the data in this scenario and there are many ways to group this data. For example, the number of students that scored above each 20 mark can be recorded.

Alternatively, the grades can be used. For example, a 90-100 all the way to F 0-59 with each category showing how many students are in each category. Histograms and frequency table are best used to show and interpret grouped data. Here is an example

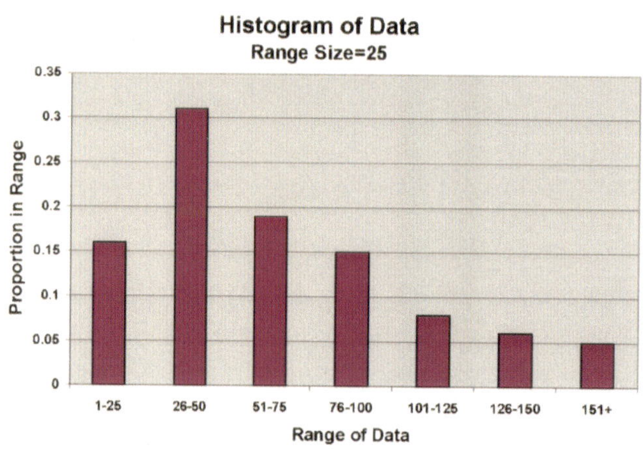

Grouping of data has the following advantages:

- Helps in improving the efficiency of estimations.
- Allows for greater balancing of statistical power of tests of the differences between strata by analyzing equal number from strata.

- Irrelevant subpopulations are ignored while the significant ones are focused on.

Class-Interval: Each class is bounded by two figures, which are called class limits. The figure-on the left side of a class is called its lower limit and that on its right is called its upper limit.

Each class is bounded by two figures, which are called class limits. The figure n the left side of a class is called its lower limit and that on its right is called its upper limit.

Types of Grouped Frequency Distribution

1. Exclusive Form (or Continuous Interval Form): A frequency distribution in which the upper limit of each class is excluded and lower limit is included, is called an exclusive form.

Example: Suppose the marks obtained by some students in an examination are given. We may consider the classes 0 – 10, 10 – 20 etc. in class 0 – 10, we include 0 and exclude 10. In class 10 – 20, we include 10 and exclude 10 and exclude 20.

2. Inclusive From (or Discontinuous Interval From): A frequency distribution in which each upper limit ad well as lower limit is included, is called an inclusive form. Thus, we have classes of the form 0 – 10, 11 – 20, 21 – 30 etc. In 0 – 10, both 0 and 10 are included.

Example 1: Given below are the marks obtained by 40 students in an examination:

3, 25, 48, 23, 17, 13, 11, 9, 46, 41, 37, 45, 10, 19, 39, 36, 34, 5, 17, 21,

39, 33, 28, 25, 12, 3, 8, 17, 48, 34, 15, 19, 32, 32, 19, 21, 28, 32, 20, 23,

Arrange the data in ascending order and present it as a grouped data in:

1. Discontinuous Interval form, taking class-intervals 1 – 10, 11 – 20, etc.
2. Continuous Interval form, taking class-intervals 1 – 10, 10 – 20, etc.

Solution: Arrange the marks in ascending order, we get:

3, 3, 5, 8, 9, 10, 11, 12, 13, 15, 17, 17, 17, 19, 19, 19, 20, 21, 21, 23

23, 25, 25, 28, 28, 32, 32, 32, 33, 34, 34, 36, 37, 39, 39, 41, 45, 46, 48, 48

We may now classify them into groups as shown below:

(i) Discontinuous Interval Form (or Inclusive Form)

Class Interval	Tally Marks	Frequency
0 – 9		5
10 – 19		11
20 – 29		9
30 – 39		10
40 – 49		5
Total		40

Note that the class 1 – 10 means, marks obtained from 1 to 10, including both.

(ii) Continuous Interval Form (or Exclusive Form)

Class Interval	Tally Marks	Frequency
0 – 10	卌	5
10 – 20	卌 卌 |	11
20 – 30	卌 ||||	9
30 – 40	卌 卌	10
40 – 50	卌	5
Total		40

Here, the class 1 – 10 means, marks obtained from 1 to 9, i.e., excluding 10.

Mean

Mean is what most people commonly refer to as an average. The mean refers to the number you obtain when you sum up a given set of numbers and then divide this sum by the total number in the set. Mean is also referred to more correctly as arithmetic mean.

$$\text{mean} = \frac{\text{sum of elements in set}}{\text{number of elements in set}}$$

Given a set of n elements from a_1 to a_n

$$a_1, a_2, a_3, a_4, \ldots a_{n-1}, a_n$$

The mean is found by adding up all the a's and then dividing by the total number, n

$$\text{mean} = \frac{a_1 + a_2 + a_3 + a_4 + \ldots + a_{n-1} + a_n}{n}$$

This can be generalized by the formula below:

$$\text{mean} = \frac{1}{n}\sum_{i=0}^{n} a_i$$

Mean Example Problems

Example

Find the mean of the set of numbers below

3, 4, - 1, 22, 14, 0, 9, 18, 7, 0, 1

Solution

The first step is to count how many numbers there are in the set, which we shall call n

$$n = 10$$

The next step is to add up all the numbers in the set

$$\text{sum} = 3 + 4 + -1 + 22 + 14 + 0 + 9 + 18 + 7 + 0 + 1$$

$$\text{sum} = 77$$

The last step is to find the actual mean by dividing the sum by n

$$\text{mean} = \frac{\text{sum}}{n}$$

$$\text{mean} = \frac{77}{10}$$

$$\text{mean} = 7.7$$

Mean can also be found for grouped data, but before we see an example on that, let us first define frequency.

Frequency in statistics means the same as in everyday use of the word. The frequency an element in a set refers to how many of that element there are in the set. The frequency can be from 0 to as many as possible. If you're told that the frequency an element a is 3, that means that there are 3 as in the set.

Example

Find the mean of the set of ages in the table below

Age (years)	Frequency
10	0
11	8
12	3
13	2
14	7

Solution

The first step is to find the total number of ages, which we shall call n. Since it will be tedious to count all the ages, we can find n by adding up the frequencies:

$$n = 0 + 8 + 3 + 2 + 7$$

$$n = 20$$

Next we need to find the sum of all the ages. We can do this in two ways: we can add up each individual age, which will be a long and tedious process; or we can use the frequency to make things faster.

Since we know that the frequency represents how many of that particular age there are, we can just multiply each age by its frequency, and then add up all these products.

$$\text{sum} = (10 \times 0) + (11 \times 8) + (12 \times 3) + (13 \times 2) + (14 \times 7)$$

$$\text{sum} = 0 + 88 + 36 + 26 + 98$$

$$\text{sum} = 248$$

The last step is to find the mean by dividing the sum by n

$$\text{mean} = \frac{248}{20}$$

$$\text{mean} = 12.4 \text{ years}$$

Population Mean vs. Sample Mean

A population mean is the true mean of the entire population of the data set while a sample mean is the mean of a small sample of the population. These different means appear frequently in both statistics and probability and should not be confused with each other.

Population mean is represented by the Greek letter μ (pronounced *mu*) while sample mean is represented by $\overline{x}$ (pronounced *x bar*). The total number of elements in a population is represented by N while the number of elements in a sample is represented by n. This leads to an adjustment in the formula we gave above for calculating the mean.

$$\text{Population mean } \mu = \frac{1}{N}\sum_{i=1}^{N} a_i$$

$$\text{Sample mean } \overline{x} = \frac{1}{n}\sum_{i=1}^{n} a_i$$

The sample mean is commonly used to estimate the population mean when the population mean is unknown. This is because they have the same expected value.

Types of Mean

Geometric Mean (GM)

The geometric mean is an average that is useful for sets of positive numbers that are interpreted according to their product and not their sum (as is the case with the arithmetic mean); e.g., rates of growth.

$$\bar{x} = \left(\prod_{i=1}^{n} x_i\right)^{\frac{1}{n}} = \left(x_1 x_2 \cdots x_n\right)^{\frac{1}{n}}$$

For example, the geometric mean of five values: 4, 36, 45, 50 and 75 is:

$$(4 \times 36 \times 45 \times 50 \times 75)^{\frac{1}{5}} = \sqrt[5]{24\,300\,000} = 30.$$

Harmonic Mean

The harmonic mean is an average which is useful for sets of numbers which are defined in relation to some unit, for example speed (distance per unit of time).

$$\bar{x} = n\left(\sum_{i=1}^{n} \frac{1}{x_i}\right)^{-1}$$

For example, the harmonic mean of the five values: 4, 36, 45, 50 and 75 is

$$\frac{5}{\frac{1}{4}+\frac{1}{36}+\frac{1}{45}+\frac{1}{50}+\frac{1}{75}} = \frac{5}{\frac{1}{3}} = 15.$$

Relationship between AM, GM and HM

AM, GM, and HM satisfy these inequalities:

$$\text{AM} \geq \text{GM} \geq \text{HM}$$

Equality holds if and only if all the elements of the given sample are equal.

Statistical Location

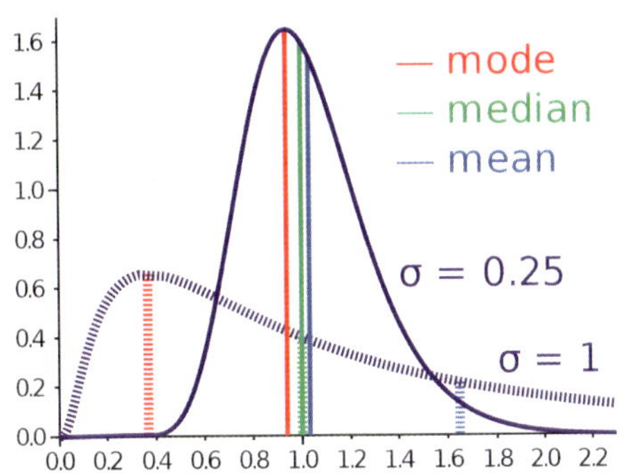

Comparison of the arithmetic mean, median and mode of two skewed (log-normal) distributions.

In descriptive statistics, the mean may be confused with the median, mode or mid-range, as any of these may be called an “average” (more formally, a measure of central tendency). The mean of a set of observations is the arithmetic average of the values; however, for skewed distributions, the mean is not necessarily the same as the middle value (median), or the most likely value (mode). For example, mean income is typically skewed upwards by a small number of people with very large incomes, so that the majorities have an income lower than the mean. By contrast, the median income is the level at which half the population is below and half is above. The mode income is the most likely income and favors the larger number of people with lower incomes. While the median and mode are often more intuitive measures for such skewed data, many skewed distributions are in fact best described by their mean, including the exponential and Poisson distributions.

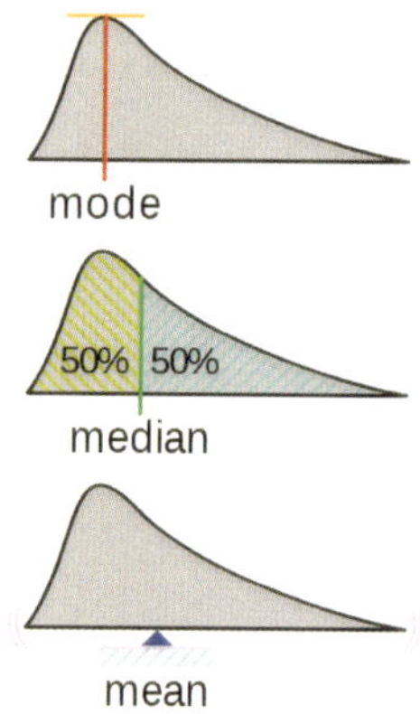

Geometric visualization of the mode, median and mean of an arbitrary probability density function.

Mean of a Probability Distribution

The mean of a probability distribution is the long-run arithmetic average value of a random variable having that distribution. In this context, it is also known as the expected value. For a discrete probability distribution, the mean is given by $\sum xP(x)$, where the sum is taken over all possible values of the random variable and $P(x)$ is the probability mass function. For a continuous distribution,the mean is $\int_{-\infty}^{\infty} xf(x)dx$,, where $f(x)$ is the probability density function. In all cases, including those in which the distribution is neither discrete nor continuous, the mean is the Lebesgue integral of the random variable with respect to its probability measure. The mean need not exist or be finite; for some probability distributions the mean is infinite $(+\infty$ or $-\infty)$, while others have no mean.

Generalized Means

Power Mean

The generalized mean, also known as the power mean or Hölder mean, is an abstraction of the quadratic, arithmetic, geometric and harmonic means. It is defined for a set of n positive numbers x_i by

$$\bar{x}(m) = \left(\frac{1}{n} \sum_{i=1}^{n} x_i^m \right)^{\frac{1}{m}}$$

By choosing different values for the parameter m, the following types of means are obtained:

$m \to \infty$	maximum of x_i
$m = 2$	quadratic mean
$m = 1$	arithmetic mean
$m \to 0$	geometric mean
$m = -1$	harmonic mean
$m \to -\infty$	minimum of x_i

F-mean

This can be generalized further as the generalized f-mean

$$\bar{x} = f^{-1}\left(\frac{1}{n}\sum_{i=1}^{n} f(x_i)\right)$$

and again a suitable choice of an invertible f will give

$f(x) = x$	arithmetic mean,
$f(x) = \frac{1}{x}$	harmonic mean,
$f(x) = x^m$	power mean,
$f(x) = \ln(x)$	geometric mean.

Weighted Arithmetic Mean

The weighted arithmetic mean (or weighted average) is used if one wants to combine average values from samples of the same population with different sample sizes:

$$\bar{x} = \frac{\sum_{i=1}^{n} w_i x_i}{\sum_{i=1}^{n} w_i}.$$

The weights w_i represent the sizes of the different samples. In other applications, they represent a measure for the reliability of the influence upon the mean by the respective values.

Truncated Mean

Sometimes a set of numbers might contain outliers, i.e., data values which are much lower or much higher than the others. Often, outliers are erroneous data caused by artifacts. In this case, one can use a truncated mean. It involves discarding given parts of the data at the top or the bottom end, typically an equal amount at each end and then taking the arithmetic mean of the remaining data. The number of values removed is indicated as a percentage of the total number of values.

Interquartile Mean

The interquartile mean is a specific example of a truncated mean. It is simply the arithmetic mean after removing the lowest and the highest quarter of values.

$$\bar{x} = \frac{2}{n} \sum_{i=\frac{n}{4}+1}^{\frac{3}{4}n} x_i$$

assuming the values have been ordered, so is simply a specific example of a weighted mean for a specific set of weights.

Mean of a Function

In some circumstances mathematicians may calculate a mean of an infinite (even an uncountable) set of values. This can happen when calculating the mean value y_{ave} of a function $f(x)$. intuitively this can be thought of as calculating the area under a section of a curve and then dividing by the length of that section. This can be done crudely by counting squares on graph paper or more precisely by integration. The integration formula is written as:

$$y_{\text{ave}}(a,b) = \frac{1}{b-a} \int_a^b f(x)dx$$

Care must be taken to make sure that the integral converges. But the mean may be finite even if the function itself tends to infinity at some points.

Mean of Angles and Cyclical Quantities

Angles, times of day, and other cyclical quantities require modular arithmetic to add and otherwise combine numbers. In all these situations, there will not be a unique mean. For example, the times an hour before and after midnight are equidistant to both midnight and noon. It is also possible that no mean exists. Consider a color wheel -- there is no mean to the set of all colors. In these situations, you must decide which mean is most useful. You can do this by adjusting the values before averaging, or by using a specialized approach for the mean of circular quantities.

Fréchet Mean

The Fréchet mean gives a manner for determining the "center" of a mass distribution on a surface or, more generally, Riemannian manifold. Unlike many other means, the Fréchet mean is defined on a space whose elements cannot necessarily be added together or multiplied by scalars. It is sometimes also known as the Karcher mean (named after Hermann Karcher).

Distribution of the Sample Mean

The arithmetic mean of a population, or population mean, is often denoted μ. The sample mean $\bar{x}$ (the arithmetic mean of a sample of values drawn from the population) makes a good estimator of the population mean, as its expected value is equal to the population mean (that is, it is an unbiased estimator). The sample mean is a random variable, not a constant, since its calculated value will randomly differ depending on which members of the population are sampled, and consequently it will have its own distribution. For a random sample of n independent observations, the expected value of the sample mean is

$$\mathrm{E}\bar{x} = \mu$$

and the variance of the sample mean is

$$\operatorname{var}(\bar{x}) = \frac{\sigma^2}{n}.$$

If the population is normally distributed, then the sample mean is normally distributed:

$$\bar{x} \sim N\left\{\mu, \frac{\sigma^2}{n}\right\}.$$

If the population is not normally distributed, the sample mean is nonetheless approximately normally distributed if n is large and $\sigma^2/n < +\infty$. This follows from the central limit theorem.

Median

The median is one of the three primary ways to find the average of statistical data. It is harder to calculate than the mode, but not as labor intensive as calculating the mean. It is the center in much the same way as finding the center of a line of people. After listing the data values in ascending order, the median is the data value with the same number of data values above it and below it.

Case One: An Odd Number of Values

Eleven batteries are tested to see how long they last. Their lifetimes, in hours, are given by 10, 99, 100, 103, 103, 105, 110, 111, 115, 130, 131. What is the median lifetime? Since there is an odd number of data values, this corresponds to a line with an odd number of people. The center will be the middle value.

There are eleven data values, so the sixth one is in the center. Therefore median battery life is the sixth value in this list, or 105 hours. Note that the median is one of the data values.

Case Two: An Even Number of Values

Twenty cats are weighed. Their weights, in pounds, are given by 4, 5, 5, 5, 6, 6, 6, 7, 7, 7, 8, 8, 9, 10, 10, 10, 11, 12, 12, 13. What is the median feline weight? Since there is an even number of data values, this corresponds to the line with an even number of people. The center is between the two middle values.

In this case the center is between the tenth and eleventh data values. To find the median we calculate the mean of these two values, and obtain (7+8)/2 = 7.5. Here the median is not one of the data values.

Other Cases

The only two possibilities are to have an even or odd number of data values. So the above two examples are the only possible ways to calculate the median. Either the median will be the middle value, or the median will be the mean of the two middle values. Typically data sets are much larger than the ones that we looked at above, but the process of finding the median is the same as these two examples.

Probability Distributions

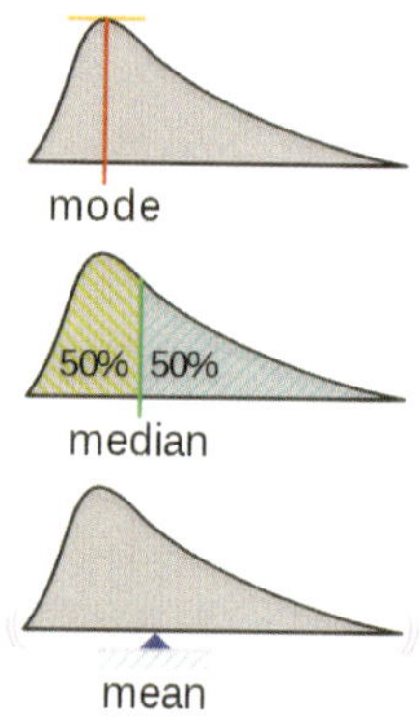

Geometric visualisation of the mode, median and mean of an arbitrary probability density function.

For any probability distribution on the real line R with cumulative distribution function F, regardless of whether it is any kind of continuous probability distribution, in particular an absolutely continuous distribution (which has a probability density function), or a discrete probability distribution, a median is by definition any real number m that satisfies the inequalities

$$P(X \leq m) \geq \frac{1}{2} \text{ and } P(X \geq m) \geq \frac{1}{2}$$

or, equivalently, the inequalities

$$\int_{(-\infty,m]} dF(x) \geq \frac{1}{2} \text{ and } \int_{[m,\infty)} dF(x) \geq \frac{1}{2}$$

In which a Lebesgue–Stieltjes integral is used. For an absolutely continuous probability distribution with probability density function f, the median satisfies

$$P(X \leq m) = P(X \geq m) = \int_{-\infty}^{m} f(x)dx = \frac{1}{2}.$$

Any probability distribution on R has at least one median, but in specific cases there may be more than one median. Specifically, if a probability distribution is zero on an interval $[a, b]$, and the cumulative distribution function at a is 1/2, any value between a and b will also be a median.

Medians of Particular Distributions

The medians of certain types of distributions can be easily calculated from their parameters; furthermore, they exist even for some distributions lacking a well-defined mean, such as the Cauchy distribution:

- The median of a symmetric unimodal distribution coincides with the mode.
- The median of a symmetric distribution which possesses a mean μ also takes the value μ.
    - The median of a normal distribution with mean μ and variance σ^2 is μ. In fact, for a normal distribution, mean = median = mode.
    - The median of a uniform distribution in the interval $[a, b]$ is $(a + b) / 2$, which is also the mean.
- The median of a Cauchy distribution with location parameter x_0 and scale parameter y is x_0, the location parameter.
- The median of a power law distribution x^{-a}, with exponent $a > 1$ is $2^{1/(a-1)}x_{min}$, where x_{min} is the minimum value for which the power law holds.
- The median of an exponential distribution with rate parameter λ is the natural logarithm of 2 divided by the rate parameter: $\lambda^{-1}\ln 2$.
- The median of a Weibull distribution with shape parameter k and scale parameter λ is $\lambda(\ln 2)^{1/k}$.

Populations

Optimality Property

The *mean absolute error* of a real variable c with respect to the random variable X is

$$|X - c|$$

Provided that the probability distribution of X is such that the above expectation exists, then m is a median of X if and only if m is a minimizer of the mean absolute error with respect to X. In particular, m is a sample median if and only if m minimizes the arithmetic mean of the absolute deviations.

More generally, a median is defined as a minimum of

$$E(|X - c| - |X|),$$

Multivariate medians (specifically, the spatial median).

This optimization-based definition of the median is useful in statistical data-analysis, for example, in k-medians clustering.

Unimodal Distributions

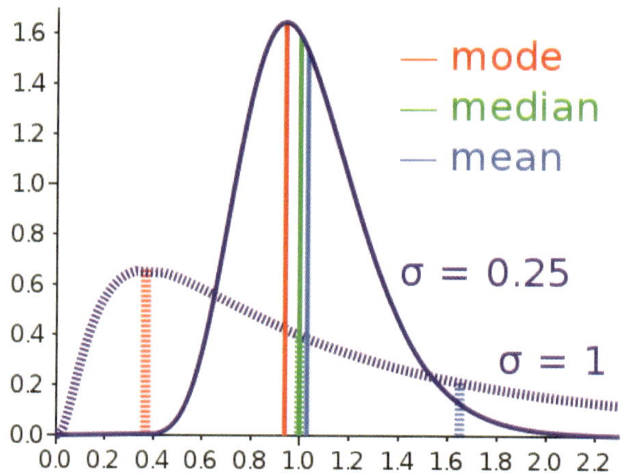

Comparison of mean, median and mode of two log-normal distributions with different skewness.

It can be shown for a unimodal distribution that the median $\tilde{X}$ and the mean $\bar{X}$ lie within $(3/5)^{1/2}$ ≈ 0.7746 standard deviations of each other. In symbols,

$$\frac{\left|\tilde{X}-\bar{X}\right|}{\sigma} \leq \left(\frac{3}{5}\right)^{\frac{1}{2}}$$

where $|\cdot|$ is the absolute value.

A similar relation holds between the median and the mode: they lie within $3^{1/2} \approx 1.732$ standard deviations of each other:

$$\frac{|\tilde{X}-\text{mode}|}{\sigma} \leq 3^{\frac{1}{2}}.$$

Inequality Relating Means and Medians

If the distribution has finite variance, then the distance between the median and the mean is bounded by one standard deviation.

This bound was proved by Mallows, who used Jensen's inequality twice, as follows. We have

$$\begin{aligned}|\mu-m| = |\mathrm{E}(X-m)| &\leq \mathrm{E}(|X-m|)\\ &\leq \mathrm{E}(|X-\mu|)\\ &\leq \sqrt{\mathrm{E}\left((X-\mu)^2\right)} = \sigma.\end{aligned}$$

The first and third inequalities come from Jensen's inequality applied to the absolute-value function and the square function, which are each convex. The second inequality comes from the fact that a median minimizes the absolute deviation function

$$a \mapsto \mathrm{E}(|X-a|).$$

This proof also follows directly from Cantelli's inequality. The result can be generalized to obtain a multivariate version of the inequality, as follows:

$$\begin{aligned}\|\mu - m\| = \|\mathrm{E}(X - m)\| &\leq \mathrm{E}\|X - m\| \\ &\leq \mathrm{E}(\|X - \mu\|) \\ &\leq \sqrt{\mathrm{E}\left(\|X - \mu\|^2\right)} = \sqrt{\mathrm{trace}\left(\mathrm{var}(X)\right)}\end{aligned}$$

where m is a spatial median, that is, a minimizer of the function $a \mapsto \mathrm{E}(\|X - a\|)$. The spatial median is unique when the data-set's dimension is two or more. An alternative proof uses the one-sided Chebyshev inequality; it appears in an inequality on location and scale parameters.

Jensen's Inequality for Medians

Jensen's inequality states that for any random variable x with a finite expectation $E(x)$ and for any convex function f

$$f[E(x)] \leq E[f(x)]$$

It has been shown that if x is a real variable with a unique median m and f is a C function then

$$f(m) \leq \mathrm{Median}[f(x)]$$

A C function is a real valued function, defined on the set of real numbers R, with the property that for any real t

$$f^{-1}\left((-\infty, t]\right) = \{x \in R \mid f(x) \leq t\}$$

is a closed interval, a singleton or an empty set.

Medians for Samples

Efficient Computation of the Sample Median

Even though comparison-sorting n items requires $\Omega(n \log n)$ operations, selection algorithms can compute the k'th-smallest of n items with only $\Theta(n)$ operations. This includes the median, which is the $\frac{n}{2}$th order statistic (or for an even number of samples, the arithmetic mean of the two middle order statistics).

Selection algorithms still have the downside of requiring $\Omega(n)$ memory, that is, they need to have the full sample (or a linear-sized portion of it) in memory. Because this, as well as the linear time requirement, can be prohibitive, several estimation procedures for the median have been developed. A simple one is the median of three rule, which estimates the median as the median of a three-element subsample; this is commonly used as a subroutine in the quicksort sorting algorithm, which uses an estimate of its input's median. A more robust estimator is Tukey's *ninther*, which is the median of three rule applied with limited recursion: if A is the sample laid out as an array, and

$$\text{med3}(A) = \text{median}(A[1], A\left[\frac{n}{2}\right], A[n]),$$

then

$$(A) = \text{med3}(\text{med3}(A[1 \ldots \frac{1}{3}n]), \text{med3}(A[\frac{1}{3}3n \ldots \frac{2}{3}n]), \text{med3}(A\frac{2}{3}n \ldots n]))$$

The remedian is an estimator for the median that requires linear time but sub-linear memory, operating in a single pass over the sample.

Sampling Distribution

The distributions of both the sample mean and the sample median were determined by Laplace. The distribution of the sample median from a population with a density function $f(x)$ is asymptotically normal with mean m and variance

$$\frac{1}{4nf(m)^2}$$

where m is the median of $f(x)$ and n is the sample size.

These results have also been extended. It is now known for the p-th quantile that the distribution of the sample p-th quantile is asymptotically normal around the p-th quantile with variance equal to

$$\frac{p(1-p)}{nf(x_p)^2}$$

where $f(x_p)$ is the value of the distribution density at the p-th quantile.

In the case of a discrete variable, the sampling distribution of the median for small-samples can be investigated as follows. We take the sample size to be an odd number $N = 2n+1$. If a given value v is to be the median of the sample then two conditions must be satisfied. The first is that at most n observations can have a value of $v-1$ or less. The second is that at most n observations can have a value of $v+1$ or more. Let i be the number of observations which have a value of $v-1$ or less and let k be the number of observations which have a value of $v+1$ or more. Then i and k both have a minimum value of 0 and a maximum of n. If an observation has a value below v, it is not relevant how far below v it is and conversely, if an observation has a value above v, it is not relevant how far above v it is. We can therefore represent the observations as following a trinomial distribution with probabilities $F(v-1)$, $f(v)$ and $1-F(v)$. The probability that the median m will have a value v is then given by

$$\Pr(m = v) = \sum_{i=0}^{n}\sum_{k=0}^{n}\frac{N!}{i!(N-i-k)!k!}[F(v-1)]^i[f(v)]^{N-i-k}[1-F(v)]^k.$$

Summing this over all values of v defines a proper distribution and gives a unit sum. In practice, the function $f(v)$ will often not be known but it can be estimated from an observed frequency distribution. An example is given in the following table where the actual distribution is not known but a sample of 3,800 observations allows a sufficiently accurate assessment of $f(v)$.

v	0	0.5	1	1.5	2	2.5	3	3.5	4	4.5	5
f(v)	0.000	0.008	0.010	0.013	0.083	0.108	0.328	0.220	0.202	0.023	0.005
F(v)	0.000	0.008	0.018	0.031	0.114	0.222	0.550	0.770	0.972	0.995	1.000

Using these data it is possible to investigate the effect of sample size on the standard errors of the mean and median. The observed mean is 3.16, the observed raw median is 3 and the observed interpolated median is 3.174. The following table gives some comparison statistics. The standard error of the median is given both from the above expression for $pr(m = v)$ and from the asymptotic approximation given earlier.

Sample size / Statistic	3	9	15	21
Expected value of median	3.198	3.191	3.174	3.161
Standard error of median (above formula)	0.482	0.305	0.257	0.239
Standard error of median (asymptotic approximation)	0.879	0.508	0.393	0.332
Standard error of mean	0.421	0.243	0.188	0.159

The expected value of the median falls slightly as sample size increases while, as would be expected, the standard errors of both the median and the mean are proportionate to the inverse square root of the sample size. The asymptotic approximation errs on the side of caution by overestimating the standard error.

In the case of a continuous variable, the following argument can be used. If a given value v is to be the median, then one observation must take the value v. The elemental probability of this is $f(v)dv$. Then, of the remaining $2n$ observations, exactly n of them must be above v and the remaining n below. The probability of this is the nth term of a binomial distribution with parameters $F(v)$ and $2n$. Finally we multiply by $2n+1$ since any of the observations in the sample can be the median observation. Hence the elemental probability of the median at the point v is given by

$$f(v)\frac{(2n)!}{n!n!}[F(v)]^n[1-F(v)]^n(2n+1)dv.$$

Now we introduce the beta function. For integer arguments α and β, this can be expressed as $B(\alpha,\beta)=(\alpha-1)!(\beta-1)!/(\alpha+\beta-1)!$. Also, we note that $f(v)=dF(v)/dv$. Using these relationships and setting both α and β equal to $(n+1)$ allows the last expression to be written as

$$\frac{[F(v)]^n[1-F(v)]^n}{B(n+1,n+1)}dF(v)$$

Hence the density function of the median is a symmetric beta distribution over the unit interval which supports $F(v)$. Its mean, as we would expect, is 0.5 and its variance is $1/(4(N+2))$. The corresponding variance of the sample median is

$$\frac{1}{4(N+2)f(m)^2}.$$

However this finding can only be used if the density function $f(v)$ is known or can be assumed. As this will not always be the case, the median variance has to be estimated sometimes from the sample data.

Estimation of Variance from Sample Data

The value of $(2f(x))^{-2}$ —the asymptotic value of $n^{-\frac{1}{2}}(\nu - m)$ where ν is the population median—has been studied by several authors. The standard "delete one" jackknife method produces inconsistent results. An alternative—the "delete k" method—where k grows with the sample size has been shown to be asymptotically consistent. This method may be computationally expensive for large data sets. A bootstrap estimate is known to be consistent, but converges very slowly (order of $n^{-\frac{1}{4}}$). Other methods have been proposed but their behavior may differ between large and small samples.

Efficiency

The efficiency of the sample median, measured as the ratio of the variance of the mean to the variance of the median, depends on the sample size and on the underlying population distribution. For a sample of size $N = 2n+1$ from the normal distribution, the efficiency for large N is

$$\frac{2}{\pi}\frac{N+2}{N}$$

The efficiency tends to $\frac{2}{\pi}$ as N tends to infinity.

Other Estimators

For univariate distributions that are *symmetric* about one median, the Hodges–Lehmann estimator is a robust and highly efficient estimator of the population median.

If data are represented by a statistical model specifying a particular family of probability distributions, then estimates of the median can be obtained by fitting that family of probability distributions to the data and calculating the theoretical median of the fitted distribution. Pareto interpolation is an application of this when the population is assumed to have a Pareto distribution.

Coefficient of Dispersion

The coefficient of dispersion (CD) is defined as the ratio of the average absolute deviation from the median to the median of the data. It is a statistical measure used by the states of Iowa, New York and South Dakota in estimating dues taxes. In symbols

$$CD = \frac{1}{n}\frac{\sum |m - x|}{m}$$

where n is the sample size, m is the sample median and x is a variate. The sum is taken over the whole sample.

Confidence intervals for a two-sample test in which the sample sizes are large have been derived by Bonett and Seier This test assumes that both samples have the same median but differ in the dispersion around it. The confidence interval (CI) is bounded inferiorly by

$$\exp\left[\log\left(\frac{t_a}{t_b}\right) - z_\alpha\left(\operatorname{var}\left[\log\left(\frac{t_a}{t_b}\right)\right]\right)^{\frac{1}{2}}\right]$$

Where t_j is the mean absolute deviation of the jth sample, var() is the variance and z_α is the value from the normal distribution for the chosen value of α: for $\alpha = 0.05$, $z_\alpha = 1.96$. The following formulae are used in the derivation of these confidence intervals

$$\operatorname{var}[\log(t_a)] = \frac{1}{n}\left[\frac{s_a^2}{t_a^2} + \left(\frac{x_a - \bar{x}}{t_a}\right)^2 - 1\right]$$

$$\operatorname{var}\left[\log\left(\frac{t_a}{t_b}\right)\right] = \operatorname{var}[\log(t_a)] + \operatorname{var}[\log(t_b)] - 2r(\operatorname{var}[\log(t_a)]\operatorname{var}[\log(t_b)])^{\frac{1}{2}}$$

where r is the Pearson correlation coefficient between the squared deviation scores

$$d_{ia} = | x_{ia} - \bar{x}_a | \text{ and } d_{ib} = | x_{ib} - \bar{x}_b |$$

a and b here are constants equal to 1 and 2, x is a variate and s is the standard deviation of the sample.

Multivariate Median

The univariate median, when the sample or population had one-dimension. When the dimension is two or higher, there are multiple concepts that extend the explanation of the univariate median; each such multivariate median agrees with the univariate median when the dimension is exactly one.

Medoid

Let be a set of points in a space with a distance function d. Medoid is defined as

$$x_{\text{medoid}} \quad \operatorname{argmin}_{y \in \{x_1, x_2, \ \ , x \ \}} \sum d(y, x_i).$$

The medoid is often used in clustering using the k-medoid algorithm.

Marginal Median

The marginal median is defined for vectors defined with respect to a fixed set of coordinates. A marginal median is defined to be the vector whose components are univariate medians. The marginal median is easy to compute, and its properties were studied by Puri and Sen.

Spatial Median

For N vectors in a normed vector space, a spatial median minimizes the average distance

$$a \mapsto \frac{1}{N}\sum_n (\| x_n - a \|),$$

where x_n and a are vectors. The spatial median is unique when the data-set's dimension is two or more and the norm is the Euclidean norm (or another strictly convex norm). The spatial median is also called the L1 median, even when the norm is Euclidean. Other names are used especially for finite sets of points: geometric median, Fermat point (in mechanics), or Weber or Fermat-Weber point (in geographical location theory). In the special case where the norm is an L1-norm, then the spatial median and the marginal median are the same.

More generally, a spatial median is defined as a minimizer of

$$a \mapsto \frac{1}{N}\sum_n (\| x_n - a \| - \| x_n \|);$$

This general definition is convenient for defining a spatial median of a population in a finite-dimensional normed space, for example, for distributions without a finite mean. Spatial medians are defined for random vectors with values in a Banach space.

The spatial median is a robust and highly efficient estimator of a central tendency of a population.

Other Multivariate Medians

An alternative generalization of the spatial median in higher dimensions that does not relate to a particular metric is the centerpoint.

Other Median-related Concepts

Interpolated Median

When dealing with a discrete variable, it is sometimes useful to regard the observed values as being midpoints of underlying continuous intervals. An example of this is a Likert scale, on which opinions or preferences are expressed on a scale with a set number of possible responses. If the scale consists of the positive integers, an observation of 3 might be regarded as representing the interval from 2.50 to 3.50. It is possible to estimate the median of the underlying variable. If, say, 22% of the observations are of value 2 or below and 55.0% are of 3 or below (so 33% have the value 3), then the median m is 3 since the median is the smallest value of x for which $f(x)$ is greater than a half. But the interpolated median is somewhere between 2.50 and 3.50. First we add half of the interval width w to the median to get the upper bound of the median interval. Then we subtract that proportion of the interval width which equals the proportion of the 33% which lies above the 50% mark. In other words, we split up the interval width pro rata to the numbers of observations. In this case, the 33% is split into 28% below the median and 5% above it so we subtract 5/33 of the interval width from the upper bound of 3.50 to give an interpolated median of 3.35. More formally, if the values $f(x)$ are known, the interpolated median can be calculated from

$$m_{\text{int}} = m + w\left[\frac{1}{2} - \frac{F(m) - \frac{1}{2}}{f(m)}\right].$$

Alternatively, if in an observed sample there are k scores above the median category, scores in it and j scores below it then the interpolated median is given by

$$m_{\text{int}} = m + \frac{w}{2}\left[\frac{k-i}{j}\right].$$

Pseudo-median

For univariate distributions that are *symmetric* about one median, the Hodges–Lehmann estimator is a robust and highly efficient estimator of the population median; for non-symmetric distributions, the Hodges–Lehmann estimator is a robust and highly efficient estimator of the population *pseudo-median*, which is the median of a symmetrized distribution and which is close to the population median. The Hodges–Lehmann estimator has been generalized to multivariate distributions.

Variants of Regression

The Theil–Sen estimator is a method for robust linear regression based on finding medians of slopes.

Median Filter

In the context of image processing of monochrome raster images there is a type of noise, known as the salt and pepper noise, when each pixel independently becomes black (with some small probability) or white (with some small probability), and is unchanged otherwise (with the probability close to 1). An image constructed of median values of neighborhoods (like 3×3 square) can effectively reduce noise in this case.

Cluster Analysis

In cluster analysis, the k-medians clustering algorithm provides a way of defining clusters, in which the criterion of maximising the distance between cluster-means that is used in k-means clustering, is replaced by maximising the distance between cluster-medians.

Median–median Line

This is a method of robust regression. The idea dates back to Wald in 1940 who suggested dividing a set of bivariate data into two halves depending on the value of the independent parameter x: a left half with values less than the median and a right half with values greater than the median. He suggested taking the means of the dependent y and independent x variables of the left and the right halves and estimating the slope of the line joining these two points. The line could then be adjusted to fit the majority of the points in the data set.

Nair and Shrivastava in 1942 suggested a similar idea but instead advocated dividing the sample into three equal parts before calculating the means of the subsamples. Brown and Mood in 1951 proposed the idea of using the medians of two subsamples rather the means. Tukey combined these ideas and recommended dividing the sample into three equal size subsamples and estimating the line based on the medians of the subsamples.

Median-unbiased Estimators

Any *mean*-unbiased estimator minimizes the risk (expected loss) with respect to the squared-error loss function, as observed by Gauss. A *median*-unbiased estimator minimizes the risk with respect to the absolute-deviation loss function, as observed by Laplace. Other loss functions are used in statistical theory, particularly in robust statistics.

The theory of median-unbiased estimators was revived by George W. Brown in 1947:

An estimate of a one-dimensional parameter θ will be said to be median-unbiased if, for fixed θ, the median of the distribution of the estimate is at the value θ; i.e., the estimate underestimates just as often as it overestimates. This requirement seems for most purposes to accomplish as much as the mean-unbiased requirement and has the additional property that it is invariant under one-to-one transformation.

Further properties of median-unbiased estimators have been reported. Median-unbiased estimators are invariant under one-to-one transformations.

There are methods of construction median-unbiased estimators that are optimal (in a sense analogous to minimum-variance property considered for mean-unbiased estimators). Such constructions exist for probability distributions having monotone likelihood-functions. One such procedure is an analogue of the Rao–Blackwell procedure for mean-unbiased estimators: The procedure holds for a smaller class of probability distributions than does the Rao—Blackwell procedure but for a larger class of loss functions.

Geometric Median

The geometric median of a discrete set of sample points in a Euclidean space is the point minimizing the sum of distances to the sample points. This generalizes the median, which has the property of minimizing the sum of distances for one-dimensional data, and provides a central tendency in higher dimensions. It is also known as the 1-median, spatial median, Euclidean minisum point or Torricelli point.

The geometric median is an important estimator of location in statistics, where it is also known as the L_1 estimator. It is also a standard problem in facility location, where it models the problem of locating a facility to minimize the cost of transportation.

The special case of the problem for three points in the plane (that is, $m = 3$ and $n = 2$ in the definition below)is sometimes also known as Fermat's problem; it arises in the construction of minimal Steiner trees, and was originally posed as a problem by Pierre de Fermat and solved by Evangelista Torricelli. Its solution is now known as the Fermat point of the triangle formed by the three sample points. The geometric median may in turn be generalized to the problem of minimizing the sum of *weighted* distances, known as the Weber problem after Alfred We-ber's discussion of the problem in his 1909 book on facility location. Some sources instead call Weber's problem the Fermat–Weber problem, but others use this name for the un weighted geometric median problem.

Formally, for a given set of m points $x_1, x_2, \ldots, x_m$ with each $x_1 \in \mathbb{R}^n$, the geometric median is de-fined as

$$= \underset{y \in \mathbb{R}^n}{\operatorname{argmin}} \sum_{i=1}^{m} \|x_i - y\|_2$$

Note that *argmin* means the value of the argument y which minimizes the sum. In this case, it is the point y from where the sum of all Euclidean distances to the x_i's is minimum.

Properties

- For the 1-dimensional case, the geometric median coincides with the median. This is be-cause the univariate median also minimizes the sum of distances from the points.
- The geometric median is unique whenever the points are not collinear.
- The geometric median is equivariant for Euclidean similarity transformations, includ-ing translation and rotation. This means that one would get the same result either by transforming the geometric median, or by applying the same transformation to the sample data and finding the geometric median of the transformed data. This property follows from the fact that the geometric median is defined only from pairwise distanc-es, and doesn't depend on the system of orthogonal Cartesian coordinates by which the sample data is represented. In contrast, the component-wise median for a multivariate dataset is not in general rotation invariant, nor is it independent of the choice of coor-dinates.
- The geometric median has a breakdown point of 0.5. That is, up to half of the sample data may be arbitrarily corrupted, and the median of the samples will still provide a robust esti-mator for the location of the uncorrupted data.

Lemma:

(a) Let $x_1, \ldots, x_k \in \mathbb{H}$ and let $x*$ be their geometric median. *Fix* $\alpha \in (0, \frac{1}{2})$ and assume that $z \in \mathbb{H}$ is such that $\|x_* - z\| > C_\alpha r$, where

$$C_\alpha = (1-\alpha)\sqrt{\frac{1}{1-2\alpha}}$$

and r > 0. Then there exists a subset $J \subseteq \{1, \ldots, k\}$ of cardinality $|J| > \alpha k$ αk such that for all $j \in J, \|x_j - z\| > r$.

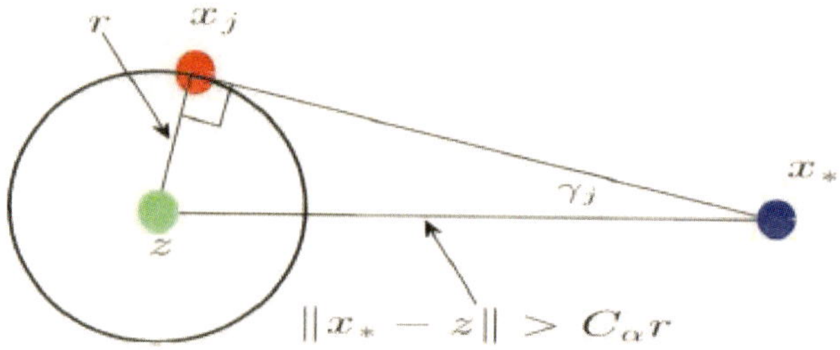

Figure: Geometric

(b) For general Banach spaces, the claim holds with a constant $C_\alpha = \frac{2(1-\alpha)}{1-2\alpha}$.

Proof: (a) Assume that the implication is not true. Without loss of generality, it means that $\|x_i - z\| \le r, i = 1, \ldots, \lfloor (1-\alpha)k \rfloor + 1.$

Consider the directional derivative

$$DF(x_*; z - x_*):\ \lim_{t \searrow 0} \frac{F(x_* + t(z - x_*)) - F(x_*)}{t}$$

at the point x_* in direction $z - x_*$. Since x_* minimizes F over $\mathbb{H}$ $DF(x_*; z - x_*) \ge 0$. On the other hand, it is easy to see that

$$\frac{DF(x_*; z - x_*)}{\|z - x_*\|} = -\sum_{j: x_j \ne x_*} \frac{\langle x_j - x_*, z - x_* \rangle}{\|x_j - x_*\| \|z - x_*\|} + \sum_{j=1}^{k} I\{x_j = x_*\}.$$

For $j = 1, \ldots, \lfloor (1-\alpha)k \rfloor + 1$ and $\gamma_j = \arccos(\frac{\langle x_j - x_*, z - x_* \rangle}{\|x_j - x_*\| \|z - x_*\|})$, y have

$$\frac{\langle x_j - x_*, z - x_* \rangle}{\|x_j - x_*\| \|z - x_*\|} = \cos(\gamma_j) > \sqrt{1 - \frac{1}{C_\alpha^2}},$$

while $\frac{\langle x_j - x_*, z - x_* \rangle}{\|x_j - x_*\| \|z - x_*\|} \ge -1$ for $j > \lfloor (1-\alpha)k \rfloor + 1$. This yields

$$\frac{DF(x_*; z - x_*)}{\|z - x_*\|} < -(1-\alpha)k\sqrt{1 - \frac{1}{C_\alpha^2}} + \alpha k \le 0$$

Whenever $C_\alpha \ge (1-\alpha)\sqrt{\frac{1}{1-2\alpha}}$, which leads to a contradiction.

Remark:

(1) Notice that in a Hilbert space, the geometric median $x_* = med(x_1, ..., x_k)$ always belongs to the convex hull of $\{x_1, ..., x_k\}$. Indeed, if x_* coincides with one of x_j's, there is nothing to prove. Otherwise, since for any $v \in \mathbb{H}$ we have $DF(x_*;v) \geq 0$ and $DF(x_*;-v) \geq 0$, , it follows from $\frac{DF(x_*;z-x_*)}{\|z-x_*\|} = -\sum_{j:x_j \neq x_*} \frac{\langle x_j - x_*, z - x_* \rangle}{\|x_j - x_*\|\|z - x_*\|} + \sum_{j=1}^{k} I\{x_j = x_*\}$. that $\sum_{j=1}^{k} \frac{x_j - x_*}{\|x_j - x_*\|} = 0$, which yields the result.

(2) In general Banach spaces, it might be convenient to consider

$$\hat{x}_* := \underset{y \in co(x_1,...,x_k)}{\arg\min} \sum_{j=1}^{k} \|y - x_j\|,$$

Where $co(x_1,...,x_k)$ is the convex hull of $\{x_1,...,x_k\}$.

Mode

Mode is the value which occurs most frequently in a set of observations. Simply put, it is the number which is repeated most, i.e. the number with the highest frequency. In the field of statistics, it is an important tool to interpret data in a relevant manner. Now it is possible for the data set to be multimodal (has more than one mode) which means more than one observation has the same number of frequencies.

Example: Let us find the Mode of the following data

4, 89, 65, 11, 54, 11, 90, 56

Here in these varied observations the most occurring number is 11, hence the Mode = 11

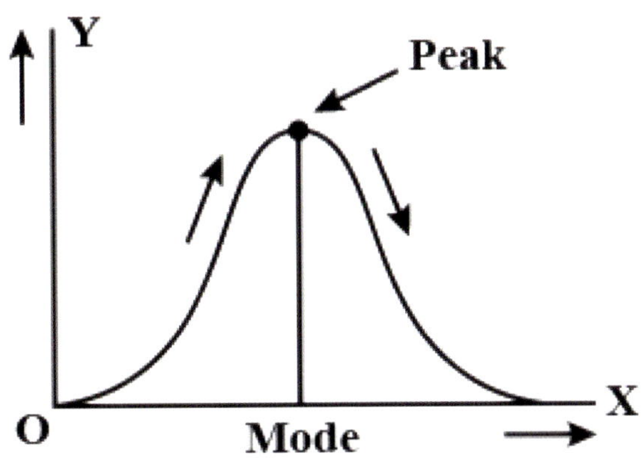

Mode of Grouped Data

As we know that Mode is the most frequently occurring number of a data set. This is easily recognizable in an ungrouped dataset, but if the data set is presented in class intervals, this can get a bit tricky. So how can we calculate Mode of grouped data?

Steps to be followed to calculate the Mode are,

1. Create a table with two columns
2. In column 1 write your class intervals
3. In column 2 write the corresponding frequencies
4. Locate the maximum frequency denoted by f_m
5. Determine the class corresponding to f_m , this will be your Modal class
6. Calculate the Mode using given formula

$$\text{Mode} = L + \frac{f_m - f_1}{(2f_m - f_1 - f_2)} \times h$$

Where,

L = lower limit of Modal Class

f_m = frequency of modal class

h = width of modal class

f_1 = frequency of pre modal class

f_2 = frequency of post modal class

Relation between Mean, Median and Mode

There is an inter-relation between the measures of central tendency. Professor Karl Pearson has suggested an empirical relationship between Mean, Median, and Mode. Via this equation, if the values of two measures are known we can find the third measure. The equation is as follows

Mean – Mode = 3 [Mean – Median]

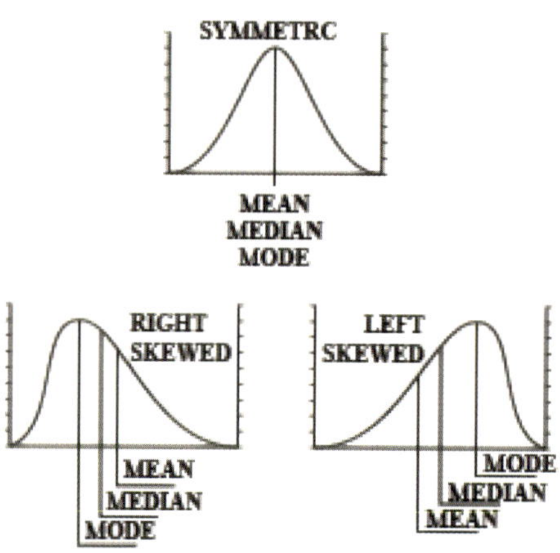

Example: If the difference between the mode and median is 2, then find the difference between the median and mean.

Solution: We know, Mean – Mode = 3 [Mean – Median]

Mode = 3 x Median – 2 x Mean

Mode – Median = 2 x (Median – Mean)

2 =2 x (Median – Mean)

Median – Mean = 1

Example: The pre modal class and the post modal class rectangles have the same height. If the modal class is 40-60, then what is the mode?

Solution: The modal class as given is 40-60

So, L = 40 and h = 20

Frequency of Modal Class = f_m

Frequency of pre modal class = f_{m-1} = Frequency of post modal class

$$\text{Mode} = L + \frac{f_m - f_1}{(2f_m - f_1 - f_2)} \times h$$

$$= L + \frac{f_m - f_m - f_1}{(2f_m - f_m - 1 - f_m - 1)} \times h$$

$$= 40 + \frac{f_m - f_m - 1_1}{(f_m - f_m - 1)2}$$

$$= 40 + 10 = 50$$

Use

Unlike mean and median, the concept of mode also makes sense for “nominal data” (i.e., not consisting of numerical values in the case of mean, or even of ordered values in the case of median). For example, taking a sample of Korean family names, one might find that “Kim” occurs more often than any other name. Then “Kim” would be the mode of the sample. In any voting system where a plurality determines victory, a single modal value determines the victor, while a multi-modal outcome would require some tie-breaking procedure to take place.

Unlike median, the concept of mode makes sense for any random variable assuming values from a vector space, including the real numbers (a one-dimensional vector space) and the integers (which can be considered embedded in the reals). For example, a distribution of points in the plane will typically have a mean and a mode, but the concept of median does not apply. The median makes sense when there is a linear order on the possible values. Generalizations of the concept of median to higher-dimensional spaces are the geometric median and the centerpoint.

Uniqueness and Definedness

For some probability distributions, the expected value may be infinite or undefined, but if defined, it is unique. The mean of a (finite) sample is always defined. The median is the value such that the fractions not exceeding it and not falling below it are each at least 1/2. It is not necessarily unique, but never infinite or totally undefined. For a data sample it is the "halfway" value when the list of values is ordered in increasing value, where usually for a list of even length the numerical average is taken of the two values closest to "halfway". Finally, as said before, the mode is not necessarily unique. Certain pathological distributions (for example, the Cantor distribution) have no defined mode at all. For a finite data sample, the mode is one (or more) of the values in the sample.

Properties

Assuming definedness, and for simplicity uniqueness, the following are some of the most interesting properties.

- All three measures have the following property: If the random variable (or each value from the sample) is subjected to the linear or affine transformation which replaces X by $aX+b$, so are the mean, median and mode.
- Except for extremely small samples, the mode is insensitive to "outliers" (such as occasional, rare, false experimental readings). The median is also very robust in the presence of outliers, while the mean is rather sensitive.
- In continuous unimodal distributions the median often lies between the mean and the mode, about one third of the way going from mean to mode. In a formula, median $\approx$ (2 $\times$ mean + mode)/3. This rule, due to Karl Pearson, often applies to slightly non-symmetric distributions that resemble a normal distribution, but it is not always true and in general the three statistics can appear in any order.
- For unimodal distributions, the mode is within $\sqrt{3}$ standard deviations of the mean, and the root mean square deviation about the mode is between the standard deviation and twice the standard deviation.

Example for a Skewed Distribution

An example of a skewed distribution is personal wealth: Few people are very rich, but among those some are extremely rich. However, many are rather poor.

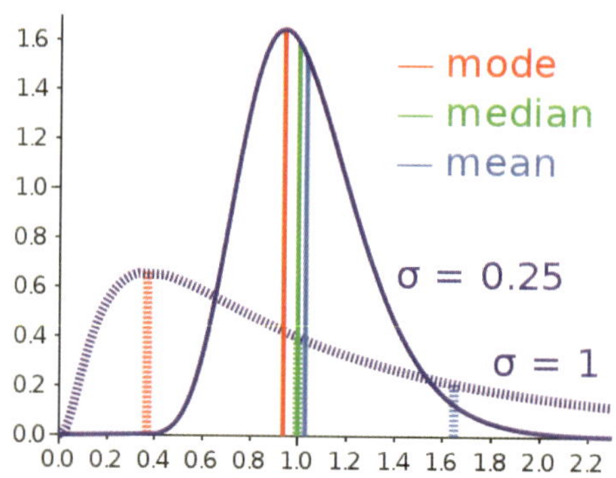

Comparison of mean, median and mode of two log-normal distributions with different skewness.

A well-known class of distributions that can be arbitrarily skewed is given by the log-normal distribution. It is obtained by transforming a random variable X having a normal distribution into random variable $Y = e^X$. Then the logarithm of random variable Y is normally distributed, hence the name.

Taking the mean μ of X to be 0, the median of Y will be 1, independent of the standard deviation σ of X. This is so because X has a symmetric distribution, so its median is also 0. The transformation from X to Y is monotonic, and so we find the median $e^0 = 1$ for Y.

When X has standard deviation σ = 0.25, the distribution of Y is weakly skewed. Using formulas for the log-normal distribution, we find:

$$\begin{aligned} \text{mean} &= e^{\mu+\sigma^2/2} &&= e^{0+0.25^2/2} &&\approx 1.032 \\ \text{mode} &= e^{\mu-\sigma^2} &&= e^{0-0.25^2} &&\approx 0.939 \\ \text{median} &= e^{\mu} &&= e^{0} &&= 1 \end{aligned}$$

Indeed, the median is about one third on the way from mean to mode.

When X has a larger standard deviation, σ = 1, the distribution of Y is strongly skewed. Now

$$\begin{aligned} \text{mean} &= e^{\mu+\sigma^2/2} &&= e^{0+1^2/2} &&\approx 1.649 \\ \text{mode} &= e^{\mu-\sigma^2} &&= e^{0-1^2} &&\approx 0.368 \\ \text{median} &= e^{\mu} &&= e^{0} &&= 1 \end{aligned}$$

Here, Pearson's rule of thumb fails.

Van Zwet Condition

Van Zwet derived an inequality which provides sufficient conditions for this inequality to hold. The inequality

$$\text{Mode} \leq \text{Median} \leq \text{Mean}$$

holds if

$$F(\text{Median} - x) + F(\text{Median} + x) \geq 1$$

for all x where F() is the cumulative distribution function of the distribution.

Unimodal Distributions

It can be shown for a unimodal distribution that the median $\tilde{X}$ and the mean $\bar{X}$ lie within $(3/5)^{1/2}$ ≈ 0.7746 standard deviations of each other. In symbols,

$$\frac{|\tilde{X} - \bar{X}|}{\sigma} \leq (3/5)^{1/2}$$

Where $|\cdot|$ is the absolute value.

A similar relation holds between the median and the mode: they lie within $3^{1/2} \approx 1.732$ standard deviations of each other:

$$\frac{|\tilde{X} - \text{mode}|}{\sigma} \leq 3^{1/2}$$

Advantages and Disadvantages of the Mode:

Advantages

- It is easy to understand and simple to calculate.
- It is not affected by extremely large or small values.
- It can be located just by inspection in ungrouped data and discrete frequency distribution.
- It can be useful for qualitative data.
- It can be computed in an open-end frequency table.
- It can be located graphically.

Disadvantages

- It is not well defined.
- It is not based on all the values.
- It is stable for large values so it will not be well defined if the data consists of a small number of values.
- It is not capable of further mathematical treatment.
- Sometimes the data has one or more than one mode, and sometimes the data has no mode at all.

Mean Log Deviation

In statistics and econometrics, the mean log deviation (MLD) is a measure of income inequality. The MLD is zero when everyone has the same income, and takes on larger positive values as incomes become more unequal, especially at the high end.

The MLD of household income has been defined as

$$\text{MLD} = \frac{1}{N}\sum_{i=1}^{N} \ln \frac{\overline{x}}{x_i}$$

where N is the number of households, x_i is the income of household i, and $\overline{x}$ is the mean of x_i. Naturally the same formula can be used for positive variables other than income and for units of observation other than households.

Equivalent definitions are

$$\text{MLD} = \frac{1}{N}\sum_{i=1}^{N}(\ln\bar{x} - \ln x_i) = \ln\bar{x} - \overline{\ln x}$$

where $\overline{\ln x}$ is the mean of ln(x). The last definition shows that MLD is nonnegative, since $\ln\bar{x} \geq \overline{\ln x}$ by Jensen's inequality.

MLD has been called "the standard deviation of ln(x)", (SDL) but this is not correct. The SDL is

$$\text{SDL} = \sqrt{\frac{1}{N}\sum_{i=1}^{N}(\ln x_i - \overline{\ln x})^2}$$

and this is not equal to the MLD. For example, for the standard lognormal distribution, MLD = 1/2 but SDL = 1.

Standardized Mean Difference

The first group has a distribution F_1 with mean μ_1 and variance σ_1^2 and the second group has a distribution F_2 with mean μ_2 and variance σ_2^2. The covariance between these two groups is σ_{12}. Let random variables P_1 and P_2 denote random values in these two groups, respectively. Suppose we have one random sample (with sample size n_1, sample mean X_1, and sample standard deviation s_1) from group P_1 and another independent random sample (with n_2, X_2 and s_2) from group P_2. Let $N=n_1+n_2$.

The t-test for testing mean difference has been widely used for the comparison of two groups. This t-test is based on t-statistic that has the following t distributions. Under the null hypothesis of $H_0 : \mu_1 = \mu_2$ (or equivalently $H_0 : \mu_D = \mu_1 - \mu_2 = 0$),

$$t\text{-statistic} = \begin{cases} \dfrac{\bar{X}_1 - \bar{X}_2}{\sqrt{(\frac{1}{n_1} + \frac{1}{n_2})(\frac{n_1 - 1}{N-2}s_1^2 + \frac{n_2 - 1}{N-2}s_2^2)}} \sim t(N-2), & \text{if equal variance;} \\ \dfrac{\bar{X}_1 - \bar{X}_2}{\sqrt{\frac{s_1^2}{n_1} + \frac{s_2^2}{n_2}}} \sim t(v), & \text{if not equal variance;} \end{cases}$$

Where v is the integer part of

$$\frac{(s_1^2/n_1 + s_2^2/n_2)^2}{(s_1^2/n_1)^2 / (n_1 - 1) + (s_2^2/n_2)^2 / (n_2 - 1)}$$

The z-score method is based on normal approximation. When n_1 and n_2 are both large,

$$z\text{-score} = \frac{\bar{X}_1 - \bar{X}_2}{\sqrt{\frac{s_1^2}{n_1} + \frac{s_2^2}{n_2}}} \sim N(0,1)$$

under the null hypothesis.

In the situation with equal sample size in the two samples, namely $n_1 = n_2 = \frac{N}{2}$,

$$t - statistic = \frac{\bar{X}_1 - \bar{X}_2}{\sqrt{\frac{2}{N}(s_1^2 + s_2^2}} \sim t(v)$$

Where v is N–2 if equal variance and is the integer part of $\frac{N-2}{2}\frac{(s_1^2+s_2^2)^2}{s_1^4+s_2^4}$ if not equal variance; and

$$z\text{-score} = \frac{\bar{X}_1 - \bar{X}_2}{\sqrt{\frac{2}{N} + (s_1^2 + s_2^2)}} \sim N(0,1).$$

$$\text{Equations}\, t\text{-statistic} = \begin{cases} \dfrac{\bar{X}_1 - \bar{X}_2}{\sqrt{(\frac{1}{n_1} + \frac{1}{n_2})(\frac{n_1 - 1}{N-2} s_1^2 + \frac{n_2 - 1}{N-2} s_2^2)}} \sim t(N-2), & \text{if equal variance;} \\ \dfrac{\bar{X}_1 - \bar{X}_2}{\sqrt{\frac{s_1^2}{n_1} + \frac{s_2^2}{n_2}}} \sim t(v), & \text{if not equal variance;} \end{cases},$$

$$\frac{(s_1^2/n_1 + s_2^2/n_2)^2}{(s_1^2/n_1)^2 / (n_1 - 1) + (s_2^2/n_2)^2 / (n_2 - 1)}$$

Especially $z\text{-score} = \frac{\bar{X}_1 - \bar{X}_2}{\sqrt{\frac{s_1^2}{n_1} + \frac{s_2^2}{n_2}}} \sim N(0,1)$, $z\text{-score} = \frac{\bar{X}_1 - \bar{X}_2}{\sqrt{\frac{s_1^2}{n_1} + \frac{s_2^2}{n_2}}} \sim N(0,1)$ clearly indicate that sample size strongly affects the t-statistic and z-score. That is, sample size and magnitude of difference are indistinguishable in the statistical significance (i.e., p-value from a t-test or z-score). As a result, we may obtain a significant result in an experiment with very small magnitude of difference but large sample size while we may obtain a non significant result in an experiment with large magnitude of difference but small sample size. Thus, the p-values are hardly comparable from various experiments with different sample sizes. Largely due to this feature of statistical significance, the use of the t-statistic for comparison of two groups has been intensively criticized in medical and social sciences. Thus, effect size has been pro-posed for group comparisons.

There are various types of effect sizes. One popular type is standardized mean difference, namely $d = \frac{\mu_1 - \mu_2}{\sigma}$. The widely used estimate is

$$\text{Cohen's } d = \frac{\bar{X}_1 - \bar{X}_2}{\sqrt{\frac{1}{N-2}((n_1 - 1)s_1^2 + (n_2 - 1)s_2^2)}}.$$

When there are treatment and control groups, Glass suggested using sample standard deviation in the control group. That is, assuming the second group is the control group,

$$\text{Glass's } \delta = \frac{\bar{X}_1 - \bar{X}_2}{s_2}.$$

The two estimates in Equations $\text{Cohen's } d = \frac{\bar{X}_1 - \bar{X}_2}{\sqrt{\frac{1}{N-2}((n_1 - 1)s_1^2 + (n_2 - 1)s_2^2)}}$. and $\text{Glass's } \delta = \frac{\bar{X}_1 - \bar{X}_2}{s_2}$.

are all based on the assumption of equal variances in the two groups. Standardized mean difference has rarely been investigated in the situation where the two groups have un-equal variances although Cohen proposes the definition of

$$d = \frac{\mu_1 - \mu_2}{\sqrt{\frac{\sigma_1^2 + \sigma_2^2}{2}}}$$

for standardized mean difference in this situation.

Strictly Standardized Mean Difference

Recently, SSMD has been proposed to measure the magnitude of difference between two groups and has been explored for both quality control and hit selection in RNA interference high-throughput screening experiments. SSMD (denoted as β) is defined as the ratio of mean to standard deviation of the difference between two groups, namely

$$\beta = \frac{\mu D}{\sigma D} = \frac{\mu_1 - \mu_2}{\sqrt{\sigma_1^2 + \sigma_2^2 - 2\sigma_{12}}}.$$

If the two groups are independent,

$$\beta = \frac{\mu_1 - \mu_2}{\sqrt{\sigma_1^2 + \sigma_2^2}}.$$

If the two independent groups have equal variances σ^2, then

$$\beta = \frac{\mu_1 - \mu_2}{\sqrt{2\sigma^2}}.$$

In the situation where the two groups are independent (i.e., $\sigma_{12} = 0$). Zhang derived maximum-likelihood estimate (MLE) and method-of-moment (MM) estimate of unpaired SSMD when the two groups have normal distributions with unequal variances. They are

$$\hat{\beta}_{\text{MLE}} = \frac{\bar{X}_1 - \bar{X}_2}{\sqrt{\frac{n_1 - 1}{n_1} + \frac{n_2 - 1}{n_2} s_2^2}},$$

$$\hat{\beta}_{MM} = \frac{\bar{X}_1 - \bar{X}_2}{\sqrt{S_1^2 + S_2^2}},$$

The confidence interval of β may be obtained using its asymptotical normality of $\hat{\beta}_{\text{MLE}}$. When the two groups have normal distributions with equal variance, the uniformly minimal variance unbiased estimate (UMVUE) of SSMD is, for $n_1 \geq 2, n_2 \geq 2$,

$$\hat{\beta}_{\text{UMVUE}} = \frac{\bar{X}_1 - \bar{X}_2}{\sqrt{\frac{2}{K}((n_1 - 1)s_1^2 + (n_2 - 1)s_2^2)}},$$

where $K = 2 \cdot \left(\frac{\Gamma\left(\frac{N-2}{2}\right)}{\Gamma\left(\frac{N-3}{2}\right)} \right)^2$. A good approximation of K is N – especially when $N \geq 5$. The confidence interval of β may be obtained using a noncentral distribution $T = \frac{(\bar{X}_1 - \bar{X}_2) / \sqrt{\frac{1}{n_1} + \frac{1}{n_2}}}{\sqrt{((n_1 - 1)s_1^2 + (n_2 - 1)s_2^2)/(N-2)}} \sim$ noncentral $t(N-2, \frac{\sqrt{2}}{\sqrt{\frac{1}{n_1} + \frac{1}{n_2}}} \beta)$. The MLE and MM estimates of β in the situation of equal variances are

$$\hat{\beta}_{\text{MLE}} = \frac{\bar{X}_1 - \bar{X}_2}{\sqrt{\frac{2}{N}((n_1 - 1)s_1^2 + (n_2 - 1)s_2^2)}},$$

and

$$\hat{\beta}_{\text{MM}} = \frac{\bar{X}_1 - \bar{X}_2}{\sqrt{\frac{2}{N-2}((n_1 - 1)s_1^2 + (n_2 - 1)s_2^2)}},$$

respectively. In the situation where the two groups are correlated (i.e., $\sigma_{12} \neq 0$), one strategy to estimate SSMD is first to estimate $\sigma_{12}, \mu_1, \mu_2, \sigma_1^2, \sigma_2^2$, and then to estimate βusing the method-of-moment method. A more commonly used strategy is first to obtain paired observations from the two groups and then to estimate SSMD based on the paired observations. Based on a paired difference D with population mean μ_D and variance σ_D^2, SSMD is $B = \frac{\mu D}{\sigma D}$. Suppose we observe n pairs of observations from P_1 and P_2, respectively. That is, we have n paired differences. Let $\bar{D}$ and s_D be the sample mean and sample standard deviation of the n paired differences. When the paired differences are normally distributed, the MM, MLE, and UMVUE of SSMD are

$$\hat{\beta}_{\mathrm{MM}} = \frac{\bar{D}}{s_D},$$

$$\hat{\beta}_{\mathrm{MLE}} = \sqrt{\frac{n}{n-1}}\frac{\bar{D}}{s_D},$$

and

$$\hat{\beta}_{\mathrm{UMVUE}} = \frac{\Gamma(\frac{n-1}{2})}{\Gamma(\frac{n-2}{2})}\sqrt{\frac{2}{n-1}}\frac{\bar{D}}{s_D},$$

respectively. The confidence interval of βcan be obtained using the noncentral distribution $\frac{\sqrt{n}\bar{D}}{s_D} \sim$ noncentral $t(n-1, \sqrt{n}\beta)$.

Comparing SSMD With *t*-Statistic

From Equations $t-statistic = \frac{\bar{X}_1 - \bar{X}_2}{\sqrt{\frac{2}{N}(s_1^2 + s_2^2}} \sim t(v)$, $z\text{-score} = \frac{\bar{X}_1 - \bar{X}_2}{\sqrt{\frac{2}{N} + (s_1^2 + s_2^2)}} \sim N(0,1)$ and $\hat{\beta}_{\mathrm{UMVUE}} = \frac{\bar{X}_1 - \bar{X}_2}{\sqrt{\frac{2}{K}((n_1-1)s_1^2 + (n_2-1)s_2^2)}}$, in the situations with equal sample sizes, the relationship between t-statistic (or z-score) and SSMD estimate can be as simple as

$$\sqrt{\frac{N}{2}}\hat{\beta}_{\mathrm{MM}} = \sqrt{\frac{N}{2}}\sqrt{\frac{N-2}{K}}\hat{\beta}_{\mathrm{UMVUE}},$$

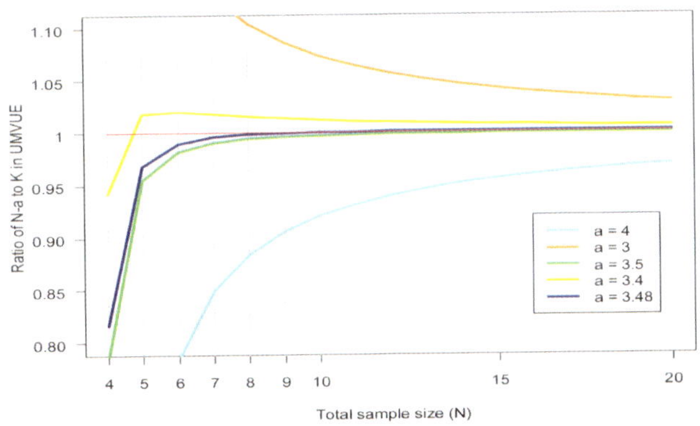

Find an approximation in the form of N – a for the constant Kin the UMVUE of SSMD.

which indicates that t-statistic and associated p-value are affected not only by the magnitude of difference but also by sample size. In general, when there is even any tiny mean difference, larger sample size (N) leads to large absolute value of t-statistic and smaller p-value. In fact, as $N \to \infty$, t-statistic value goes to ∞ or $-\infty$ and p-value goes to 0 in above Figure. By contrast, as sample size in-creases, the SSMD estimate goes to the true SSMD value in probability in above Figure. Sample size can only impact the precision of $\hat{\beta}$ in representing β. Thus, we can still have the benefit of increasing sample size in an experiment: in-creasing the precision of SSMD estimation. When there are unequal sample sizes in the two groups, the relation-ship between t-statistic and SSMD estimate can be com-plicate; however, the moving trends of SSMD estimate, t-statistic and p-value are still the same as illustrated in below figure.

In the traditional methods of testing mean difference μ_D using t-statistic or z-score, one may also use both the point estimate and confidence interval of μ_D. However, they will lead to the same misleading results as the use of p-value of testing mean difference. This is because, as $N \to \infty$, the point estimate and the upper and low bounds of the confidence interval all approach the population value of μ_D in above Figure, and it is well-known that the mean difference μ_D alone cannot represent the magnitude of difference between two groups.

In this example, two studies are conducted to investigate two treatments for reducing blood pressure. Suppose the blood pressures in people without treatment (i.e., controls), with Treatment 1 and with Treatment 2 have normal distributions $N(126,4)$, $N(118,16)$, and $N(120,2)$, respectively. The interest is the difference between with a treatment and without a treatment. In Study 1, there are 800 people without treatment and 800 with Treatment 1; in Study 2, there are 100 people without treatment and 100 with Treatment 2. Based on these two studies, the p-value for testing no mean difference between Treatment 1 and control is 6 ×10−42; the mean difference between Treatment 1 and control is estimated to be −8.10 and its 95% confidence interval is (−9.22, −6.99). By contrast, the p-value for testing no mean difference between Treatment 2 and control is 2×10−21; the mean difference between Treatment 2 and control is estimated to be −5.88 and its 95% confidence interval is (−6.88, −4.88). If one makes a judgment based on p-value, the estimate of mean difference and its confidence interval, one would easily make a conclusion that Treatment 1 is better than Treatment 2 in reducing blood pressure. This is because Treatment 1 has a much smaller p-value and a larger amount of reduction in blood pressure on average and because the CI upper bound for Treatment 1 is smaller than the CI lower bound for Treatment 2. However, the comparison of distributions clearly reveals that Treatment 2 is better than Treatment 1 even though Treatment 2 has a smaller amount of reduction on average in below Figure.

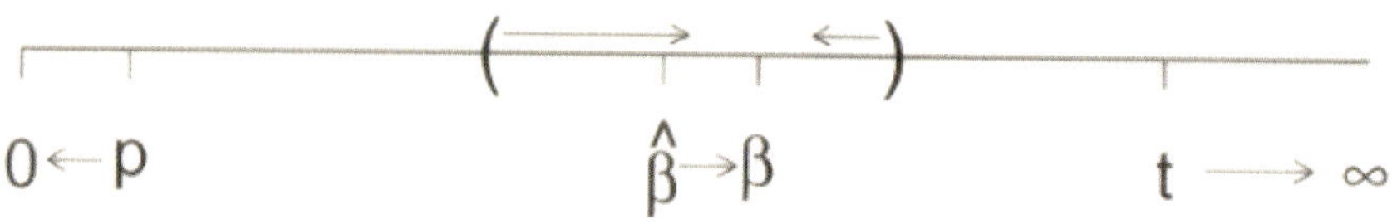

A graph showing the moving trends of SSMD estimate $\hat{\beta}$, the bounds of its confidence interval (CI), t-statistic and p-value as sample size N increases. The parentheses denote the CI bounds. β denotes the true value of SSMD. As $N \to \infty$, t-statistic goes to ∞ or $-\infty$, p-value goes to zero; however, the estimate $\hat{\beta}$ and the lower and upper bounds of CI all go to β.

Using Equation $\hat{\beta}_{MLE} = \frac{\bar{X}_1 - \bar{X}_2}{\sqrt{\frac{n_1 - 1}{n_1} + \frac{n_2 - 1}{n_2} s_2^2}}$, the MLE estimate and 95% confidence interval of SSMD are −0.506 and(−0.579,−0.433), respectively, for Treatment 1and are −1.308 and (−1.582,−1.0346), respectively, for Treatment 2, which correctly indicate that Treatment2 has a large effect whereas Treatment 1 only has a medium effect in reducing blood pressure. The use of d+-probability also reaches a similar result. For Treatment 1, the d+-probability is estimated to beo.69, which means that the chance that the people with Treatment 1 have blood pressure less than those without population distributions of blood pressures in people treated with two treatments, Treatment 1 and Treatment 2, and without treatment(i.e., Control), respectively.

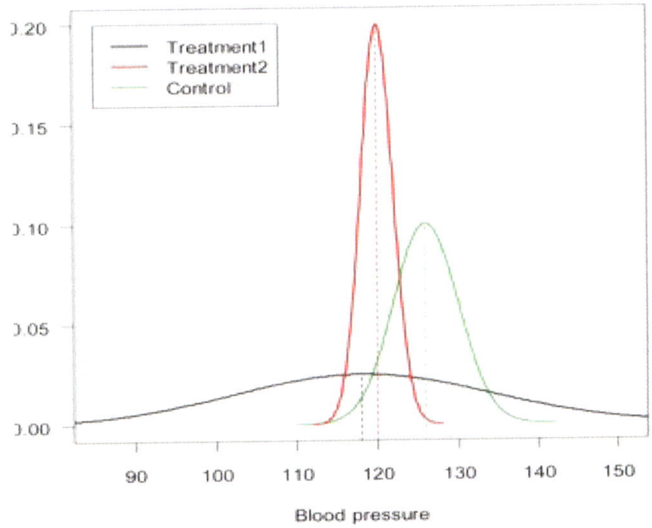

A treatment is 69%. By contrast, for Treatment 2, the d+-probability is estimated to be 0.90, which means that the chance that the people with Treatment 2 have blood pressure less than those without a treatment is90%. Thus, the d+-probability indicates that people will have a much larger chance of reducing blood pressure by taking Treatment 2 than by taking Treatment 1. All these indicate that SSMD and d+-probability provide a potentially better alternative to evaluate treatment effect than p-value and confidence interval of mean difference in traditional approaches.

Comparing SSMD With Standardized Mean Difference

In the situation where the two groups do not have correlation, the relationship between Cohen's *d* and SSMD is

$$d = \sqrt{2}\beta.$$

The definition of SSMD is applicable in not only independent groups but also correlated groups. For two correlated groups, we can either estimate the covariance σ^{12} between two groups, or use paired sample to estimate SSMD directly based on paired differences. In addition, SSMD can be extended to thc comparison involving more than two groups. Therefore, SSMD has a broader utility than Co-hen's din two-group comparisons.

SSMD has two clear and meaningful interpretations when it is used to assess the magnitude of difference between two groups. The first interpretation is that it is the ratio of mean to standard deviation of a random variable representing the difference between two groups, and the second interpretation is that it reflects the probability that a random value from the first group is larger than a random value from the second group, namely d+-probability. When the data are normally distributed in both groups.

$$d^{+} - probability = \Phi(\beta),$$

Where Φ (·)is the cumulative density function of the standard normal distribution. When the data are not normally distributed, there is still a relationship between d+-probability and β Zhang. Because of clear and meaningful interpretations of SSMD, we can construct meaningful and interpretable SSMD-based criteria for classifying the magnitude of difference between two groups, namely $|\beta| \geq 1.645$ for "extra-large", $1.645 > |\beta| \geq 1$ for "large," $1 > |\beta| > 0.25$ for "medium, and $|\beta| \leq 0.25$ for "small" under normality assumption (Zhang 2009). Under normality assumption, these SSMD-based criteria can be interpreted using probability as follows. An "extra-large effect" means that the probability of a value from the first group being greater than a value from the second group is greater than 0.95 in the positive direction of the difference, and is less than 0.05 in the negative direction. A "large effect" means that this probability (i.e., d+-probability) is between 0.84 and 0.95 in the positive direction, and is between 0.05 and 0.16 in the negative direction. A "medium effect" means that this probability is between 0.6 and 0.84 in the positive direction, and is between 0.16 and 0.40 in the negative direction. A "small effect" means that this probability is between 0.5 and 0.6 in the positive direction, and is between 0.40 and 0.50 in the negative direction.

Based on the relationship between SSMD and standardized mean difference, the original and probability meanings of SSMD also give clear interpretations to Co-hen's criterion: Cohen's d being 0.20 for small effects, 0.50 for medium effects, and 0.80 for large effects. Co-hen's d=0.20, 0.50, 0.80 corresponds to SSMD =0.1414, 0.3536, 0.5656 which indicates a small effect based on Cohen's criterion is also a small effect based on SSMD; however, both medium and large effects based on Cohen's criterion are medium effects based on SSMD. Under normality assumption, Cohen's d=0.20, 0.50, 0.80 indicates that d+-probability = 0.556, 0.638, 0.714.These values of d+-probability provide a probability interpretation to Cohen's criteria.

References

- Piché, Robert (2012). Random Vectors and Random Sequences. Lambert Academic Publishing. ISBN 978-3659211966
- Difference-between-grouped-data-and-ungrouped-data, words-language: differencebetween.net, Retrieved 26 May 2018
- Mallows, Colin (August 1991). "Another comment on O'Cinneide". The American Statistician. 45 (3): 257. doi: 10.1080/00031305.1991.10475815.
- Grouped-and-ungrouped-data: tutorvista.com, Retrieved 14 April 2018
- Derek Bissell (1994). Statistical Methods for Spc and Tqm. CRC Press. pp. 26–. ISBN 978-0-412-39440-9. Retrieved 25 February 2013.
- Shao, J.; Wu, C. F. (1989). "A General Theory for Jackknife Variance Estimation". Ann. Stat. 17 (3): 1176–1197. doi:10.1214/aos/1176347263. JSTOR 2241717

- What-is-the-median-3126370: thoughtco.com, Retrieved 26 June 2018
- Bonett, DG; Seier, E (2006). "Confidence interval for a coefficient of dispersion in non-normal distributions". Biometrical Journal. 48 (1): 144–148. doi:10.1002/bimj.200410148
- Nair, K. R.; Shrivastava, M. P. (1942). "On a Simple Method of Curve Fitting". Sankhyā: The Indian Journal of Statistics. 6 (2): 121–132. JSTOR 25047749
- Geometric-Median-339287567: scribd.com, Retrieved 09 May 2018
- David J. Sheskin (27 August 2003). Handbook of Parametric and Nonparametric Statistical Procedures: Third Edition. CRC Press. pp. 7–. ISBN 978-1-4200-3626-8. Retrieved 25 February 2013

3

Statistical Inference

Statistical inference is a data analysis method that is used for the deduction of properties of an underlying probability distribution. This chapter closely examines the vital aspects of statistical inference, such as frequentist inference, Bayesian inference, fiducial inference, decision theory, etc.

Statistical inference is the act of using observed data to infer unknown properties and characteristics of the probability distribution from which the observed data have been generated. The set of data that is used to make inferences is called sample.

Samples

In the simplest possible case, we observe the realizations $x_1,\ldots,x_n$ of n independent random variables $X_1,\ldots,X_n$ having a common distribution function $F_X(x)$ and we use the observed realizations to infer some characteristics of $F_X(x)$. With a slight abuse of language, we sometimes say "*an* independent realizations of a random variable X" instead of saying "the realizations of independent random variables $X_1,\ldots,X_n$ having a common distribution function $F_X(x)$".

Example: The lifetime of a certain type of electronic device is a random variable X, whose distribution function $F_X(x)$ is unknown. Suppose we independently observe the lifetimes of components. Denote these realizations by $x_1,x_2,\ldots,x_{10}$. We are interested in the expected value of X, which is an unknown characteristic of $F_X(x)$. We infer from the data, and estimate $E[X]$ with the sample mean

$$\bar{x} = \frac{1}{10}\sum_{i=1}^{10} x_i$$

In this simple example the observed data $x_1,x_2,\ldots,x_{10}$ constitute our sample and $E[X]$ is the quantity about which we are making a statistical inference.

While in the simplest case $X_1,\ldots,X_n$ are independent random variables, more complicated cases are possible. For example:

1. $X_1,\ldots,X_n$ are not independent;
2. $X_1,\ldots,X_n$ are random vectors having a common joint distribution function $F_X(x)$;
3. $X_1,\ldots,X_n$ do not have a common probability distribution.

Is there a definition of sample that generalizes all of the above special cases? Fortunately, there is one and it is extremely simple.

- A sample ξ is the realization of a random vector Ξ.

The distribution function of Ξ, denoted by $F_\Xi(\xi)$, is the unknown distribution function that constitutes the object of inference.

Therefore, "sample" is just a synonym of "realization of a random vector". The following examples show how this general definition accommodates the special cases mentioned above.

Example When we observe n realizations $x_1, \ldots, x_n$ of n independent random variables $X_1, \ldots, X_n$ having a common distribution function $F_X(x)$, the sample is the n-dimensional vector $\xi = [x_1 \ldots x_n]$, which is a realization of the random vector $\Xi[X_1 \ldots X_n]$. The joint distribution function of Ξ is

$$F_{\Xi}(\xi) = F_X(x_1) \bullet \ldots \bullet F_X(x_n)$$

Example When we observe n realizations $x_1, \ldots, x_n$ of n random variables $X_1, \ldots, X_n$ that are not independent but have a common distribution function $F_X(x)$, the sample is again the n-dimensional vector $\xi = [x_1 \ldots x_n]$, which is a realization of the random vector $\Xi = [X_1 \ldots X_n]$. However, in this case the joint distribution function $F_{\Xi}(\xi)$ can no longer be written as the product of the distribution functions of $X_1, \ldots, X_n$.

Example When we observe n realizations $x_1 \ldots x_n$ of independent K-dimensional random vectors $X_1, \ldots, X_n$ having a common joint distribution function $F_X(x)$, the sample is the nK – dimensional vector $\xi = [x_1 \ldots x_n]$, which is a realization of the random vector $\Xi[X_1 \ldots X_n]$. The joint distribution function of Ξ is

$$F_{\Xi}(\xi) = F_X(x_1) \bullet \ldots \bullet F_X(x_n)$$

Example When we observe n realizations $x_1, \ldots, x_n$ of n independent K-dimensional random vectors $x_1, \ldots, x_n$ having different joint distribution functions $F_{X_1}(x_1), \ldots, F_{Xn}(x_n)$, the sample is the nK-dimensional vector $\xi = [x_1 \ldots x_n]$, which is a realization of the random vector $\Xi = [X_1 \ldots X_n]$. The joint distribution function of Ξ is

When the sample is made of n realizations $x_1, \ldots, x_n$ of n random variables (or random vectors):

$$\xi = [x_1 \ldots x_n]$$

Then we say that the sample has size n (or that the sample size is n). An individual realization x_1 is referred to as an observation from the sample.

Statistical Models

The sample ξ is used to infer some characteristics of $F_{\Xi}(\xi)$ that are not fully known by the statistician. The properties and the characteristics of $F_{\Xi}(\xi)$ that are already known (or are assumed to be known) before observing the sample are called a model for Ξ. In mathematical terms, a model for Ξ is a set of joint distribution functions to which $F_{\Xi}(\xi)$ is assumed to belong.

- Let the sample ξ be a realization of a l-dimensional random vector Ξ having joint distribution function $F_{\Xi}(\xi)$. Let ψ be the set of all l-dimensional joint distribution functions:
  $\psi = \{F : \mathbb{R}^l \to \mathbb{R}_+ \textit{ such that F is a joint distribution function}\}$

A subset $\Phi \subseteq \psi$ is called a statistical model (or a model specification or, simply, a model) for Ξ. If $F_{\Xi} \in \Phi$ the model is said to be correctly specified (or well-specified). Otherwise, if $F_{\Xi} \notin \Phi$ the model is said to be mis-specified.

Example, Suppose our sample is made of n realizations $x_1,\ldots,x_n$ of n random variables $X_1,\ldots,X_n$. Assume that the n random variables are mutually independent and that they have a common distribution function $F_X(x)$. The sample is the n-dimensional vector $\xi=[x_1 \ldots x_n]$. ψ is the set of all possible distribution functions of the random vector $\Xi=[X_1 \ldots X_n]$. Recalling the definition of marginal distribution function and the characterization of mutual independence, we can define the statistical model Φ as follows:

$\Phi=\{F\in\psi:$ all the marginals of F are equal and F is equal to the product of its marginals$\}$

Example, Take the example above and drop the assumption that the n random variables $X_1,\ldots,X_n$ are mutually independent. The statistical model Φ is now:

$$\Phi=\{F\in\psi: \textit{all the marginal of } F \textit{ are equal}\}$$

Parametric Models

A model Φ for Ξ is called a parametric model if the joint distribution functions belonging to Φ are put into correspondence with a set Θ of real vectors.

- Let Φ be a model for Ξ. Let $\Theta\subseteq\mathbb{R}^p$ be a set of p-dimensional real vectors. Let $\gamma(\theta)$ be a correspondence that associates a subset of Φ to each $\theta\in\Theta$. The triple (Φ,Θ,γ) is a parametric model if and only if

$$\Phi=\bigcup_{\theta\in\Theta}\gamma(\theta)$$

The set Φ is called parameter space. A vector $\theta\in\Theta$ is called a parameter.

Therefore, in a parametric model every element of Φ is put into correspondence with at least one parameter θ.

When $\gamma(\theta)$ associates to each parameter a unique joint distribution function (i.e., when $\gamma(\theta)$ is a function) the parametric model is called a parametric family.

- Let (Φ,Θ,γ) be a parametric model. If γ is a function from Θ to Φ, then the parametric model is called a parametric family. In this case, the joint distribution function associated to a parameter θ is denoted by $F_\Xi(\xi;\theta)$.

When each distribution function is associated with only one parameter, the parametric family is said to be identifiable.

- (Φ,Θ,γ) be a parametric family. If γ is one-to-one (i.e., each distribution function F is associated with only one parameter), then the parametric family is said to be identifiable.

Statistical Inferences

Now we can define that A statistical inference is a statement about the unknown distribution function $F_\Xi(\xi)$, based on the observed sample ξ and the statistical model Φ. Statistical inferences are often chosen among a set of possible inferences and take the form of model restrictions. Given a subset of the original model $\Phi_R\subset\Phi$, a model restriction can be either an inclusion restriction:

$$F_{\Xi} \in \Phi_R$$

or an exclusion restriction:

$$F_{\Xi} \notin \Phi_R$$

The following are common kinds of statistical inferences:

1. In hypothesis testing, a restriction $F_{\Xi} \in \Phi_R$ is proposed and the choice is between two possible statements:
   a. reject the restriction;
   b. do not reject the restriction.
2. In estimation, a restriction $F_{\Xi} \in \Phi_R$ must be chosen among a set of possible restrictions.
3. In Bayesian inference, the observed sample ξ is used to update the subjective probability that a restriction $F_{\Xi} \in \Phi_R$ is true.

Frequentist Inference

Frequentist Statistics tests whether an event (hypothesis) occurs or not. It calculates the probability of an event in the long run of the experiment (i.e the experiment is repeated under the same conditions to obtain the outcome).

Here, the sampling distributions of fixed size are taken. Then, the experiment is theoretically repeated infinite number of times but practically done with a stopping intention. For example, this perform an experiment with a stopping intention in mind that will stop the experiment when it is repeated 1000 times or we can see minimum 300 heads in a coin toss.

Now, we'll understand frequentist statistics using an example of coin toss. The objective is to estimate the fairness of the coin. Below is a table representing the frequency of heads:

no. of tosses	no. of heads	difference
10	4	-1
50	25	0
100	44	-6
500	255	5
1000	502	2
5000	2533	33
10000	5067	67

We know that probability of getting a head on tossing a fair coin is 0.5. No. of heads represents the actual number of heads obtained. Difference is the difference between 0.5*(No. of tosses) - no. of heads.

An important thing is to note that, though the difference between the actual number of heads and expected number of heads(50% of number of tosses) increases as the number of tosses are increased, the proportion of number of heads to total number of tosses approaches 0.5 (for a fair coin).

This experiment presents us with a very common flaw found in frequentist approach i.e. *Dependence of the result of an experiment on the number of times the experiment is repeated.*

The Inherent Flaws in Frequentist Statistics

20th century saw a massive upsurge in the *frequentist statistics* being applied to numerical models to check whether one sample is different from the other, a parameter is important enough to be kept in the model and variousother manifestations of hypothesis testing. But *frequentist statistics* suffered some great flaws in its design and interpretation which posed a serious concern in all real life problems. For example:

- *p*-values measured against a sample (fixed size) statistic with some stopping intention changes with change in intention and sample size. i.e If two persons work on the same data and have different stopping intention, they may get two different p- values for the same data, which is undesirable.

  For example: Person A may choose to stop tossing a coin when the total count reaches 100 while B stops at 1000. For different sample sizes, we get different t-scores and different p-values. Similarly, intention to stop may change from fixed number of flips to total duration of flipping. In this case too, we are bound to get different *p-values*.

- 2- Confidence Interval (C.I) like p-value depends heavily on the sample size. This makes the stopping potential absolutely absurd since no matter how many persons perform the tests on the same data, the results should be consistent.

- 3- Confidence Intervals (C.I) are not probability distributions therefore they do not provide the most probable value for a parameter and the most probable values.

Hypothesis tests and confidence intervals are constructed under the assumption that p is not random, but just unknown.

Let $n = 40$ (sample size) and

$$\hat{p} = \frac{22}{40} = 0.55$$

the sample proportion of successes, this estimates the unknown "population proportion" p of successes (in fact the success probability).

a) Confidence interval. We obtain a confidence interval of level C

$$[\hat{p} - m, \hat{p} + m]$$

with margin of error

$$m = z_{\alpha}^{*}\sqrt{\frac{\hat{p}(1-\hat{p})}{n}}$$

where z^*_α is the upper α-quantile of the standard normal random variable Z

$$\Pr(Z > z^*_\alpha) = \alpha$$

with $\alpha = (1 - C)/2 = 0.025$ for confidence level $C = 0.95$ (95%). From the normal table $z^*_\alpha = 1.96$, and

$$m = 1.96 \cdot \sqrt{\frac{0.55 \cdot (1 - 0.55)}{40}} = 0.154$$

so the confidence interval is

$$[0.55 - 0.154, 0.55 + 0.154] = [0.396, 0.704].$$

This interval includes the values 0.4,...,0.7, the values were also the posterior probabilities in the Bayesian model are concentrated.

Here we used the normal approximation to the standardized sample proportion

$$\frac{\hat{p} - p}{\sqrt{p(1-p)/n}} \approx Z$$

for large n. Use of a large sample sample approximation is not specific for frequentiest inference; is can also be used in a Bayesian framework.

b) Hypothesis test. Suppose we set up a null hypothesis H_0: $p = p_0$ and an alternative $H_a : p > p_0$. Here we are "trying to reject the null hypothesis", that is trying to collect statistically significant evidence for the claim that the treatment is better than a certain threshold. For instance, p_0 might be the known effectiveness of another treatment, or it might be the probability of success (no relapse) with no treatment. Let us consider $p_0 = 0.3$. We will be using a one sided Z-test. The test statistic is, with $p_0 = 0.3$

$$\hat{Z} = \frac{\hat{p} - p_0}{\sqrt{p_0(1-p_0)/n}}$$

and we reject at the $\alpha = 0.05$ (5%) significance level if

$$\hat{Z} > z^*_\alpha$$

Where $\Pr(Z > z^*_\alpha) = \alpha = 0.05$, thus $z^*_\alpha = 1.645$. We find

$$P = \Pr(Z > \hat{Z}) = \Pr(Z > 3.45) = 0.0003.$$

Again we used the normal approximation (0.1) at the value $p = p_0 = 0.3$. This should be justified by checking the success-failure condition:

$$np_0 = 12 > 10,$$
$$n(1 - p_0) = 28 > 10.$$

Bayesian Inference

Bayesian inference is a statistical method of inductive reasoning based on the reassessment of competing hypotheses in the presence of new evidence. Conceptually similar to the scientific method itself, formal Bayesian techniques are increasingly used in Artificial Intelligence and Brain models for decision-making under uncertainty Bayes' theorem shows the relation between two conditional probabilities that are the reverse of each other. This theorem is named after Reverend Thomas Bayes, and is also referred to as Bayes' law or Bayes' rule. Bayes' theorem expresses the conditional probability, or 'posterior probability', of an event A after B is observed in terms of the 'prior probability' of A, prior probability of B, and the conditional probability of B given A. Bayes' theorem is valid in all common interpretations of probability. The two (related) examples below should be sufficient to introduce Bayes' theorem.

Bayes' Theorem, Example 1

Bayes' theorem provides an expression for the conditional probability of A given B, which is equal to

$$\Pr(A \mid B) = \frac{\Pr(B \mid A)\Pr(A)}{\Pr(B)}$$

For example, suppose one asks the question: what is the probability of going to Hell, conditional on consorting (or given that a person consorts) with Laplace's Demon. By replacing A with Hell and B with Consort, the question becomes

$$\Pr(\text{Hell} \mid \text{Consort}) = \frac{\Pr(\text{Consort} \mid \text{Hell})\Pr(\text{Hell})}{\Pr(\text{Consort})}$$

Note that a common fallacy is to assume that $\Pr(A|B) = \Pr(B|A)$, which is called the conditional probability fallacy.

Bayes' Theorem, Example 2

Another way to state Bayes' theorem is

$$\Pr(A_i|B) = \frac{\Pr(B|A_i)\Pr(A_i)}{\Pr(B|A_i)\Pr(A_i) + \ldots + \Pr(B|A_n)\Pr(A_n)}$$

Let's examine our burning question, by replacing Ai with Hell or Heaven, and replacing B with Consort

- $\Pr(A_1) = \Pr(\text{Hell})$
- $\Pr(A_2) = \Pr(\text{Heaven})$
- $\Pr(B) = \Pr(\text{Consort})$
- $\Pr(A_1|B) = \Pr(\text{Hell}|\text{Consort})$

- $\Pr(A_2|B) = \Pr(\text{Heaven}|\text{Consort})$
- $\Pr(B|A_1) = \Pr(\text{Consort}|\text{Hell})$
- $\Pr(B|A_2) = \Pr(\text{Consort}|\text{Heaven})$

Laplace's Demon was conjured and asked for some data.

Data

- 6 people consorted out of 9 who went to Hell.
- 5 people consorted out of 7 who went to Heaven.
- 75% of the population goes to Hell.
- 25% of the population goes to Heaven.

Now, Bayes' theorem is applied to the data. Four pieces are worked out as follows

- $\Pr(\text{Consort}|\text{Hell}) = 6/9 = 0.666$
- $\Pr(\text{Consort}|\text{Heaven}) = 5/7 = 0.714$
- $\Pr(\text{Hell}) = 0.75$
- $\Pr(\text{Heaven}) = 0.25$

Finally, the desired conditional probability Pr(Hell|Consort) is calculated using Bayes' theorem

- $\Pr(\text{Hell}|\text{Consort}) = \dfrac{0.666(0.75)}{0.666(0.75) + 0.714(0.25)}$
- $\Pr(\text{Hell}|\text{Consort}) = 0.737$

The probability of someone consorting with Laplace's Demon and going to Hell is 73.7%, which is less than the prevalence of 75% in the population. According to these findings, consorting with Laplace's Demon does not increase the probability of going to Hell.

Model-Based Bayesian Inference

The basis for Bayesian inference is derived from Bayes' theorem. Here is Bayes' theorem, equation $\Pr(A|B) = \dfrac{\Pr(B|A)\Pr(A)}{\Pr(B)}$, again

$$\Pr(A|B) = \frac{\Pr(B|A)\Pr(A)}{\Pr(B)}$$

Replacing B with observations y, A with parameter set Θ, and probabilities Pr with densities p (or sometimes π or function f), results in the following.

$$p(\Theta|y) = \frac{p(y|\Theta)p(\Theta)}{p(y)}$$

Where $p(y)$ will be discussed below, $p(\Theta)$ is the set of prior distributions of parameter set Θ before y is observed, $p(y \mid \Theta)$ is the likelihood of y under a model, and $p(\Theta \mid y)$ is the joint posterior distribution, sometimes called the full posterior distribution, of parameter set Θ that expresses uncertainty about parameter set Θ after taking both the prior and data into account. Since there are usually multiple parameters, Θ represents a set of j parameters.

$$\Theta = \theta_1, \ldots, \theta_j$$

The denominator

$$p(y) = \int p(y \mid \Theta) p(\Theta) d\Theta$$

Defines the "marginal likelihood" of y, or the "prior predictive distribution" of y, and may be set to an unknown constant c. The prior predictive distribution4 indicates what y should look like, given the model, before y has been observed. Only the set of prior probabilities and the model's likelihood function are used for the marginal likelihood of y. The presence of the marginal likelihood of y normalizes the joint posterior distribution, $p(\Theta|y)$, ensuring it is a proper distribution and integrates to one.

By replacing $p(y)$ with c, which is short for a 'constant of proportionality', the model-based formulation of Bayes' theorem becomes

$$p(\Theta \mid y) = \frac{p(y \mid \Theta) p(\Theta)}{c}$$

By removing c from the equation, the relationship changes from 'equals' (=) to 'proportional to' ($\propto$)[5]

$$p(\Theta \mid y) \propto p(y \mid \Theta) p(\Theta)$$

This form can be stated as the unnormalized joint posterior being proportional to the likelihood times the prior. However, the goal in model-based Bayesian inference is usually not to summarize the unnormalized joint posterior distribution, but to summarize the marginal distributions of the parameters. The full parameter set Θ can typically be partitioned into

$$\Theta = \{\Phi, \wedge\}$$

where Φ is the sub-vector of interest, and Λ is the complementary sub-vector of Θ, often referred to as a vector of nuisance parameters. In a Bayesian framework, the presence of nuisance parameters does not pose any formal, theoretical problems. A nuisance parameter is a parameter that exists in the joint posterior distribution of a model, though it is not a parameter of interest. The marginal posterior distribution of φ, the parameter of interest, can simply be written as

$$p(\phi \mid y) = \int p(\phi, \wedge \mid y) d\wedge$$

In model-based Bayesian inference, Bayes' theorem is used to estimate the unnormalized joint posterior distribution, and finally the user can assess and make inferences from the marginal posterior distributions.

Components of Bayesian Inference

The components of Bayesian inference are

1. $p(\Theta)$ is the set of prior distributions for parameter set Θ, and uses probability as a means of quantifying uncertainty about Θ before taking the data into account.
2. $p(y|\Theta)$ is the likelihood or likelihood function, in which all variables are related in a full probability model.
3. $p(\Theta|y)$ is the joint posterior distribution that expresses uncertainty about parameter set Θ after taking both the prior and the data into account. If parameter set Θ is partitioned into a single parameter of interest φ and the remaining parameters are considered nuisance parameters, then $p(\varphi|y)$ is the marginal posterior distribution.

Inference Over Exclusive and Exhaustive Possibilities

If evidence is simultaneously used to update belief over a set of exclusive and exhaustive propositions, Bayesian inference may be thought of as acting on this belief distribution as a whole.

General Formulation

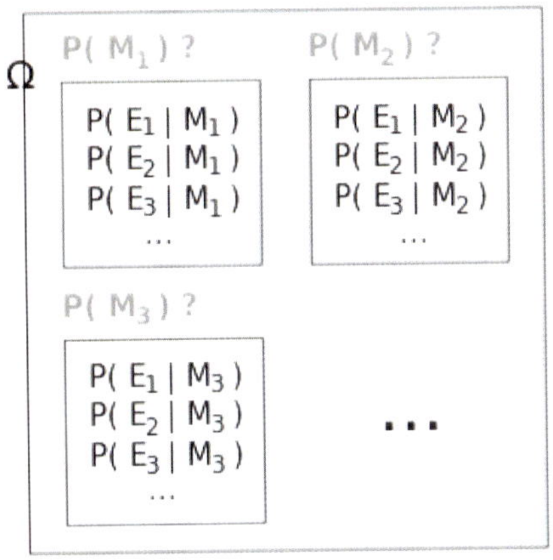

Diagram illustrating event space Ω in general formulation of Bayesian inference. Although this diagram shows discrete models and events, the continuous case may be visualized similarly using probability densities.

Suppose a process is generating independent and identically distributed events E_n, but the probability distribution is unknown. Let the event space Ω represent the current state of belief for this process. Each model is represented by event M_m. The conditional probabilities $P(E_n | M_m)$ are specified to define the models. $P(M_m)$ is the degree of belief in M_m. Before the first inference step, $\{P(M_m)\}$ is a set of *initial prior probabilities*. These must sum to 1, but are otherwise arbitrary.

Suppose that the process is observed to generate $E \in \{E_n\}$. For each $M \in \{M_m\}$, the prior $P(M)$ is updated to the posterior $P(M | E)$. From Bayes' theorem:

$$P(M | E) = \frac{P(E | M)}{\sum_m P(E | M_m) P(M_m)} \cdot P(M)$$

Upon observation of further evidence, this procedure may be repeated.

Multiple Observations

For a sequence of independent and identically distributed observations $\mathrm{E}=(e_1,\ldots,e_n),$, it can be shown by induction that repeated application of the above is equivalent to

$$P(M\mid \mathrm{E})=\frac{P(\mathrm{E}\mid M)}{\sum_m P(\mathrm{E}\mid M_m)P(M_m)}\cdot P(M)$$

Where

$$P(\mathrm{E}\mid M)=\prod_k P(e_k\mid M).$$

For a sequence where the conditional independence of the observations cannot be guaranteed, Rachael Bond, Yang-Hui He, and Thomas Ormerod have shown from quantum mechanics that

$$P(M_\alpha\mid E_1\cap E_2\ldots\cap E_m)=\frac{\sum_{i,j}\sqrt{E_{i\alpha}E_{j\alpha}c_{ij}^{\alpha}}}{\sum_{i,j,\beta}\sqrt{E_{i\beta}E_{j\beta}c_{ij}^{\beta}}}$$

such that

$$P(M_1\mid E_1\cap E_2)=\frac{\frac{P(E_1\mid M_1)P(E_2\mid M_1)}{P(E_1\mid M_2)P(E_2\mid M_2)}+P(E_1\mid M_1)+P(E_2\mid M_1)}{\frac{P(E_1\mid M_1)P(E_2\mid M_1)}{P(E_1\mid M_2)P(E_2\mid M_2)}+P(E_1\mid M_1)+P(E_2\mid M_1)+\frac{P(E_1\mid M_2)P(E_2\mid M_2)}{P(E_1\mid M_1)P(E_2\mid M_1)}+P(E_1\mid M_2)+P(E_2\mid M_2)}$$

Parametric Formulation

By parameterizing the space of models, the belief in all models may be updated in a single step. The distribution of belief over the model space may then be thought of as a distribution of belief over the parameter space. The distributions in this topic are expressed as continuous, represented by probability densities, as this is the usual situation. The technique is however equally applicable to discrete distributions.

Let the vector θ span the parameter space. Let the initial prior distribution over θ be $p(\theta\mid\alpha),$, where á is a set of parameters to the prior itself, or *hyperparameters*. Let $\mathrm{E}=(e_1,\ldots,e_n)$ be a sequence of independent and identically distributed event observations, where all e_i are distributed as $p(e\mid\theta)$ for some θ. Bayes' theorem is applied to find the posterior distribution over è :

$$p(\theta\mid \mathrm{E},\alpha)=\frac{p(\mathrm{E}\mid\theta,\alpha)}{p(\mathrm{E}\mid\alpha)}\cdot p(\theta\mid\alpha)=\frac{p(\mathrm{E}\mid\theta,\alpha)}{\int p(\mathrm{E}\mid\theta,\alpha)p(\theta\mid\alpha)d\theta}\cdot p(\theta\mid\alpha)$$

Where

$$p(\mathrm{E} \mid \theta, \alpha) = \prod_k p(e_k \mid \theta)$$

Mathematical Properties

Interpretation of Factor

$\frac{P(E \mid M)}{P(E)} > 1 \Rightarrow P(E \mid M) > P(E)$. That is, if the model were true, the evidence would be more likely than is predicted by the current state of belief. The reverse applies for a decrease in belief. If the belief does not change, $\frac{P(E \mid M)}{P(E)} = 1 \Rightarrow P(E \mid M) = P(E)$. That is, the evidence is independent of the model. If the model were true, the evidence would be exactly as likely as predicted by the current state of belief.

Cromwell's Rule

If $P(M) = 0$ then $P(M \mid E) = 0$. If $P(M) = 1$, then $P(M \mid E) = 1$. This can be interpreted to mean that hard convictions are insensitive to counter-evidence.

The former follows directly from Bayes' theorem. The latter can be derived by applying the first rule to the event "not M" in place of "M", yielding "if $1 - P(M) = 0$, then $1 - P(M \mid E) = 0$", from which the result immediately follows.

Asymptotic Behavior of Posterior

Consider the Behavior of a belief distribution as it is updated a large number of times with independent and identically distributed trials. For sufficiently nice prior probabilities, the Bernstein-von Mises theorem gives that in the limit of infinite trials, the posterior converges to a Gaussian distribution independent of the initial prior under some conditions firstly outlined and rigorously proven by Joseph L. Doob in 1948, namely if the random variable in consideration has a finite probability space. The more general results were obtained later by the statistician David A. Freedman who published in two seminal research papers in 1963 and 1965 when and under what circumstances the asymptotic Behavior of posterior is guaranteed. His 1963 paper treats, like Doob, the finite case and comes to a satisfactory conclusion. However, if the random variable has an infinite but countable probability space (i.e., corresponding to a die with infinite many faces) the 1965 paper demonstrates that for a dense subset of priors the Bernstein-von Mises theorem is not applicable. In this case there is almost surely no asymptotic convergence. Later in the 1980s and 1990s Freedman and Persi Diaconis continued to work on the case of infinite countable probability spaces. To summarise, there may be insufficient trials to suppress the effects of the initial choice, and especially for large (but finite) systems the convergence might be very slow.

Conjugate Priors

In parameterized form, the prior distribution is often assumed to come from a family of distributions called conjugate priors. The usefulness of a conjugate prior is that the corresponding posterior distribution will be in the same family, and the calculation may be expressed in closed form.

Estimates of Parameters and Predictions

It is often desired to use a posterior distribution to estimate a parameter or variable. Several methods of Bayesian estimation select measurements of central tendency from the posterior distribution.

For one-dimensional problems, a unique median exists for practical continuous problems. The posterior median is attractive as a robust estimator.

If there exists a finite mean for the posterior distribution, then the posterior mean is a method of estimation.

$$\tilde{\theta} = \mathrm{E}[\theta] = \int \theta p(\theta \mid \mathrm{X}, \alpha) d\theta$$

Taking a value with the greatest probability defines maximum *a posteriori* (MAP) estimates:

$$\{\theta_{\mathrm{MAP}}\} \subset \arg\max_{\theta} p(\theta \mid \mathrm{X}, \alpha).$$

There are examples where no maximum is attained, in which case the set of MAP estimates is empty.

There are other methods of estimation that minimize the posterior *risk* (expected-posterior loss) with respect to a loss function, and these are of interest to statistical decision theory using the sampling distribution ("frequentist statistics").

The posterior predictive distribution of a new observation $\tilde{x}$ (that is independent of previous observations) is determined by

$$p(\tilde{x} \mid \mathrm{X}, \alpha) = \int p(\tilde{x}, \theta \mid \mathrm{X}, \alpha) d\theta = \int p(\tilde{x} \mid \theta) p(\theta \mid \mathrm{X}, \alpha) d\theta.$$

Examples

Probability of a Hypothesis

Suppose there are two full bowls of cookies. Bowl #1 has 10 chocolate chip and 30 plain cookies, while bowl #2 has 20 of each. Our friend Fred picks a bowl at random, and then picks a cookie at random. We may assume there is no reason to believe Fred treats one bowl differently from another, likewise for the cookies. The cookie turns out to be a plain one. How probable is it that Fred picked it out of bowl #1?

Intuitively, it seems clear that the answer should be more than a half, since there are more plain cookies in bowl #1. The precise answer is given by Bayes' theorem. Let H_1 correspond to bowl #1, and H_2 to bowl #2. It is given that the bowls are identical from Fred's point of view, thus $P(H_1) = P(H_2)$, and the two must add up to 1, so both are equal to 0.5. The event E is the observation of a plain cookie. From the contents of the bowls, we know that $P(E \mid H_1) = 30/40 = 0.75$ and $P(E \mid H_2) = 20/40 = 0.5$. Bayes' formula then yields

$$P(H_1|E) = \frac{P(E|H_1)P(H_1)}{P(E|H_1)P(H_1) + P(E|H_2)P(H_2)}$$
$$= \frac{0.75 \times 0.5}{0.75 \times 0.5 + 0.5 \times 0.5}$$
$$= 0.6$$

Before we observed the cookie, the probability we assigned for Fred having chosen bowl #1 was the prior probability, $P(H_1)$, which was 0.5. After observing the cookie, we must revise the probability to $P(H_1|E)$, which is 0.6.

Making a Prediction

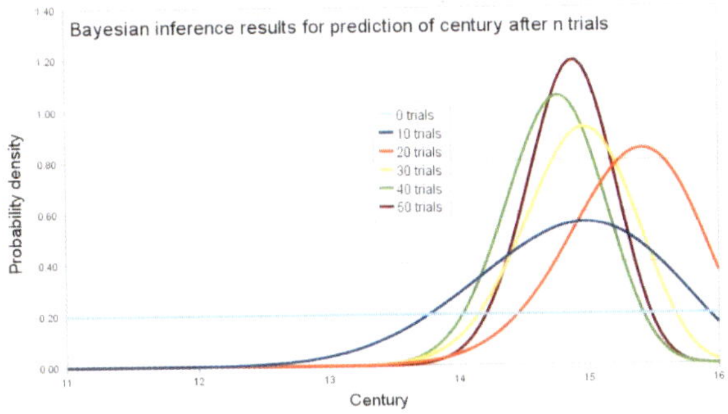

Example results for archaeology example. This simulation was generated using c=15.2.

An archaeologist is working at a site thought to be from the medieval period, between the 11th century to the 16th century. However, it is uncertain exactly when in this period the site was inhabited. Fragments of pottery are found, some of which are glazed and some of which are decorated. It is expected that if the site were inhabited during the early medieval period, then 1% of the pottery would be glazed and 50% of its area decorated, whereas if it had been inhabited in the late medieval period then 81% would be glazed and 5% of its area decorated. How confident can the archaeologist be in the date of inhabitation as fragments are unearthed?

The degree of belief in the continuous variable C (century) is to be calculated, with the discrete set of events $\{GD, G\bar{D}, \bar{G}D, \bar{G}\bar{D}\}$ as evidence. Assuming linear variation of glaze and decoration with time, and that these variables are independent,

$$P(E = GD | C = c) = (0.01 + \frac{0.81 - 0.01}{16 - 11}(c - 11))(0.5 - \frac{0.5 - 0.05}{16 - 11}(c - 11))$$

$$P(E = G\bar{D} | C = c) = (0.01 + \frac{0.81 - 0.01}{16 - 11}(c - 11))(0.5 + \frac{0.5 - 0.05}{16 - 11}(c - 11))$$

$$P(E = \bar{G}D | C = c) = ((1 - 0.01) - \frac{0.81 - 0.01}{16 - 11}(c - 11))(0.5 - \frac{0.5 - 0.05}{16 - 11}(c - 11))$$

$$P(E = \bar{G}\bar{D} | C = c) = ((1 - 0.01) - \frac{0.81 - 0.01}{16 - 11}(c - 11))(0.5 + \frac{0.5 - 0.05}{16 - 11}(c - 11))$$

Assume a uniform prior of $f_C(c) = 0.2$, , and that trials are independent and identically distributed. When a new fragment of type e is discovered, Bayes' theorem is applied to update the degree of belief for each c :

$$f_C(c \mid E = e) = \frac{P(E = e \mid C = c)}{P(E = e)} f_C(c) = \frac{P(E = e \mid C = c)}{\int_{11}^{16} P(E = e \mid C = c) f_C(c) dc} f_C(c)$$

A computer simulation of the changing belief as 50 fragments are unearthed is shown on the graph. In the simulation, the site was inhabited around 1420, or $c = 15.2$. By calculating the area under the relevant portion of the graph for 50 trials, the archaeologist can say that there is practically no chance the site was inhabited in the 11th and 12th centuries, about 1% chance that it was inhabited during the 13th century, 63% chance during the 14th century and 36% during the 15th century. Note that the Bernstein-von Mises theorem asserts here the asymptotic convergence to the "true" distribution because the probability space corresponding to the discrete set of events $\{GD, G\bar{D}, \bar{G}D, \bar{G}\bar{D}\}$ is finite.

Applications

Computer Applications

Bayesian inference has applications in artificial intelligence and expert systems. Bayesian inference techniques have been a fundamental part of computerized pattern recognition techniques since the late 1950s. There is also an ever-growing connection between Bayesian methods and simulation-based Monte Carlo techniques since complex models cannot be processed in closed form by a Bayesian analysis, while a graphical model structure *may* allow for efficient simulation algorithms like the Gibbs sampling and other Metropolis–Hastings algorithm schemes. Recently-Bayesian inference has gained popularity among the phylogenetics community for these reasons; a number of applications allow many demographic and evolutionary parameters to be estimated simultaneously.

As applied to statistical classification, Bayesian inference has been used in recent years to develop algorithms for identifying e-mail spam. Applications which make use of Bayesian inference for spam filtering include CRM, DSPAM, Bogofilter, SpamAssassin, SpamBayes, Mozilla, XEAMS, and others. Spam classification is treated in more detail in the topic on the naive Bayes classifier.

Solomonoff's Inductive inference is the theory of prediction based on observations; for example, predicting the next symbol based upon a given series of symbols. The only assumption is that the environment follows some unknown but computable probability distribution. It is a formal inductive framework that combines two well-studied principles of inductive inference: Bayesian statistics and Occam's Razor. Solomonoff's universal prior probability of any prefix p of a computable sequence x is the sum of the probabilities of all programs (for a universal computer) that compute something starting with p. Given some p and any computable but unknown probability distribution from which x is sampled, the universal prior and Bayes' theorem can be used to predict the yet unseen parts of x in optimal fashion.

In the Courtroom

Bayesian inference can be used by jurors to coherently accumulate the evidence for and against a defendant, and to see whether, in totality, it meets their personal threshold for 'beyond a reasonable doubt'. Bayes' theorem is applied successively to all evidence presented, with the posterior from one stage becoming the prior for the next. The benefit of a Bayesian approach is that it gives the juror an unbiased, rational mechanism for combining evidence. It may be appropriate to explain Bayes' theorem to jurors in odds form, as betting odds are more widely understood than probabilities. Alternatively, a logarithmic approach, replacing multiplication with addition, might be easier for a jury to handle.

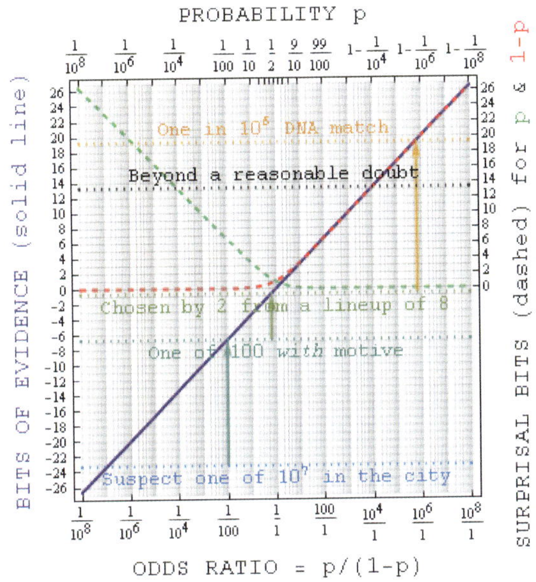

Adding up evidence.

If the existence of the crime is not in doubt, only the identity of the culprit, it has been suggested that the prior should be uniform over the qualifying population. For example, if 1,000 people could have committed the crime, the prior probability of guilt would be 1/1000.

The use of Bayes' theorem by jurors is controversial. In the United Kingdom, a defence expert witness explained Bayes' theorem to the jury in *R v Adams*. The jury convicted, but the case went to appeal on the basis that no means of accumulating evidence had been provided for jurors who did not wish to use Bayes' theorem. The Court of Appeal upheld the conviction, but it also gave the opinion that "To introduce Bayes' Theorem, or any similar method, into a criminal trial plunges the jury into inappropriate and unnecessary realms of theory and complexity, deflecting them from their proper task."

Gardner-Medwin argues that the criterion on which a verdict in a criminal trial should be based is *not* the probability of guilt, but rather the *probability of the evidence, given that the defendant is innocent* (akin to a frequentist p-value). He argues that if the posterior probability of guilt is to be computed by Bayes' theorem, the prior probability of guilt must be known. This will depend on the incidence of the crime, which is an unusual piece of evidence to consider in a criminal trial. Consider the following three propositions:

A. The known facts and testimony could have arisen if the defendant is guilty

B. The known facts and testimony could have arisen if the defendant is innocent

C. The defendant is guilty.

Gardner-Medwin argues that the jury should believe both A and not-B in order to convict. A and not-B implies the truth of C, but the reverse is not true. It is possible that B and C are both true, but in this case he argues that a jury should acquit, even though they know that they will be letting some guilty people go free.

Bayesian Epistemology

Bayesian epistemology is a movement that advocates for Bayesian inference as a means of justifying the rules of inductive logic.

Karl Popper and David Miller have rejected the idea of Bayesian rationalism, i.e. using Bayes rule to make epistemological inferences: It is prone to the same vicious circle as any other justificationist epistemology, because it presupposes what it attempts to justify. According to this view, a rational interpretation of Bayesian inference would see it merely as a probabilistic version of falsification, rejecting the belief, commonly held by Bayesians, that high likelihood achieved by a series of Bayesian updates would prove the hypothesis beyond any reasonable doubt, or even with likelihood greater than 0.

Other

- The scientific method is sometimes interpreted as an application of Bayesian inference. In this view, Bayes' rule guides (or should guide) the updating of probabilities about hypotheses conditional on new observations or experiments. The Bayesian inference has also been applied to treat stochastic scheduling problems with incomplete information.
- Bayesian search theory is used to search for lost objects.
- Bayesian inference in phylogeny.
- Bayesian tool for methylation analysis.
- Bayesian approaches to brain function investigate the brain as a Bayesian mechanism.
- Bayesian inference in ecological studies.
- Bayesian inference is used to estimate parameters in stochastic chemical kinetic models.

Bayes and Bayesian Inference

The problem considered by Bayes in Proposition 9 of his essay, "An Essay towards solving a Problem in the Doctrine of Chances", is the posterior distribution for the parameter a (the success rate) of the binomial distribution.

Fiducial Inference

R. A. Fisher proposed a statistical method for obtaining from observed data a probability distribution concerning a parameter value; he called the distribution a *fiducial probability distribution.* Fisher denied the equivalence and in his subsequent theoretical papers developed and extended fiducial probability.

As an example, consider a sample of independent measurements, $X_1, \ldots, X_n$, on a physical characteristic μ and suppose that the measurement error is normally distributed with mean 0 and known variance σ_0^2. Fisher requires that fiducial inference be based on the simplest statistic containing all the information about the parameter, in this case, on the sample mean X. The expression W = X — μ, involving the variable X and the characteristic (*μ*,has a known distribution: normal with mean 0 and variance σ_0^2/n. In an application of the method, the value of X is obtained and substituted in the expression *W* = *X* — *μ;* the expression is solved for μ in terms of *W,* giving *μ* = *X* - *W;* the fiducial distribution of μ then derives from the known distribution of W: μ is normal with mean X and variance σ_0^2/n. A 95 per cent fiducial interval is.

Fisher claimed that a fiducial probability statement has the same meaning as an ordinary probability statement. In the example, suppose that the 95 per cent fiducial interval as calculated in a specific application is 163.9 ± 0.8. The fiducial statement is that there is 95 per cent probability that the unknown value of μ lies in this interval. The interval 163.9 ± 0.8 is also a 95 per cent confidence interval, but as a confidence interval its interpretation is different. Confidence methods and fiducial methods do not, however, always lead to the same numerical results.

The proponents of the confidence method claim that in this context probability statements concerning μ cannot be made; the value of μ is something that exists: either it *is* in the interval or it is *not,* and we don't know which.

The proponents of the fiducial method reply that probabilities concerning realized values are commonplace: in the play of card games, for example, a player may observe his own hand and perhaps other cards and make a probability statement concerning the distribution of cards in the concealed hands.

The rejoinder is that μ did not arise from a random process such as the card shuffling and dealing. The relevance of this rejoinder is perhaps the key element to criticisms of the fiducial method.

In more complex problems the fiducial method may give a result different than the confidence method. In one prominent problem mentioned below, the Behrens-Fisher problem, the fiducial method gives an answer where confidence methods have not yet produced an entirely satisfactory result.

Fisher considered a statistic, T, obtained by the maximum likelihood method for estimating a parameter, θ Let F(T, o) be the cumulative distribution function for the statistic *T.* The probability density function for T is obtained by differentiating with respect to T: f(T, θ) = <9F(T, o)/<9T. Correspondingly, the fiducial density function for 6, given an observed value for T, is obtained by differentiating with respect to *θ: g*(*θ*, T) = <9F(T, o)/<9o.

Fisher illustrated the method with the correlation coefficient *r* of a sample from a bivariate normal distribution with population correlation coefficient *p*.

For more complex problems, Fisher proposed the use of a *pivotal quantity*, $W = h(T, \theta)$, a function of the statistic T and the parameter θ that has a fixed known distribution regardless of the value of θ For the first example, $W = X - \mu$. is a pivotal quantity. In an application, the observed value of the statistic T is substituted in the expression $W = h(T,\theta)$; the parameter θ is expressed in terms of W; and the fiducial distribution of θ is obtained from the known distribution of *W*.

Fisher's original method for obtaining a fiducial distribution is a special case of the pivotal method. As a function of a continuous statistic, *T*, the cumulative distribution function $W = F(T, \theta)$ has a uniform distribution on the interval (0, 1); this relationship is called that of the *probability integral transformation*.The fiducial density of θ for fixed T is obtained from the uniform distribution of W, in the same way as the density of *T* for fixed θ is obtained by differentiation.

As a second example, consider a random sample, $X_1, \ldots, X_n$, from a normal distribution with mean μ and variance σ^2, both unknown, and suppose that interest centers on the parameter μ. The quantity, using the sample mean $\bar{X}$ and sample standard deviation s_x, has a known distribution, the t-distribution on $n - 1$ degrees of freedom. In an application, the values of X and s_x are substituted and the parameter is solved for: This equation gives a fiducial distribution for μ that is of t-distribution form $(n - 1$ degrees of freedom), located at *X* and scaled by the factor.

The Behrens-Fisher problem is an extension of this example. Consider a first random sample, $X_1, \ldots, X_n$, from a normal distribution with mean μ_y and variance σ^2_x and a second independent random sample, $Y_1, \ldots Y_m$, from a normal distribution with mean μ_y and variance σ^2_Y. The Behrens-Fisher problem concerns inference about the parameter difference, $\mu_x - \mu_y$. The fiducial method gives a distribution described by for μ_x and a distribution described by for $/\mu_y$, . (Here t_1, and t_2 are independent *t* variables with n — 1, m — 1 degrees of freedom.) The fiducial distribution for $\mu_x - \mu_y$ is that of the difference, some percentage points are given in Fisher and Yates.

For many problems involving normal and chisquare distributions, the fiducial distribution has the form of a Bayesian posterior distribution as based on a prior distribution with uniformity characteristics.

Fisher considers a wide range of statistical problems and derives the corresponding fiducial distributions.

Consider the example involving a sample of measurements, $X_1, \ldots, X_n$ on a physical quantity, *p*. Let *me* be a variable describing the error introduced by the measuring instrument: in the example, *e* is normally distributed with mean 0 and variance σ^2_0. A measurement, X_I can then be expressed in the form $X_1 = \mu + e_i$. Correspondingly, the sample mean takes the form $\bar{X} = \mu + \bar{e}_i$, where $\bar{e}$ is normally distributed with mean 0 and variance σ^2_0. Now consider an application and suppose there is *no information* concerning μ With no information concerning μ, there is no information concerning *e* other than that describing its distribution. Probability statements can then be made concerning the unknown *e*just as the card player makes statements concerning the realized but un-revealed cards in his opponents' hands.

Suppose the normal distribution of $\bar{e}$ with variance σ^2_0/n gives a 95 per cent probability for $\bar{e}$ lying in the interval 0.0 ± 0.8. Then, *with an observed* x =163.9, the probability statement concerning $\bar{e}$ is equivalent to the statement that μ is in the interval 163.9 ± 0.8 with probability 95 per cent.

This analysis involving error variables applies to many of Fisher's examples, and it extends to other problems. The name structural probability has been introduced to distinguish it in cases where the method conflicts with the fiducial method. None of the criticisms mentioned concerning fiducial probability apply to structural probability.

Algorithmic "Solomonoff" Probability

Algorithmic "Solomonoff" Probability (AP) assigns to objects an *a priori* probability that is in some sense universal. This prior distribution has theoretical applications in a number of areas, including inductive inference theory and the time complexity analysis of algorithms. Its main drawback is that it is not computable and thus can only be approximated in practice.

Bayes, Occam and Epicurus

In an inductive inference problem there is some observed data $D=x_1, x_2,\ldots$ and a set of hypotheses $H=h_1,h_2,\ldots$, one of which may be the true hypothesis generating D . The task is to decide which hypothesis, or hypotheses, are the most likely to be responsible for the observations. An elegant solution to this problem is Bayes' rule,

$$P(h|D)=\frac{P(D|h)P(h)}{P(D)}.$$

To compute the relative probabilities of different hypotheses $P(D)$ can be dropped as it is an independent constant. Furthermore, for a given hypothesis h it is often straight forward to compute $P(D|h)$. A conceptually more difficult problem is to assign the prior probability $P(h)$. Indeed, how can one assign a probability to a hypothesis h *before* observing any data? A theoretically sound solution to this problem was not known until R.J. Solomonoff founded algorithmic probability theory in the 1960s.

Algorithmic probability rests upon two philosophical principles. From Bayes' rule it is clear that in order to keep consistent hypotheses what is needed is $\forall h : P(h) > 0$.

Occam's razor states that when inferring causes *entities should not be multiplied beyond necessity*. This is widely understood to mean: Among all hypotheses consistent with the observations, choose the simplest. In terms of a prior distribution over hypotheses, this is the same as giving simpler hypotheses higher a priori probability, and more complex ones lower probability.

Using Turing's model of universal computation, Solomonoff produced a universal prior distribution that unifies these two principles. This work was later generalized and extended by a number of researchers, in particular L. A. Levin who formalized the initial intuitive approach in a rigorous mathematical setting and supplied the basic theorems of the field.

Discrete Universal A Priori Probability

Consider an unknown process producing a binary string of one hundred 1s. The probability that such a string is the result of a uniform random process, such as fair coin flips, is just 2−100≈10−30, like that of any other string of the same length. Intuitively, we feel that there is a difference between a string that we can recognize and distinguish, and the vast majority of strings that are indistinguishable to us. In fact, we feel that a far more likely explanation is that some deterministic algorithmic simple process has generated this string. Presently we do this using the lucky confluence of combinatorics, information theory and theory of algorithms. It is clear that, in a world with computable processes, patterns which result from simple processes are relatively likely, while patterns that can only be produced by very complex processes are relatively unlikely. Formally, a computable process that produces a string x is a program p that when executed on a universal Turing machine U produces the string x as output.

As p is itself a binary string, we can define the discrete universal a priori probability $m(x)$ to be the probability that the output of a universal prefix Turing machine U is x when provided with fair coin flips on the input tape. Formally,

$$m(x) := \sum_{p:U(p)=x} 2^{-\ell(p)},$$

Where the sum is over all halting programs p for which U outputs the string x. As U is a *prefix* universal Turing machine the set of valid programs forms a prefix-free set and thus the sum is bounded due to Kraft's inequality. In fact, the boundedness of the sum should be an obvious consequence from realizing that the above expression is indeed a probability, which hinges on the fact that a prefix machine is one with only {0,1} input symbols and no end-of-input marker of any kind. It is easy to see that this distribution is strongly related to Kolmogorov complexity in that $m(x)$ is at least the maximum term in the summation, which is$-K(x)$. The Coding Theorem of L.A. Levin states that equality also holds in the other direction in the sense that $m(x) = \Theta(2^{-K(x)})$, or equivalently,

$$-\log m(x) = K(x) + O(1).$$

As every string can be computed by at least one program on U, it follows that m assigns a non-zero probability to all computable hypotheses and thus this distribution respects Epicurus' principle. Furthermore, if we measure the complexity of x by its Kolmogorov complexity, we see that the simplest strings have the highest probability under $m(x)$, while complex ones have a correspondingly low probability. In this way m also formalises the principle of Occam's razor.

Unfortunately, it is impossible in general to know whether any given program p will halt when run on U, due to the halting problem. This means that the sum can not be computed, it can only be approximated from below. Stated technically, the function $m()$ is only lower semi-computable. Furthermore, m is not a proper probability measure but rather a semi-measure as $\sum_x m(x) < 1$.

The universal a priori probability m has many remarkable properties. A theorem of L.A. Levin states that among the lower semi-computable semi-measures it is the largest such function in the following sense: If p is a lower semi-computable semi-measure, then there is a constant c such that

$cm(x) \geq p(x)$ for all x. In fact, we can take $c = 2^{K(p)}$ with $K(p)$ the prefix Kolmogorov complexity of the distribution p(defined as the least complexity of a prefix Turing machine lower semi-computing the distribution). This justifies calling m a universal distribution.

Thus, the same mathematical object m arises in three different ways:

- The family of lower semi-computable semi-measures contains an element that multiplicatively dominates every other element: a `universal' lower semi-computable semi-measure m;
- The probability mass function defined as the probability that the universal prefix machine U outputs x when the input is provided by fair coin flips, is the `a priori' probability m; and
- The probability $m(x)$ has $k(x)$ as its associated Shannon-Fano code ($m(x) = 2^{-K(x)+O(1)}$).

When three very different definitions (or properties) turn out to define the same object, then this is usually taken in Mathematics to mean that this object is objectively important.

Continuous Universal a Priori Probability

The universal distribution m is defined in a discrete domain, its arguments are finite binary strings. For applications such as the prediction of growing sequences it is necessary to define a similar distribution on infinite binary sequences. This leads to the universal semi-measure M defined as the probability that the output of a monotone universal Turing machine U starts with x when provided with fair coin flips on the input tape. Formally,

$$m(x) = \sum_{p:U(p)=x*} 2^{-\ell(p)},$$

where the sum is over all minimal programs p for which U's output starts with x (denoted by *).

Like m, it can be shown that M is only a semi-computable semi-measure. Due to a theorem of L.A. Levin it is the largest such function: If μ is a lower semi-computable semi-measure, then there is a constant c such that $cM(x) \geq \mu(x)$ for all x.

The semi-measure M has similar remarkable properties to m, for example, $-logM(x) = K(x) - O(log\ell(x)) = Km(x) - O(log\ell(x))$, where $Km(x) := min\{\ell(p) : U(p) = x*\}$ is the monotone complexity. Also M divided by the uniform measure $2^{-\ell(x)}$ is the maximal semicomputable supermartingale. Additionally, if the infinite binary sequences are distributed according to a computable measure μ,then the predictive distribution $M(x_{n+1} | x_1 \ldots x_n) := M(x_1 \ldots x_{n+1}) / M(x_1 \ldots x_n)$ converges rapidly to $\mu(x_{n+1} | x_1 \ldots x_n) := \mu(x_1 \ldots x_{n+1}) / \mu(x_1 \ldots x_n)$ with μ-probability 1. Hence, M predicts almost as well as does the true distribution μ.

Applications

Algorithmic probability has a number of important theoretical applications; some of them are listed below.

"Solomonoff" Induction

Solomonoff developed a quantitative formal theory of induction based on universal Turing machines (to compute, quantify and assign codes to all quantities of interest) and algorithmic complexity (to define what simplicity/complexity means). The important Prediction Error Theorem is due to Solomonoff: Let

$$Sn = \sum_{\ell(x)=n-1} \mu(x)(M(0\,|\,x) - \mu(0\,|\,x))^2,$$

be the expected squared error made in the n-th prediction of the next bit of an infinite binary sequence $x = x_1x_2x_3\ldots$ in the ensemble of infinite binary sequences distributed according to computable measure μ when predicting according to the fixed semi-measure M. Then,

$$\sum_n S_n \le K(\mu)(\ln 2)/2,$$

that is, the total summed expected squared prediction error is bounded by a constant, and therefore, if smooth enough, typically decreases faster than $1/n$. In particular it implies $M(x_{n+1}\,|\,x_1\ldots x_n) \to \mu(x_{n+1|}x_1\ldots x_n)$ with μ-probability 1, stated above. This is remarkable as it means that Solomonoff's inductive inference system will learn to correctly predict *any* computable sequence with only the absolute minimum amount of data. It would thus, in some sense, be the perfect universal prediction algorithm, if only it were computable.

AIXI

It can also be shown that M leads to excellent predictions and decisions in more general stochastic environments. Essentially *AIXI* is a generalisation of Solomonoff induction to the reinforcement learningsetting, that is, where the agent's actions can influence the state of the environment. AIXI can be proven to be, in some sense, an optimal reinforcement learning agent embedded in an arbitrary unknown environment. Like Solomonoff induction, the AIXI agent is not computable and thus can only be approximated in practice.

Instead of defining an optimal general agent, it is possible to turn things around and define a universal performance measure for agents acting in computable environments, known as universal intelligence.

Expected Time/Space Complexity of Algorithms under the Universal Distribution

For every algorithm whatsoever, the m-average-case complexity is always of the same order of magnitude as the worst-case complexity for computational resources such as time and storage space. This result of Li and Vitanyi means that, if the inputs are distributed according to the universal distribution, rather than according to the uniform distribution as is customary, then the expected case is as bad as the worst case. For example, the sorting procedure Quicksort has worst-case running time n^2, but expected running time $n\log n$ on a list of n keys for permutations drawn at random from uniformly distributed permutations of n keys. But if the permutations are distributed according to the universal distribution, that is, according to Occam's dictum that the simple permutations have the highest probabilities, as is often the case in practice, then the expected running time rises to the worst-case of n^2. Experiments appear to confirm this theoretical prediction.

PAC Learning using the Universal Distribution in the Learning Phase

Using m as the sampling distribution in the learning phase in PAC-learning properly increases the learning power for the family of discrete concept classes under computable distributions, and similarly for PAC learning of continuous concepts over computable measures by using M . This model of Li and Vitanyi is akin to using a teacher in the learning phase who gives the simple examples first.

Halting Probability

A formally related quantity is the probability that U halts when provided with fair coin flips on the input tape (i.e., that a random computer program will eventually halt). This halting probability $\Omega = \sum_x m(x)$ is also known as *Chaitin's constant* or «the number of wisdom», and has numerous remarkable mathematical properties. For example, it can be used to quantify Goedel's Incompleteness Theorem.

Decision Theory

In statistics, Decision theory is a set of quantitative methods for reaching optimal decisions.

A solvable decision problem must be capable of being tightly formulated in terms of initial conditions and choices or courses of action, with their consequences. In general, such consequences are not known with certainty but are expressed as a set of probabilistic outcomes. Each outcome is assigned a "utility" value based on the preferences of the decision maker. An optimal decision, following the logic of the theory, is one that maximizes the expected utility. Thus, the ideal of decision theory is to make choices rational by reducing them to a kind of routine calculation.

Decision Types

The decision problems can be classified into five types and they are:

1. Decision Making Under Certainty

There are a few problems where the decision maker gets almost complete information so that he knows all the facts about the state of nature and again which state of nature would occur and also the consequences of the state of nature. In such a situation, the problem of decision making is simple because the decision maker has only to choose the strategy which will give him maximum pay-off in terms of utility.

In cases where the strategy rows are normally very large and it is impossible even to list them, the technique of operational research like linear and nonlinear programming and geometric programming would have to be used to achieve the optimal strategy.

2. Decision Making Under Risk

A problem of this kind arises when the state of nature is unknown, but based on the objective or empirical evidence, we can possibly assign probabilities to various states of nature. In a number of problems on the basis of historical data and past experience, we are able to assign probabilities to various states of nature. In such cases, the pay-off matrix is of immense help for reaching an optimal decision by assigning probabilities to various states of nature.

3. Decision Making Under Uncertainty

The process of making decision under conditions of uncertainty takes place when there is hardly any knowledge about states of nature and no objective information about their probabilities of occurrence. In such cases of absence of historical data and relative frequency, the probability of the occurrence of the particular state of nature cannot be indicated.

Such situations arise when a new product is introduced or a new plant is set up. Of course, even in such cases some market surveys are conducted and relevant information is gathered though it is not generally sufficient to indicate a probability figure for the occurrence of a particular state of nature.

4. Decision Making Under Partial Information

This type of situation is somewhere between the conditions of risk and conditions of uncertainty. As regards conditions of risk, we have seen that the probability of the occurrence of various states of nature are known as the basis of past experience, and in conditions of uncertainty, there is no such data available. But many situations arise where there is partial availability of data. In such circumstances, we can say that decision making is done on the basis of partial information.

5. Decision Making Under Conflict

A condition of conflict is supposed to occur when we are dealing with rational opponent rather than the state of nature. The decision maker, therefore, has to choose a strategy taking into consideration the action or counter-action of his opponent. Brand competition, military weapons, market place, etc. are problems which come under this category. The strategy choice is done as the basis of game theory where a decision maker anticipates the action of the opponent and then determines his own strategy.

Logical Decision Framework

The main purpose of studying decision theory is to put the problem into a suitable logical framework. It includes identification of the problem. Personal perception and innovativeness are two essential things for the identification of the problem, and then generating alternative course of action and finally evolving criteria for evaluating the different alternatives to arrive at the best choice of action.

The basic components of a decision situation are the following:

1. Acts

There are many alternative courses of action in any decision problem. But only some relevant alternatives need be considered. For instance, the business firm may decide to market its goods within the state or within the country or beyond the boundaries of the country. Here, there are three alternatives. There may be more such alternatives. The final choice of any one will depend upon the payoffs from each strategy.

2. States of Nature

There are those possible events or the states of nature which are uncertain but are vital for the choice of any one of the alternative acts. For example, the radio dealer does not know how many radios he will be able to sell. There is an element of uncertainty about it and for this reason he cannot decide how many radios to buy. This uncertainty is known as the state of nature or the state of the world.

3. Outcomes

There is an outcome of the combination of each of the likely acts and possible states of nature. This is otherwise known as conditional value. The outcome has not much significant unless we calculate the pay-offs in terms of monetary gain or loss for each outcome. Thus outcome refers to the result of the combination of an act and each of the states of nature.

4. Pay-off

The pay-off deals with the monetary gain or loss from each of the outcomes. It can be also in terms of cost-saving or lime-saving but the expression of pay-off should always be in quantitative terms to help precise analysis. Therefore where the value of output is expressed directly in terms of gain expressed in money it is called pay-off. The calculation of pay-off or utility of each outcome has to be carefully done.

5. Expected Values of Each Act

In practical business situation, there is risk and uncertainty. In the case of risk, the probability of each state of nature is known, and in uncertainty, it is unknown. Therefore, each likely outcome of an act has to be appraised with reference to the probability of occurrence.

The expected value of a given act can be calculated by the following formula:

$$\sum_{j=1}^{n} PO_{ij} = P_1O_{i1} + P_2O_{i2} + \ldots\ldots\ldots P_nO_{in}.$$

Where P_1 to P_n refers to event probabilities of events E_1 to E_n and O_{ij}, the pay-offs of the outcome with the combination of each event and act. The expected value of each alternative is thus calculated with reference to probability assigned to each state of nature.

Choice of Decision Criteria

The nature of the decision criteria would depend upon the type of the decision situation as follows:

1. Under Conditions of Certainty

Under this condition; there is one pay-off for each strategy. The pay-off represents the degree of achievements of the objective, hence the largest pay-off is chosen and the corresponding strategy is selected.

2. Under Conditions of Risk

Under condition of risk, there would be more than one state of nature but the probabilities of their occurrence are known on the basis of past experience. The strategy which gives the maximum pay-off is selected.

3. Under Conditions of Uncertainty

Under conditions of uncertainty, we do not have a set of probabilities for the state of nature. Therefore, for each alternative only pay-offs or utilities are known. But nothing is known about the likelihood of each state of nature. The problem becomes more complex and the personality of the decision maker plays an important role in the selection of the strategy.

The following criteria are generally followed:

(1) Maximin Criterion

The strategy which gives the highest minimum pay-off will be chosen. The basic rationale behind this criterion is that pessimism is not irrational under the state of uncertainty. The idea is to avoid the worst. In this criterion the motive of self preservation is considered.

(2) Maximax Criterion

If the decision maker is an optimist by nature, he would always think that the state of nature would be the best from his point of view. He will find out the expected pay-off of all the strategies and pick up the strategy which gives the maximum pay-off out of all the strategies. He will always think that the state of nature would be favourable.

(3) Minimax Regret Criterion

When the criterion is in terms of cost or regret then the decision maker would choose the strategy in which the maximum regret or the cost is the lowest. The regrets have to be calculated for each act with reference to the best pay-off of the various alternative acts.

(4) Laplace Criterion

Under this criterion when under conditions of uncertainty there is complete ignorance about the probability of the occurrences of state of nature, it is assumed that the probability of the occurrence of each state of nature is the same. After this, the strategy which maximises the expected pay-off is chosen.

(5) Subjective Expected Utility Criterion

Under this criterion not only the knowledge gathered from past experience but also the judgment of the decision maker is taken into account in assigning probabilities to the states of nature. In this criterion, therefore, the expected value will be calculated by taking into account the posterior probabilities in regard to the state of nature in place of prior probabilities given.

References

- Aster, Richard; Borchers, Brian, and Thurber, Clifford (2012). Parameter Estimation and Inverse Problems, Second Edition, Elsevier. ISBN 0123850487, ISBN 978-0123850485

- Statistical-inference, fundamentals-of-statistics: statlect.com, Retrieved 31 March 2018
- Algorithmic-probability: scholarpedia.org, Retrieved 19 April 2018
- Edwards, Ward (1982). Daniel Kahneman, Paul Slovic and Amos Tversky, ed. "Judgment under uncertainty: Heuristics and biases". Science. 185 (4157): 1124–1131. Bibcode:1974Sci...185.1124T. doi:10.1126/science.185.4157.1124. PMID 17835457.
- Bayesian-statistics-beginners-simple-english: analyticsvidhya.com, Retrieved 11 July 2018
- Decision-theory-statistics: britannica.com, Retrieved 20 May 2018
- Phillips, L. D.; Edwards, Ward (October 2008). "Chapter 6: Conservatism in a Simple Probability Inference Task (Journal of Experimental Psychology (1966) 72: 346-354)". In Jie W. Weiss; David J. Weiss. A Science of Decision Making:The Legacy of Ward Edwards. Oxford University Press. p. 536. ISBN 978-0-19-532298-9.
- Fiducial-inference, applied-and-social-sciences-magazines, social-sciences: encyclopedia.com, Retrieved 30 June 2018

4

Probability

Probability refers to the likelihood for the occurrence of an event. It is a number that has value between 0 and 1. The concepts and axioms of probability are of significant use in statistics. An elaborate study of the varied principles of probability has been provided in this chapter, which includes topics that cover the diverse axioms of probability, random variable, conditional probability, probability distributions, etc.

Probability is a branch of mathematics that deals with calculating the likelihood of a given event's occurrence, which is expressed as a number between 1 and 0. An event with a probability of 1 can be considered a certainty: for example, the probability of a coin toss resulting in either "heads" or "tails" is 1, because there are no other options, assuming the coin lands flat. An event with a probability of .5 can be considered to have equal odds of occurring or not occurring: for example, the probability of a coin toss resulting in "heads" is .5, because the toss is equally as likely to result in "tails." An event with a probability of 0 can be considered an impossibility: for example, the probability that the coin will land (flat) without either side facing up is 0, because either "heads" or "tails" must be facing up. A little paradoxical, probability theory applies precise calculations to quantify uncertain measures of random events.

In its simplest form, probability can be expressed mathematically as: the number of occurrences of a targeted event divided by the number of occurrences plus the number of failures of occurrences (this adds up to the total of possible outcomes):

$$p(a) = p(a)/[p(a) + p(b)]$$

Calculating probabilities in a situation like a coin toss is straightforward, because the outcomes are mutually exclusive: either one event or the other must occur. Each coin toss is an *independent* event; the outcome of one trial has no effect on subsequent ones. No matter how many consecutive times one side lands facing up, the probability that it will do so at the next toss is always .5 (50-50). The mistaken idea that a number of consecutive results (six "heads" for example) makes it more likely that the next toss will result in a "tails" is known as the *gambler's fallacy*, one that has led to the downfall of many a bettor.

Probability theory had its start in the 17th century, when two French mathematicians, Blaise Pascal and Pierre de Fermat carried on a correspondence discussing mathematical problems dealing with games of chance. Contemporary applications of probability theory run the gamut of human inquiry, and include aspects of computer programming, astrophysics, music, weather prediction, and medicine.

For example, a merchant notices one day that 5 out of 50 visitors to her store make a purchase. The next day, 20 out of 50 visitors make a purchase. The two relative frequencies (5/50 or 0.10 and 20/50 or 0.40) differ. However, summing results over many visitors, she might find that the probability that a visitor makes a purchase gets closer and closer to 0.20.

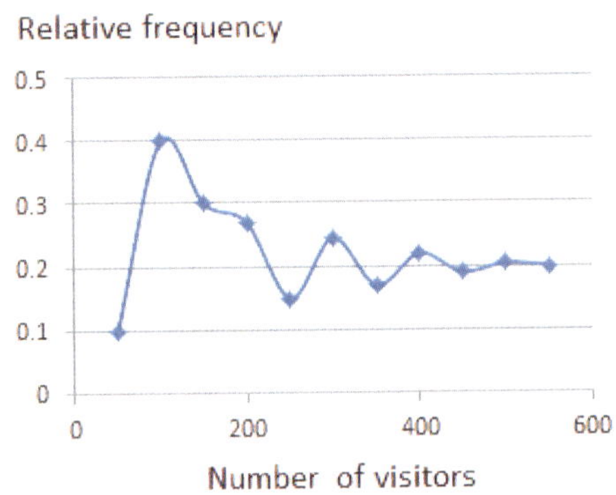

The scatterplot above shows the relative frequency of purchase as the number of trials (in this case, the number of visitors) increases. Over many trials, the relative frequency converges toward a stable value (0.20), which can be interpreted as the probability that a visitor to the store will make a purchase.

The idea that the relative frequency of an event will converge on the probability of the event, as the number of trials increases, is called the law of large numbers.

Applications of Probability and Statistics

Cell phone transmission over a noisy medium has a probability of error. Number of extra bits sent depend on statistics of the noise.

Weather prediction based on past observations makes use of probabilities.

Speech recognition is based on determining the most likely (highest probability) spoken words based on statistics of past speech. The statistics may be speaker specific.

Image compression standards like jpeg make use of unequal probabilities of pixel intensities.

Example: Testing the effect of a new drug

Problem: A neurologist wants to determine if a certain drug slows down the progress of Alzheimer's disease.

It is known that Alzheimer's disease results in abnormal enlargement of the ventricles (a compartment of the brain) as it progresses.

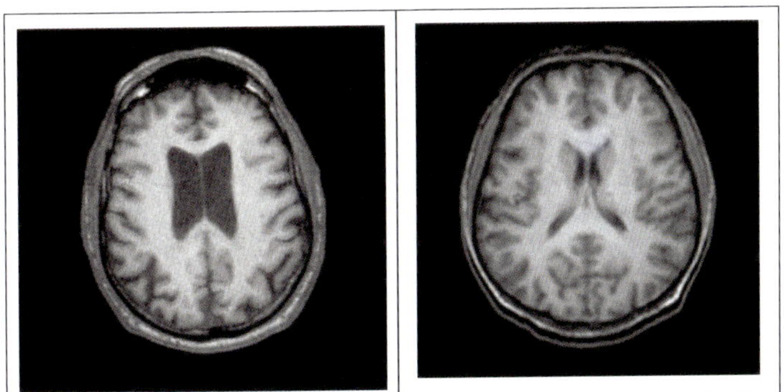

Alzheimer's patient | Normal brain

Measuring Ventricular Volume

The neurologist works together with an engineer who specializes in digital image processing to develop a computer program that automatically measures ventricular volume from Magnetic Resonance Images (MRI).

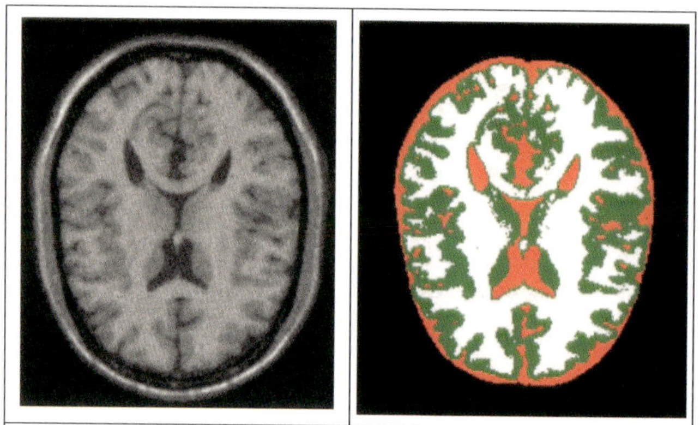

MRI	Partitioned image

Clinical Trial

The neurologist then recruits 20 Alzheimer's patients into a clinical trial.

He randomly assigns the 20 patients into 2 groups of 10. One group will take the drug that is being tested for 6 months while the other group will take the placebo (no drug) for 6 months.

At the beginning of the 6 months all patients have a MRI taken and their ventricular volume is measured.

This is repeated at the end of the 6 months which allows us to compute the change in ventricular volume for each patient.

Data

The following ventricular volume changes (in units of ml) are recorded

Drug	4.5	3.5	7.8	-1.1	5.8	7.2	6.7	6.2	4.6	6.5
Placebo	10.5	9.6	7.4	7.6	10.6	6.4	11.6	11.0	6.8	8.5

Do you think the drug was effective in slowing down Alzheimer's disease?

How can we start to look at this data.

in a more visual way?

in a more quantitative way?

Looking at Data Visually

Lets make a plot of the data:

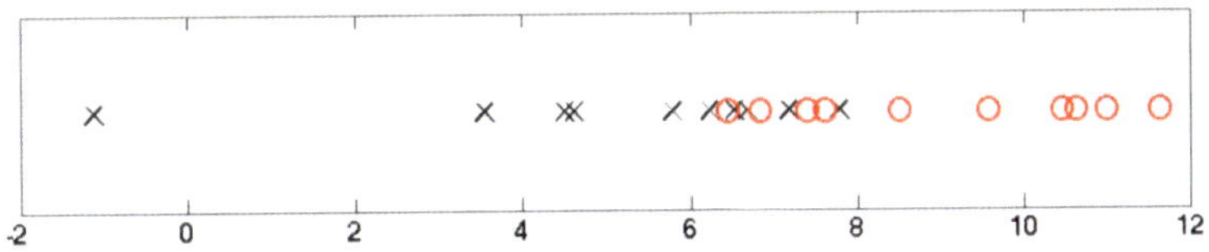

Crosses correspond to "drug" sample

Circles correspond to "placebo" sample

Does this help in forming an opinion?

Does it directly allow us to make a formal decision on whether the drug is effective?

Towards more Quantitative Analysis

If you had to summarize both samples with a single number what would you choose?

Answer: Sample mean

Let's take the "drug" sample:

Drug	4.5	3.5	7.8	-1.1	5.8	7.2	6.7	6.2	4.6	6.5

What is the sample mean?

How about the "placebo" sample:

Placebo	10.5	9.6	7.4	7.6	10.6	6.4	11.6	11.0	6.8	8.5

What is the sample mean?

Let's change our data a bit.

Assume the following data was recorded instead:

Drug	8.6	5.6	-0.4	6.7	5.9	11.9	10.2	1.0	-1.9	4.1
Placebo	7.3	6.7	12.3	18.2	6.0	4.2	7.8	13.3	9.1	5.1

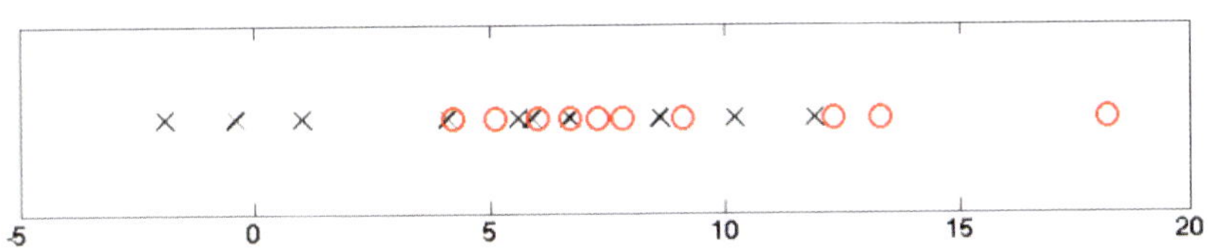

Do you think the drug was effective in slowing down Alzheimer's disease in this case?

Compute the Sample Means

"Drug" sample:

Drug	8.6	5.6	-0.4	6.7	5.9	11.9	10.2	1.0	-1.9	4.1

Sample mean = 5.17

"Placebo" sample:

Placebo	7.3	6.7	12.3	18.2	6.0	4.2	7.8	13.3	9.1	5.1

Sample mean =9.0

Observation: These are the exact same sample means as before.

Conclusion: Sample mean by itself is not sufficient to describe the data.

Question: So what changed?

Measuring Variability

Sample variance is a measure of variability.

Sample variances in our examples:

	Drug	Placebo
Dataset 1	6.61	3.59
Dataset 2	20.55	18.99

Sample standard deviations:

	Drug	Placebo
Dataset 1	2.57	1.89
Dataset 2	4.53	4.36

There is greater variability in the second dataset.

Large variability can hide the difference between two samples.

So even though the sample means remained the same, we are less sure about what to conclude about dataset 2.

Axioms of Probability

One strategy in mathematics is to start with a few statements, then build up more mathematics from these statements. The beginning statements are known as axioms. An axiom is typically something that is mathematically self-evident. From a relatively short list of axioms, deductive logic is used to prove other statements, called theorems or propositions.

The area of mathematics known as probability is no different. Probability can be reduced to three axioms. This was first done by the mathematician Andrei Kolmogorov. The handful of axioms that are underlying probability can be used to deduce all sorts of results. But what are these probability axioms?

In order to understand the axioms for probability, we must first discuss some basic points. We suppose that we have a set of outcomes called the sample space S. This sample space can be thought of as the universal set for the situation that we are studying. The sample space is comprised of subsets called events $E_1, E_2, \ldots, E_n$.

We also assume that there is a way of assigning a probability to any event E. This can be thought of as a function that has a set for an input, and a real number as an output. The probability of the event E is denoted by $P(E)$.

Axiom One

The first axiom of probability is that the probability of any event is a nonnegative real number. This means that the smallest that a probability can ever be is zero and that it cannot be infinite. The set of numbers that we may use are real numbers. This refers to both rational numbers, also known as fractions, and irrational numbers that cannot be written as fractions.

One thing to note is that this axiom says nothing about how large the probability of an event can be. The axiom does eliminate the possibility of negative probabilities. It reflects the notion that smallest probability, reserved for impossible events, is zero.

Axiom Two

The second axiom of probability is that the probability of the entire sample space is one. Symbolically we write $P(S) = 1$. Implicit in this axiom is the notion that the sample space is everything possible for our probability experiment and that there are no events outside of the sample space.

By itself, this axiom does not set an upper limit on the probabilities of events that are not the entire sample space. It does reflect that something with absolute certainty has a probability of 100%.

Axiom Three

The third axiom of probability deals with mutually exclusive events. If E_1 and E_2 are mutually exclusive, meaning that they have an empty intersection and we use U to denote the union, then

$$P(E_1 U E_2) = P(E_1) + P(E_2).$$

The axiom actually covers the situation with several (even countably infinite) events, every pair of which are mutually exclusive. As long as this occurs, the probability of the union of the events is the same as the sum of the probabilities:

$$P(E_1 U E_2 U \ldots U E_n) = P(E_1) + P(E_2) + \ldots + E_n$$

Consequences

From the Kolmogorov axioms, one can deduce other useful rules for calculating probabilities.

The Probability of the Empty set

$$P(\varnothing) = 0.$$

In some cases, $\varnothing$ is not the only event with probability 0.

Monotonicity

$$\text{if} \quad A \subseteq B \quad \text{then} \quad P(A) \leq P(B).$$

If A is a subset of, or equal to B, then the probability of A is less than, or equal to the probability of B.

Numeric Bound

It immediately follows from the monotonicity property that

$$0 \leq P(E) \leq 1 \qquad \forall E \in F.$$

Proof

The proofs of these properties are both interesting and insightful. They illustrate the power of the third axiom, and its interaction with the remaining two axioms. When studying axiomatic probability theory, many deep consequences follow from merely these three axioms. In order to verify the monotonicity property, we set $E_1 = A$ and $E_2 = B \setminus A$, where $A \subseteq B$ and $E_i = \varnothing$ for $i \geq 3$. It is easy to see that the sets E_i are pairwise disjoint and $E_1 \cup E_2 \cup \cdots = B$. Hence, we obtain from the third axiom that

$$P(A) + P(B \setminus A) + \sum_{i=3}^{\infty} P(E_i) = P(B).$$

Since the left-hand side of this equation is a series of non-negative numbers, and since it converges to $P(B)$ which is finite, we obtain both $P(A) \leq P(B)$ and $P(\varnothing) = 0$. The second part of the statement is seen by contradiction: if $P(\varnothing) = a$ then the left hand side is not less than infinity

$$\sum_{i=3}^{\infty} P(E_i) = \sum_{i=3}^{\infty} P(\varnothing) = \sum_{i=3}^{\infty} a = \begin{cases} 0 & \text{if } a = 0, \\ \infty & \text{if } a > 0. \end{cases}$$

If $a > 0$ then we obtain a contradiction, because the sum does not exceed $P(B)$ which is finite. Thus, $a = 0$. We have shown as a byproduct of the proof of monotonicity that $P(\varnothing) = 0$.

Further Consequences

Another important property is:

$$P(A \cup B) = P(A) + P(B) - P(A \cap B).$$

This is called the addition law of probability, or the sum rule. That is, the probability that A *or* B will happen is the sum of the probabilities that A will happen and that B will happen, minus the probability that both A *and* B will happen. The proof of this is as follows:

Firstly,

$$P(A \cup B) = P(A) + P(B \setminus A) \text{ (by Axiom 3)}$$

So,

$$P(A \cup B) = P(A) + P(B \setminus (A \cap B)) \text{ (by : } P(B \setminus A) = P(B \setminus (A \cap B))\text{)}.$$

Also,

$$P(B) = P(B \setminus (A \cap B)) + P(A \cap B)$$

and eliminating $P(B \setminus (A \cap B))$ from both equations gives us the desired result.

An extension of the addition law to any number of sets is the inclusion–exclusion principle.

Setting B to the complement A^c of A in the addition law gives

$$P\left(A^c\right) = P(\Omega \setminus A) = 1 - P(A)$$

That is, the probability that any event will *not* happen (or the event's complement) is 1 minus the probability that it will.

Simple Example: Coin Toss

Consider a single coin-toss, and assume that the coin will either land heads (H) or tails (T) (but not both). No assumption is made as to whether the coin is fair.

We may define:

$$\Omega = \{H, T\}$$

$$F = \{\varnothing, \{H\}, \{T\}, \{H, T\}\}$$

Kolmogorov's axioms imply that:

$$P(\varnothing) = 0$$

The probability of *neither* heads *nor* tails, is 0.

$$P(\{H, T\}) = 1$$

The probability of *either* heads *or* tails is 1.

$$P(\{H\}) + P(\{T\}) = 1$$

The sum of the probability of heads and the probability of tails is 1.

Events

The set of all the possible outcomes is called the sample space of the experiment and is usually denoted by S. Any subset E of the sample space S is called an event. Here are some examples.

Example 1 Tossing a coin. The sample space is $S = \{H, T\}$. $E = \{H\}$ is an event.

Example 2 Tossing a die. The sample space is $S = \{1, 2, 3, 4, 5, 6\}$. E = {2, 4, 6} is an event, which can be described in words as "the number is even".

Example 3 Tossing a coin twice. The sample space is $S = \{HH, HT, TH, T\,T\}$. $E = \{HH, HT\}$ is an event, which can be described in words as" the first toss results in a Heads.

Example 4 Tossing a die twice. The sample space is $S = \{(i, j): i, j = 1, 2, \ldots, 6\}$, which contains 36 elements. "The sum of the results of the two toss is equal to 10" is an event.

Example 5 Tossing a die twice. The sample space is $S = \{(i, j) : i, j = 1, 2, \ldots, 6\}$, which contains 36 elements. "The sum of the results of the two toss is equal to 10" is an event.

Example 5 Choosing a point from the interval (0, 1). The sample space is $S = (0, 1)$. $E = (1/3, 1/2)$ is an event.

Example 6 Measuring the lifetime of a light bulb. The sample space is S = $[0, \infty)$. E = $[90, \infty)$ is an event.

Example 7 Keeping on tossing a coin until one gets a Heads. The sample space of this experiment is $S = \{H, T\,H, T\,T\,H, T\,T\,T\,H, \ldots\}$. $E = \{H, T\,H\}$ is an event.

Suppose that E is an event. We say that the event E "occurs" if the outcome of the experiment is contained in E.

Since events are simply subsets of the sample space, we can talk about various set theoretic operations on events. In the following, $E, F, G, E_i, = 1, 2, \ldots$ are events.

Types of Events

Simple Event

If the event E has only one sample point of a sample space, it is called a simple event or an Elementary Event. It is an event that consists of exactly one outcome. Let us understand this with an example. Say you throw a die, the possibility of 2 appearing on the die is a simple event and is given by $E = \{2\}$

Compound Event

As opposed to a simple event, if there is more than one sample point on a sample space, such an event is called Compound Event. It involves combining two or more events together and finding the probability of such a combination of events.

For example, let us take another example. When we throw a die, the possibility of an even number appearing is a compound event, as there is more than one possibility, there are three possibilities i.e. E = {2,4,6}

Certain Event

Just as the name suggests, an event which is sure to occur in any given experiment is a certain event. The probability of this type of event is 1

Impossible Event

On the other hand, when an event cannot occur i.e. there is no chance of the event occurring it is said to be an impossible event. The probability of this event is 0. Like the probability that the card you drew from a deck is both red and black is an impossible event.

Equally likely Events

When the outcomes of an experiment are equally likely to happen, they are called equally likely events. Like during a coin toss you are equally likely to get heads or tails.

Complimentary Events

For an event E the non-occurrence of the event is called its complimentary event. Basically complimentary events are events that cannot occur at the same time. So when a die is thrown getting an odd face and an even face are complementary events.

Mutually Exclusive Events

Two events are said to be mutually exclusive events when both cannot occur at the same time. Mutually exclusive events always have a different outcome. Two simple events are always mutually exclusive, whereas two compound events may or may not be.

If A and B are two events, then

$$\left(A \cap B\right) = \emptyset$$

$$P\left(A \cap B\right) = 0$$

$$P\,(A \cup B) = P\,(A) + P\left(B\right)$$

Example: Three coins are tossed simultaneously

P is the event of getting at least 2 heads Q is the event of getting no heads R is the event of getting heads on the second coin Which of the pairs is mutually exclusive?

Solution: n (S) = 2 x 2 x 2 = 8

n (P) = HHT, HTH, THH, HHH = 4

n (Q) = TTT = 1

n (R) = THT, HHH, HHT, THH = 4

So Q & R and P & R are mutually exclusive as they have nothing in their intersection.

Event Relations

Let us consider the example of throwing a fair die. The sample space of this experiment is

S = {1 , 2, 3, 4, 5, 6}

If A be the event of getting an even number, then the sample points 2, 4 and 6 are favourable to the event A.

The remaining sample points 1, 3 and 5 are not favourable to the event A. Therefore, these will occur when the event A will not occur.

In an experiment, the outcomes which are not favourable to the event A are called complement of A and defined as follows:

The outcomes favourable to the complement of an event A consists of all those outcomes which are not favourable to the event A, and are denoted by 'not' A or by $\bar{A}$.

Event 'A or B'

Let us consider the example of throwing a die. A is an event of getting a multiple of 2 and B be another event of getting a multiple of 3.

The outcomes 2, 4 and 6 are favourable to the event A and the outcomes 3 and 6 are favourable to the event B.

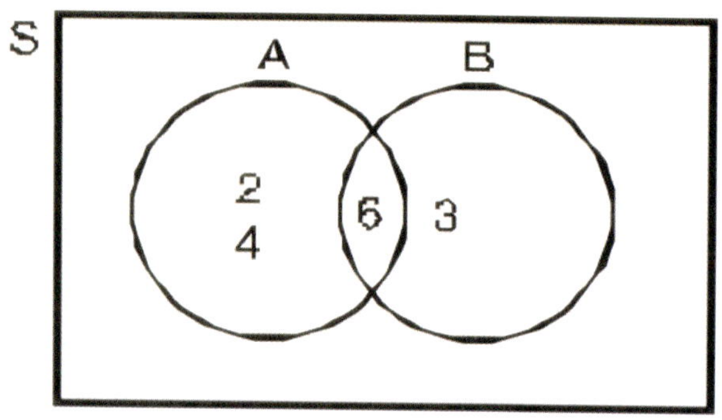

The happening of event A or B is

$A \cup B$ = {2, 3, 4, and 6}

Again, if A be the event of getting an even number and B is another event of getting an odd number, then

A = {2, 4, 6},

B = {1, 3, 5}

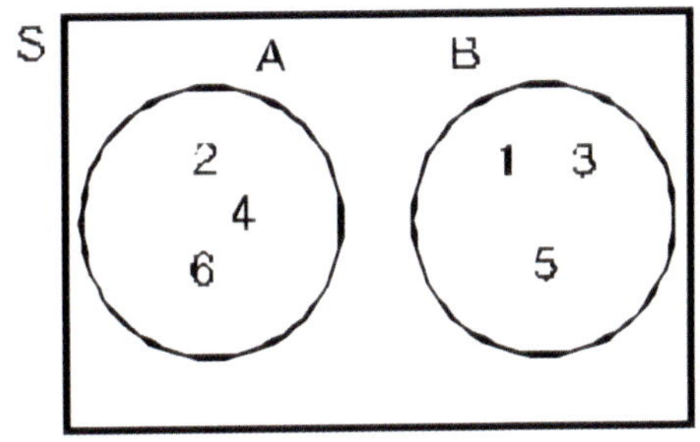

$A \cup B$ = {1, 2, 3, 4, 5 and 6}

Here, it may be observed that if A and B are two events, then the event ' A or B ' ($A \cup B$) will consist of the outcomes which are either favourable to the event A or to the event B or to both the events.

Thus, the event 'A or B' occurs, if either A or B or both occur.

Event "A and B'

Recall the example of throwing a die in which A is the event of getting a multiple of 2 and B is the event of getting a multiple of 3. The outcomes favourable to A are 2, 4, 6 and the outcomes favourable to B are 3, 6.

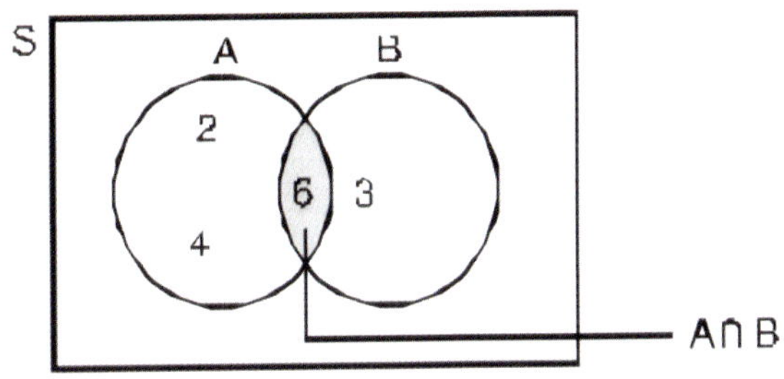

Here, we observe that the outcome 6 is favourable to both the events A and B. Draw a card from a well shuffled deck of 52 cards. A and B are two events defined as

A: a red card

B: a king

We know that there are 26 red cards and 4 kings in a deck of cards. Out of these 4 kings, two are red.

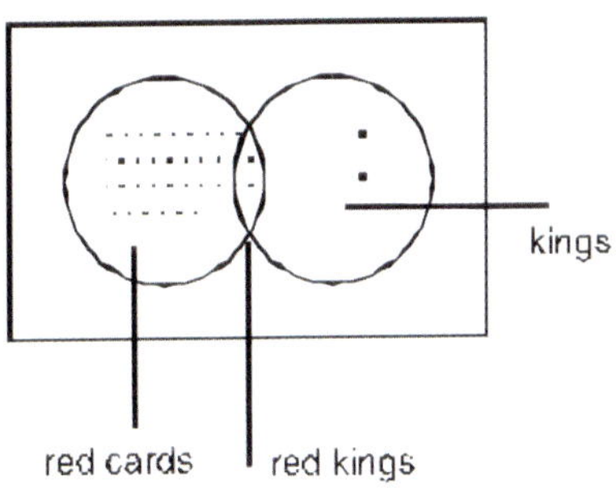

Here, we see that the two red kings are favourable to both the events.

Here, we see that the two red kings are favourable to both the events. Hence, the event ' A and B' consists of all those outcomes which are favourable to both the events A and B. That is, the event 'A and B' occurs, when both the events A and B occur simultaneously. Symbolically, it is denoted as $A \cap B$.

Event A but not B

The event A but not B shows the sample points which are in A but not in B. Event A but not B = A ∩ B' = A – A ∩ B. This event shows the unique sample points of A other than that in B. Suppose event A = {1, 3, 4, 5, 6, 7} and B' = {2, 3, 5. 6}. A ∩ B' = {1, 4, 7}.

Some Results on Probability of Events

Result: Probability of impossible event is zero.

Solution: Impossible event contains no sample points. Therefore, the certain event S and the impossible event φ are mutually exclusive

Hence, $S \cup \phi = S$

$$\Rightarrow P(\ S \cup \varphi =)\ P(S)$$

$$\text{or} \quad P(S) + P\varphi = P\ S$$

$$\Rightarrow\ P(\varphi) = 0.$$

Result: Probability of the complementary event $\bar{A}$ of A is given by

$$P(\bar{A}) = 1 - P(A)$$

Solution: A and $\bar{A}$ are disjoint events. Also

$$A \cup \bar{A} = S \ \Rightarrow \ P(A \cup \bar{A}) = P(S)$$

Using additive laws (ii) and (iii), we get

$$P(A) + P(\bar{A}) = 1 \ \Rightarrow \ P(\bar{A}) = 1 - P(A).$$

Result: Prove that $0 \le P(A) \le 1$, for any A in S.

Solution: We know that

$$A \subset S \ \Rightarrow \ P(A) \le P(S)$$

$$\Rightarrow \ P(A) \le 1.$$

We know that $P(A) \ge 0$.

Hence, $0 \le P(A) \le 1$.

Result: If A and B are any two events, then

$$P(A \cup B) = P(A) + P(B) - P(A \cap B).$$

Solution:

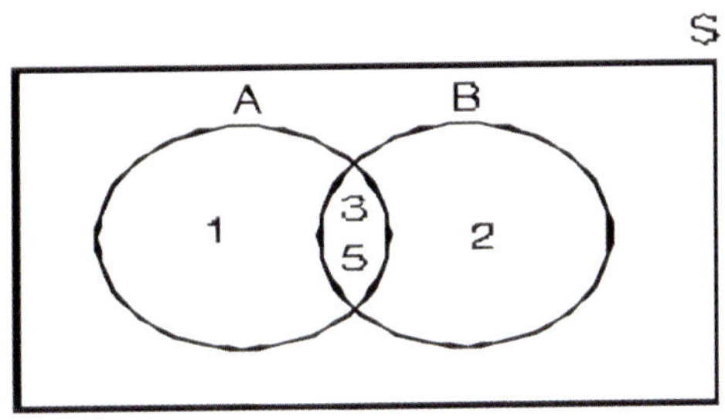

From the above figure, we can write

$$A \cup B = A \cup (\bar{A} \cap B)$$

$$\Rightarrow \quad P(A \cup B) = P[A \cup (\bar{A} \cap B)].$$

Since the events A and $(\bar{A} \cap B)$ are disjoint, therefore law (iii) gives

$$P[A \cup (\bar{A} \cap B)] = P(A) + P(\bar{A} \cap B)$$

Substituting this value in $P(A \cup B) = P[A \cup (\bar{A} \cap B)]$, we get

$$P(A \cup B) = P(A) + P(\bar{A} \cap B)$$

$$\text{or} \quad P(A \cup B) = P(A) + [P(\bar{A} \cap B) + P(A \cap B)] - P(A \cap B)$$

From figure, we see that

$$(\bar{A} \cap B) \cup (A \cap B) = B$$

$$\therefore \quad P[(\bar{A} \cap B) \cup (A \cap B)] = P(B).$$

Further, the $(\bar{A} \cap B)$ and $(A \cap B)$ are disjoint, so from additive law, we get

$$P(\bar{A} \cap B) + P(A \cap B) = P(B)$$

Substituting the above value we get,

$$P(A \cup B) = P(A) + P(B) - P(A \cap B)$$

Result: If A and B are mutually exclusive events, then

Solution: From additive law, we have

$$P(A \cup B) = P(A) + P(B) - P(A \cap B)$$

Since A and B are mutually exclusive events

Therefore $A \cap B = \phi$

$$\Rightarrow \quad (A \cap B) = P\phi = 0$$

Substituting this value in equation $P(A \cup B) = P(A) + P(B) - P(A \cap B)$, we get

$$P(A \cup B) = P(A) + P(B),$$

which is additive law for mutually exclusive events.

Multiplication Law of Probability for Independent Events

Two events A and B are said to be independent, if the occurrence or non-occurrence of one does not affect the probability of the occurrence (and hence non-occurrence) of the other.

The event of getting 'H' on first coin and the event of getting 'T' on the second coin in a simultaneous toss of two coins are independent events.

What about the event of getting 'H' on the first toss and event of getting 'T' on the second toss in two successive tosses of a coin? They are also independent events.

Let us consider the event of 'drawing an ace' and the event of 'drawing a king' in two successive draws of a card from a well-shuffled deck of cards without replacement.

Are these independent events?

No, these are not independent events, because we draw an ace in the first draw with probability $\frac{4}{52}$ Now, we do not replace the card and draw a king from the remaining 51 cards and this affect the probability of getting a king in the second draw, i.e., the probability of getting a king in the second draw without replacement will be $\frac{4}{51}$.

Note: If the cards are drawn with replacement, then the two events become independent. Is there any rule by which we can say that the events are independent?

How to find the probability of simultaneous occurrence of two independent events? If A and B are independent events, then

$$P(A \text{ and } B) = P(A) \cdot P(B)$$

Or

$$P(A \cap B) = P(A) \cdot P(B)$$

Thus, the probability of simultaneous occurrence of two independent events is the product of their separate probabilities.

The above law can be extended to more than two independent events, i.e.

$$P(A \cap B \cap C\ldots) = P(A) \cdot P(B) \cdot P(C)\ldots$$

On the other hand, if the probability of the event 'A' and 'B' is equal to the product of the probabilities of the events A and B, then we say that the events A and B are independent.

Example: A die is tossed twice. Find the probability of a number greater than 4 on each throw.

Solution: Let us denote by A, the event 'a number greater than 4' on first throw. B be the event 'a number greater than 4' in the second throw. Clearly A and B are independent events. In the first throw, there are two outcomes, namely, 5 and 6 favourable to the event A.

$$\therefore \quad P(A) = \frac{2}{6} = \frac{1}{3}$$

Similarly, $\therefore \quad P(B) = \frac{1}{3}$

Hence, $P(A \text{ and } B) = P(A) \cdot P(B)$

$$= \frac{1}{3} \cdot \frac{1}{3} = \frac{1}{9}.$$

Example: Arun and Tarun appear for an interview for two vacancies. The probability of Arun's selection is $\frac{1}{3}$ and that of Tarun's selection is $\frac{1}{5}$. Find the probability that

(a) Both of them will be selected.

(b) None of them is selected.

(c) At least one of them is selected.

(d) Only one of them is selected.

Solution: Probability of Arun's selection $= P(A) = \frac{1}{3}$

Probability of Tarun's selection $= P(T) = \frac{1}{5}$

(a) P (both of them will be selected) = P (A) P (T)

$$= \frac{1}{3} \times \frac{1}{5}$$

$$= \frac{1}{15}$$

(b) P (none of them is selected)

$$= P(\bar{A})P(\bar{T})$$

$$= \left(1 - \frac{1}{3}\right)\left(1 \frac{1}{5}\right)$$

$$= \frac{2}{3} \times \frac{4}{5} = \frac{8}{15}$$

(c) P (at least one of them is selected)

= 1- P (None of them is selected)

$$= 1 - P(\bar{A})P(\bar{T})$$

$$= 1 - \left(1 - \frac{1}{3}\right)\left(1 \frac{1}{5}\right)$$

$$= 1 - \left(\frac{2}{3} \times \frac{4}{5}\right)$$

$$= 1 - \frac{8}{15} = \frac{7}{15}$$

(d) P (only one of them is selected)

$$= P(A)P(\bar{T}) + P(\bar{A})\,P(T)$$

$$= \frac{1}{3} \times \frac{4}{5} + \frac{2}{3} \times \frac{1}{5}$$

$$= \frac{6}{15} = \frac{2}{5}$$

Example: A problem in statistics is given to three students, whose chances of solving it are $\frac{1}{2}, \frac{1}{3}$ and $\frac{1}{4}$ respectively. What is the probability that problem will be solved?

Solution: Let p_1, p_2 and p_3 be the probabilities of three persons of solving the problem

Here, $p_1 = \frac{1}{2}$, $p_2 = \frac{1}{3}$ and $p_3 \frac{1}{4}$.

The problem will be solved, if at least one of them solves the problem.

$\therefore$ P (at least one of them solves the problem)

= – 1 P (None of them solves the problem)

Now, the probability that none of them solves the problem will be

P (none of them solves the problem) $=(1-p_1)(1-p_2)(1-p_3)$

$$=\frac{1}{2}\times\frac{2}{3}\times\frac{3}{4}=\frac{1}{4}$$

When we putting above this value, we get

P (at least one of them solves the problem) $=1-\frac{1}{4}$

$$=\frac{3}{4}$$

Hence, the probability that the problem will be solved is $\frac{3}{4}$.

Example: Two balls are drawn at random with replacement from a box containing 15 red and 10 white balls. Calculate the probability that

(a) Both balls are red.

(b) First ball is red and the second is white.

(c) One of them is white and the other is red.

Solution:

(a) Let A be the event that first drawn ball is red and B be the event that the second ball drawn is red. Then as the balls drawn are with replacement,

therefore $P(A)=\frac{15}{25}=\frac{3}{5}, P(B)=\frac{3}{5}$

As A and B are independent events

Therefore P (both red) = P (A and B)

$$=P(A)\times P(B)$$

$$=\frac{3}{5}\times\frac{3}{5}=\frac{9}{25}$$

(b) Let A : First ball drawn is red.

B : Second ball drawn is white.

$$\therefore \quad P(A \text{ and } B)=P(A)\times P(B)$$

$$=\frac{3}{5}\times\frac{2}{5}=\frac{6}{25}.$$

(c) If WR denotes the event of getting a white ball in the first draw and a red ball in the second draw and the event RW of getting a red ball in the first draw and a white ball in the second draw. Then as 'RW' and WR' are mutually exclusive events, therefore

∴ P (a white and a red ball)

$$= P(WR \text{ or } RW)$$
$$= P(WR) + P(RW)$$
$$= P(W)\,P(R) + P(R)\,P(W)$$
$$= \frac{2}{5}\cdot\frac{3}{5} + \frac{3}{5}\cdot\frac{2}{5}$$
$$= \frac{6}{25} + \frac{6}{25} = \frac{12}{25}$$

Example: The odds against Manager X settling the wage dispute with the workers are 8 : 6 and odds in favor of manager Y settling the same dispute are 14 : 16.

(i) What is the chance that neither settles the dispute, if they both try independently of each other?

(ii) What is the probability that the dispute will be settled?

Solution: Let A be the event that the manager X will settle the dispute and B be the event that the manager Y will settle the dispute. Then, clearly

$$P(A) = \frac{6}{14} = \frac{3}{7},\ P(\bar{A}) = 1 - P(A) = 1 - \frac{3}{7} = \frac{4}{7}$$

$$P(B) = \frac{14}{30} = \frac{7}{15},\ P(\bar{B}) = 1 - \frac{14}{30} = \frac{16}{30} = \frac{8}{15}$$

The required probability that neither settles the dispute is given by

$$P(\bar{A} \cap \bar{B}) = P(\bar{A})\,P(\bar{B})$$

$$= \frac{4}{7}\cdot\frac{8}{15} = \frac{32}{105}$$

(Since A, B are independent, therefore, $\bar{A}, \bar{B}$ also independent)

(ii) The dispute will be settled, if at least one of the managers X and Y settles the dispute. Hence, the required probability is given by

$P(A \cup B) = P$ [At least one of X and Y settles the dispute]

$= 1 - P$ [None settles the dispute]

$$= 1 - P(\bar{A} \cap \bar{B})$$

$$= 1 - \frac{32}{105} = \frac{73}{105}$$

Example: A dice is thrown 3 times. Getting a number '5 or 6' is a success. Find the probability of getting

(a) 3 successes (b) exactly 2 successes (c) at most 2 successes (d) at least 2 successes.

Solution: Let S denote the success in a trial and F denote the ' not success' i.e. failure. Therefore,

$$P(S)=\frac{2}{6}=\frac{1}{3}$$

$$P(F)=1-\frac{1}{3}=\frac{2}{3}$$

(a) As the trials are independent, by multiplication theorem for independent events,

$$P(SSS)=P(S)\,P(S)\,P(S)$$

$$=\frac{1}{3}\times\frac{1}{3}\times\frac{1}{3}=\frac{1}{27}$$

$$P(SSF)=P(S)\,P(S)\,P(F)$$

$$=\frac{1}{3}\times\frac{1}{3}\times\frac{2}{3}=\frac{2}{27}$$

Since the two successes can occur in ${}^{3}C_{2}$ ways

$\therefore$ P (exactly two successes) $= {}^{3}C_{2}\times\frac{2}{27}=\frac{2}{9}$

(c) P (at most two successes) = – 1 P(3successes)

$$=1-\frac{1}{27}=\frac{26}{97}$$

(d) P (at least two successes) = P (exactly 2 successes) + P (3 successes)

$$=\frac{2}{9}+\frac{1}{27}=\frac{7}{27}$$

Example: A card is drawn from a pack of 52 cards so that each card is equally likely to be selected. Which of the following events are independent?

(i) A: the card drawn is a spade

B: the card drawn is an ace

(ii) A: the card drawn is black

B: the card drawn is a king

(iii) A: the card drawn is a king or a queen

B: the card drawn is a queen or a jack

Solution: (i) There are 13 cards of spade in a pack.

$$P(A) = \frac{13}{52} = \frac{1}{4}$$

There are four aces in the pack.

$$\therefore \quad P(B) = \frac{4}{52} = \frac{1}{13}$$

$A \cap B =$ {an ace of spade }

$$\therefore \quad P(A \cap B) = \frac{1}{52}$$

Now $P(A)\,P(B) = \frac{1}{4} \times \frac{1}{13} = \frac{1}{52}$

Since $P(A \cap B) = P(A).\,P(B)$

Hence, the events A and B are independent.

(ii) There are 26 black cards in a pack.

$$\therefore \quad P(A) = \frac{26}{52} = \frac{1}{2}$$

There are four kings in the pack.

$$\therefore \quad P(B) = \frac{4}{52} = \frac{1}{13}$$

$$A \cap B = \{2 \text{ black kings}\}$$

$$\therefore \quad P(A \cap B) = \frac{2}{52} = \frac{1}{26}$$

Now, $\quad P(A) \times (B) = \frac{1}{2} \times \frac{1}{13} = \frac{1}{26}$

Since $\quad P(A \cap B) = P(A)\,P(B)$

Hence, the events A and B are independent.

(iii) There are 4 kings and 4 queens in a pack of cards.

∴ Total number of outcomes favourable to the event A is 8.

$$\therefore \quad P(A) = \frac{8}{52} = \frac{2}{13}$$

Similarly, $P(B) = \frac{2}{13}$

$A \cap B = \{4 \text{ queens}\}$

$\therefore \quad P(A \cap B) = \frac{4}{52} = \frac{1}{13}$

$\therefore \quad P(A) \times P(B) = \frac{2}{13} \times \frac{2}{13} = \frac{4}{169}$

Here, $P(A \cap B) \neq P(A).\ P(B)$

Hence, the events A and B are not independent

Example: Consider a group of 36 students. Suppose that A and B are two properties that each student either has or does not have. The events are:

A. Student has blue eyes

B. Student is a male

Out of 36, there are 12 male and 24 female students and half of them in each has blue eyes. Are these events independent?

Solution: With regard to the given two properties, i.e., either has or does not have, the 36 students are distributed as follows:

	Blue eyes A	Not blue eyes $(\bar{A})$	Total
Male (B)	6	6	12
Female $(\bar{B})$	12	12	24
Total	18	18	36

If we choose a student at random, the probabilities corresponding to the events A and B are

$P(A) = \frac{18}{36} = \frac{1}{2}$

$P(B) = \frac{12}{36} = \frac{1}{3}$

$P(A \cap B) = \frac{6}{36} = \frac{1}{6}$

Also $P(A) \cdot P(B) = \frac{1}{2} \times \frac{1}{3} = \frac{1}{6}$

Here, $P(A \cap B) = P(A) \cdot P(B)$

Hence, the events A and B are independent.

Example: Suppose that we toss a coin three times and record the sequence of heads and tails. Let A be the event ‘ at most one head occurs’ and B the event ‘ both heads and tails occur’. Are these event independent?

Solution: The sample space in tossing a coin three times will be

$$S = \{HHH, HHT, HTH, HTT, THH, TTH, THT, TTT\}$$

Also, $(A \cap B) = \{TTH, THT, HTT\}$

$$\therefore \quad P(A) = \frac{4}{8} = \frac{1}{2}, P(B) = \frac{6}{8} = \frac{3}{4}, \ P(A \cap B) = \frac{3}{8}$$

Moreover, $P(A) \times P(B) = \frac{1}{2} \times \frac{3}{4} = \frac{3}{8}$

Which equals $P(A \cap B)$. Hence, A and B are independent?

Random Variable

Random variables are very important in statistics and probability and a must have if anyone is looking forward to understand probability distributions. Random Variables many a times confused with traditional variables.

Random variable is different from our traditional variable in terms of the value which it takes. It's a function which performs the mapping of the outcomes of a random process to a numeric value.

As it is subject to randomness, it takes different values.

Random variables allows us to ask questions in mathematical way.

If we flip 5 coins and want to answers questions like:

1. What is the probability of getting exactly 3 heads?
2. What is the probability of getting less than 4 heads?
3. What is the probability of getting more than 1 head?

Then our general way of writing would be:

- P(Probability of getting exactly 3 heads when we flip a coin 5 times)
- P(Probability of getting less than 4 heads when we flip a coin 5 times)
- P(Probability of getting more than 1 head when we flip a coin 5 times)

But if we use random variables to represent above questions then we would write:

1. P(X=3)
2. P(X<4)
3. P(X>1)

As we can see above random variables makes our task much easier to quantify results of any random process and apply math and perform further computation.

Suppose we have a random process/experiment of flipping a coin. One of the two possible outcomes could be either a head or a tail. So here we use X to denote random variable, which represents the outcomes of the random process.

Therefore we can write

X= 1, if the outcome is head

X= 0, if the outcome is tail

Here the random Variable X is mapping the outcomes of the random process (flipping a coin) to the numerical values (1 and 0).

The values assigned to denote head and tail can be anything it's not necessarily be 1 and 0.

To make understanding simple we have used 1 and 0.

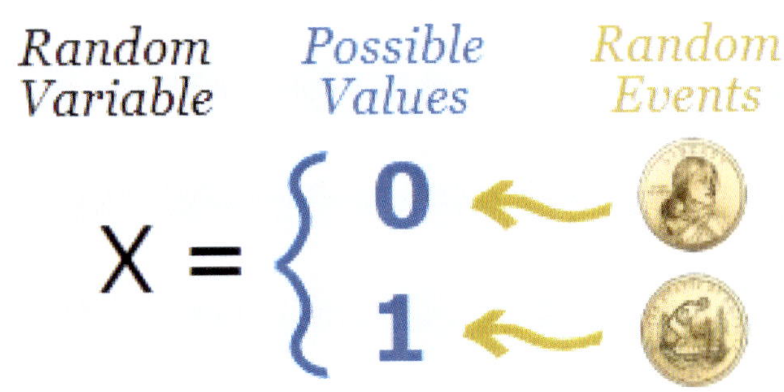

Other way of assigning numerical values to outcomes of a random process could be:

X= 100, if the outcome is head

X= 50, if the outcome is tail

Difference between Random Variables and Traditional Variables used in Algebra

Let's say variables used in algebra as x, y, z. Here x can be the number of cell phones, y = no of heads or z= no of students. A variable is nothing but an alphabetical character which represents an unknown number.

For Example:

$x + 5 = 10$

x is the variable whose value is unknown and we are trying to find its value.

After evaluating, x=5.

A Random Variable is different from the variable in algebra as it has whole set of values and it can take any of those randomly. Variable used in algebra cannot have more than a single value at a time.

If random variable X= {0, 1, 2, 3}

Then X could be 0, 1, 2 or 3 randomly where each of them might have a different probability.

We use capital letter for random variables to avoid confusion with traditional variables.

Properties of a Random Variable

- It only takes the real value.
- If X is a random variable and C is a constant, then CX is also a random variable.
- If X_1 and X_2 are two random variables, then $X_1 + X_2$ and $X_1 X_2$ are also random.
- For any constants C_1 and C_2, $C_1X_1 + C_2X_2$ is also random.
- |X| is a random variable.

Mean of Random Variables

In probability and statistics, we can find out the average of a random variable. The term average is the mean or the expected value or the expectation in probability and statistics. Once we have calculated the probability distribution for a random variable, we can calculate its expected value. Mean of a random variable shows the location or the central tendency of the random variable.

The expectation or the mean of a discrete random variable is a weighted average of all possible values of the random variable. The weights are the probabilities associated with the corresponding values. It is calculated as,

$$E(X) = \mu = \Sigma_i x_i p_i \quad i = 1, 2, \ldots, n$$

$$E(X) = x_1 p_1 + x_2 p_2 + \ldots + x_n p_n.$$

Properties of Mean of Random Variables

- If X and Y are random variables, then $E(X + Y) = E(X) + E(Y)$.
- If $X_1, X_2, \ldots, X_n$ are random variables, then $E(X_1 + X_2 + \ldots + X_n) = E(X_1) + E(X_2) + \ldots + E(X_n) = \Sigma_i E(X_i)$.
- For random variables, X and Y, $E(XY) = E(X)\,E(Y)$. Here, X and Y must be independent.
- If a is any constant and X is a random variable, $E[aX] = a\,E[X]$ and $E[X + a] = E[X] + a$.
- For any random variable, $X > 0$, $E(X) > 0$.
- $E(Y) \geq E(X)$ if the random variables X and Y are such that $Y \geq X$.

Variance of Random Variables

Suppose you calculated the mean or the average marks in the five tests of mathematics. You can easily see the difference of marks in each of the tests from this average marks. This difference in marks shows the variability of the possible values of the random variable. The random variable being the marks scored in the test.

The variance of a random variable shows the variability or the scatterings of the random variables. It shows the distance of a random variable from its mean. It is calculated as

$$\sigma_x^2 = \text{Var}(X) = \Sigma_i (x_i - \mu)^2 p(x_i) = E(X - \mu)^2 \text{ or, Var } X = E(X)^2 - [E(X)]^2.$$

$$E(X)^2 = \Sigma_i x_1^2 p(x_i), \text{and } [E(X)]^2 = [\Sigma_i x_1 p(x_i)]^2 = \mu^2.$$

If the value of the variance is small, then the values of the random variable are close to the mean.

Properties of Variance of Random Variables

- The variance of any constant is zero i.e, $V(a) = 0$, where a is any constant.
- If X is a random variable, and a and b are any constants, then $V(aX + b) = a^2 V(X)$.
- For any pair-wise independent random variables, $X_1, X_2, \ldots, X_n$ and for any constants

$$a_1, a_2, \ldots, a_n; V(a_1 X_1 + a_2 X_2 + \ldots + a_n X_n) = a_1^2\,V(X_1) + a_2^2\,V(X_2) + \ldots + a_n^2\,V(X_n).$$

Example: Calculate the mean and variance for a random variable, X defined as the number of tails in four tosses of a coin. Also, draw the probability distribution.

Solution: Let T represents a tail and H, a head. X denotes the number of tails in four tosses of a coin. X takes the value 0, 1, 2, 3, 4.

S. No.	Possible outcomes	Number of Tails, X	S. No.	Possible outcomes	Number of Tails, X
1	THTH	2	9	THTT	3
2	HHTH	1	10	HHTT	2
3	TTTH	3	11	TTTT	4
4	HTTH	2	12	HTTT	3
5	THHH	1	13	THHT	2
6	HHHH	0	14	HHHT	1
7	TTHH	2	15	TTHT	3
8	HTHH	1	16	HTHT	2

$P(X = 0) = 1/16$, $P(X = 1) = 4/16 = 1/4$, $P(X = 2) = 6/16 = 3/8$, $P(X = 3) = 4/16 = 1/4$, $P(X = 4) = 1/16$

The probability distribution of X is

x	0	1	2	3	4
p(x)	1/16	1/4	3/8	1/4	1/16

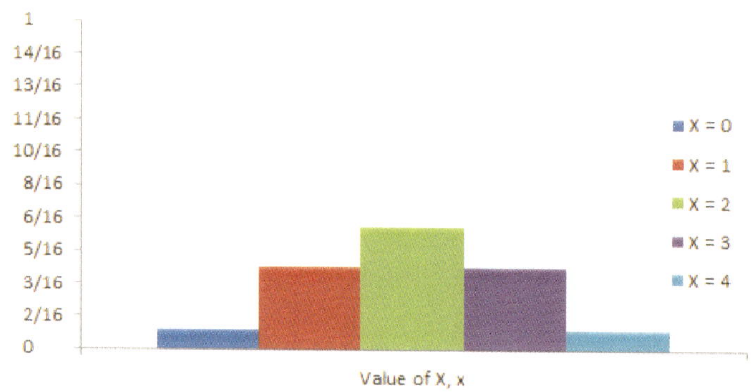

$$E(X) = \Sigma_i x_i p_i = 1 \times 1/4 + 2 \times 3/8 + 3 \times 1/4 + 4 \times 1/16 = 8/4 = 2.$$

$$E(X^2) = 1^2 \times 1/4 + 2^2 \times 3/8 + 3^2 \times 1/4 + 4^2 \times 1/16 = 1/4 + 3/2 + 9/4 + 1 = 5.$$

So, Variance of X = $V(X) = E(X^2) - [E(X)]^2 = 5 - 2^2 = 1$.

Discrete Random Variables

A random variable X is called discrete if it can assume a finite number of values. If two bulbs are selected from a certain lot, the number of defective bulbs may be 0, 1 or 2. The range of the variable is from 0 to 2 and the random variable can take some selected values in this range. The number of defective bulbs cannot be 1.1 or −1–1 or 3, etc. This random variable can take only the specific values which are 0, 1 and 2. When two dice are rolled, the total on the two dice will be 2, 3,...,12. The total on the two dice is a discrete random variable.

Discrete Probability Distribution

Suppose a discrete random variable X can assume the values $x_1, x_2, x_3, \ldots, x_n$ with corresponding probabilities $p(x_1), p(x_2), p(x_3), \ldots, p(x_n)$. The set of ordered pairs

$[x_1, p(x_1)], [x_2, p(x_2)], [x_3, p(x_3)], \ldots, [x_n, p(x_n)]$ is called the probability distribution or probability function of the random variable X. A probability distribution can be presented in a tabular form showing the values of the random variable X and the corresponding probabilities donated by $p(x_i)$ or $f(x_i)$. The above information can be collected in the form of a table as below, which is called the probability distribution of the random variable X. The probability that the random variable X will take the value x_i is denoted by $p(x_i)$ where $p(x_i) = P(X = x_i)$.

Values of random variable (x_i)	x_1	x_2	x_3	$\cdots$	x_i	$\cdots$	x_n
Probability $p(x_i)$	$p(x_1)$	$p(x_2)$	$p(x_3)$	$\cdots$	$p(x_i)$	$\cdots$	$p(x_n)$

The probability of some interval can be calculated by adding the probabilities of all points in the interval. For example,

$$P[x_1 < X < x_4] = P(X = x_2) + P(X = x_3).$$

This type of addition will not be possible in continuous random variable to find the probability of an interval. Therein we shall use integral calculus to find the probability of an interval.

The probability distribution can also be described in the form of an equation for $f(x_i)$ with a list of possible values of the random variable X. Some probability distributions in the form of equations are

(i) $f(x_i) = \frac{1}{6}$ for $x_i = 1, 2, 3, \cdots, 6$

For each value of X the probability is 1/6. It is the probability distribution when a fair die is rolled.

(ii) $f(x_i) = \binom{3}{x}\left(\frac{1}{2}\right)^3$ for $x = 0, 1, 2, 3$

Example

A digit is selected from the first 8 natural numbers. Write the probability distribution of X where X is the number of factors (divisors) of digits.

Solution

It is assumed here that the probability of selection for each digit is $\frac{1}{8}$. We can write:

Digit	Factors	Number of Factors	Probability
1	1	1	1/8
2	1, 2	2	1/8
3	1, 3	2	1/8
4	1, 2, 4	3	1/8
5	1, 5	2	1/8
6	1, 2, 3, 6	4	1/8
7	1, 7	2	1/8
8	1, 2, 4, 8	4	1/8

The information in the table above can be summarized as below in the form of a table of probability distribution:

Number of factors r. v. (x_i)	1	2	3	4	Total
$p(x_i)$	1/8	4/8	1/8	2/8	1

The probability of $X = 2$ is 4/8 which has been obtained by adding 1/8 four times. Similarly the probability of 4 is 2/8 which has been obtained by adding 1/8 two times.

The Mean and Standard Deviation of a Discrete Random Variable

The mean (also called the expected value) of a discrete random variable X is the number

$$\mu = E(X) = \Sigma x\, P(x)$$

The mean of a random variable may be interpreted as the average of the values assumed by the random variable in repeated trials of the experiment.

Example:

Find the mean of the discrete random variable X whose probability distribution is

x	−2	1	2	3.5
$P(x)$	0.21	0.34	0.24	0.21

Solution:

The formula in the definition gives

$$= (-2)\cdot 0.21 + (1)\cdot 0.34 + (2)\cdot 0.24 + (3.5)\cdot 0.21 = 1.135$$

Example:

A service organization in a large town organizes a raffle each month. One thousand raffle tickets are sold for \$1 each. Each has an equal chance of winning. First prize is \$300, second prize is \$200, and third prize is \$100. Let X denote the net gain from the purchase of one ticket.

a. Construct the probability distribution of X.

b. Find the probability of winning any money in the purchase of one ticket.

c. Find the expected value of X, and interpret its meaning.

Solution:

a. If a ticket is selected as the first prize winner, the net gain to the purchaser is the \$300 prize less the \$1 that was paid for the ticket, hence $X = 300 - 1 = 299$. There is one such ticket, so $P(299) = 0.001$. Applying the same "income minus outgo" principle to the second and third prize winners and to the 997 losing tickets yields the probability distribution:

x	299	199	99	−1
$P(x)$	0.001	0.001	0.001	0.997

b. Let W denote the event that a ticket is selected to win one of the prizes. Using the table

$$P(W) = P(299) + P(199) + P(99) = 0.001 + 0.001 = 0.003$$

c. Using the formula in the definition of expected value,

$$E(X) = 299 \cdot 0.001 + 199 \cdot 0.001 + 99.0.001 + (-1) \cdot 0.997 = -0.4$$

The negative value means that one loses money on the average. In particular, if someone were to buy tickets repeatedly, then although he would win now and then, on average he would lose 40 cents per ticket purchased.

Example:

A life insurance company will sell a $200,000 one-year term life insurance policy to an individual in a particular risk group for a premium of $195. Find the expected value to the company of a single policy if a person in this risk group has a 99.97% chance of surviving one year.

Solution:

Let X denote the net gain to the company from the sale of one such policy. There are two possibilities: the insured person lives the whole year or the insured person dies before the year is up. Applying the "income minus outgo" principle, in the former case the value of X is 195 – 0; in the latter case it is 195 – 200,000 = – 199, 805.195 – 200,000 = – 199,805. Since the probability in the first case is 0.9997 and in the second case is 1 – 0.9997 = 0.00031 – 0.9997 = 0.0003, the probability distribution for X is:

x	195	−199,805
$P(x)$	0.997	0.0003

Therefore

$$E(X) = \Sigma x\, P(x) = 195 \cdot 0.997 + (-199{,}805) \cdot 0.0003 = 135$$

Occasionally (in fact, 3 times in 10,000) the company loses a large amount of money on a policy, but typically it gains $195, which by our computation of $E(X)E(X)$ works out to a net gain of $135 per policy sold, on average.

The variance, σ^2, *of a discrete random variable X is the number*

$$\sigma^2 = \Sigma(x-\mu)^2\, P(x)$$

which by algebra is equivalent to the formula

$$\sigma^2 = [\Sigma x^2\, P(x)] - \mu^2$$

The standard deviation, *σ, of a discrete random variable X is the square root of its variance, hence is given by the formulas*

$$\sigma = \sqrt{\Sigma(x-\mu)^2 P(x)} = \sqrt{[\Sigma x^2 \, P(x)] - \mu^2}$$

The variance and standard deviation of a discrete random variable X may be interpreted as measures of the variability of the values assumed by the random variable in repeated trials of the experiment. The units on the standard deviation match those of X.

Example:

A discrete random variable X has the following probability distribution

x	-1	0	1	4
$P(x)$	0.2	0.5	a	0.1

A histogram that graphically illustrates the probability distribution is given in below Figure "Probability Distribution of a Discrete Random Variable".

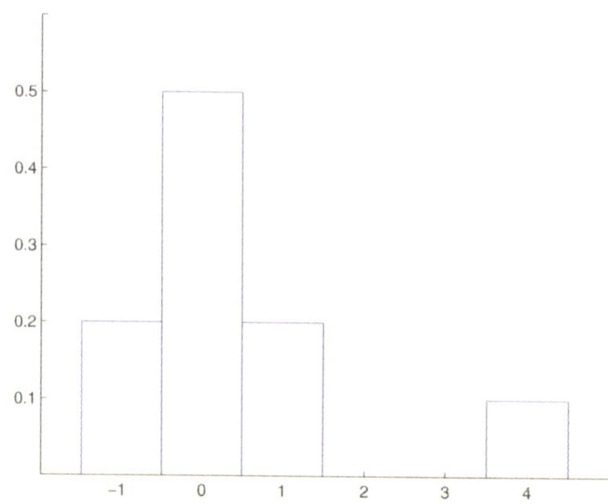

Probability Distribution of a Discrete Random Variable

Compute each of the following quantities.

a. a.

b. $P(0)$.

c. $P(X > 0)$.

d. $P(X \geq 0)$.

e. $P(X \leq -2)$.

f. The mean μ of X.

g. The variance σ^2 of X.

h. The standard deviation σ of X.

Solution:

a. Since all probabilities must add up to 1, a = 1 – (0.2 + 0.5 + 0.1) = 0.2.

b. Directly from the table, P (0) = 0.5.

c. From the table, $P(X > 0) = P(1) + P(4) = 0.2 + 0.1 = 0.3$.

d. From the table, $P(X \geq 0) = P(0) + P(1) + P(4) = 0.5 + 0.2 + 0.1 = 0.8$.

e. Since none of the numbers listed as possible values for X is less than or equal to −2, the event $X \leq -2$ is impossible, so $P(X \leq -2) = 0$.

f. Using the formula in the definition of μ,

$$\mu = \Sigma x\, P(x) = (-1)\cdot 0.2 + 0\cdot 0.5 + 1\cdot 0.2 + 4\cdot 0.1 = 0.4$$

g. Using the formula in the definition of σ^2 and the value of μ that was just computed,

$$\begin{aligned}\sigma^2 &= S(x-\mu)^2\, P(x)\\ &= (-1-0.4)^2\cdot 0.2 + (0-0.4)^2\cdot 0.5 + (1-0.4)^2\cdot 0.2 + (4-0.4)^2\cdot 0.1\\ &= 1.84\end{aligned}$$

h. Using the result of part (g), $\sigma = \sqrt{1.84} = 1.3565$.

Continuous Random Variables

A continuous random variable is a random variable which can take values measured on a continuous scale e.g. weights, strengths, times or lengths.

For any pre-determined value $x, P(X = x) = 0$, since if we measured X accurately enough, we are never going to hit the value x exactly. However the probability of some region of values near x can be non-zero.

Probability Density Function (pdf): *f*(*x*)

$$P(a \leq X \leq b) = \int_a^b f(x')dx'$$

Probability of X in the range a to b.

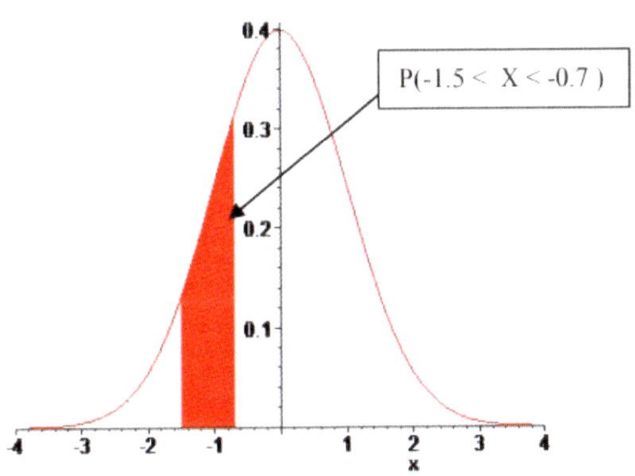

Since X has to have some value

$$\int_{-\infty}^{\infty} f(x)dx = P(-\infty < X < \infty) = 1$$

And since $0 \le P \le 1$, For a pdf $f(x) \ge 0$, for all x.

Cumulative Distribution Function

This is the probability of $X < x$.

$$F(x) \equiv P(X < x) = \int_{-\infty}^{x} f(x')dx'$$

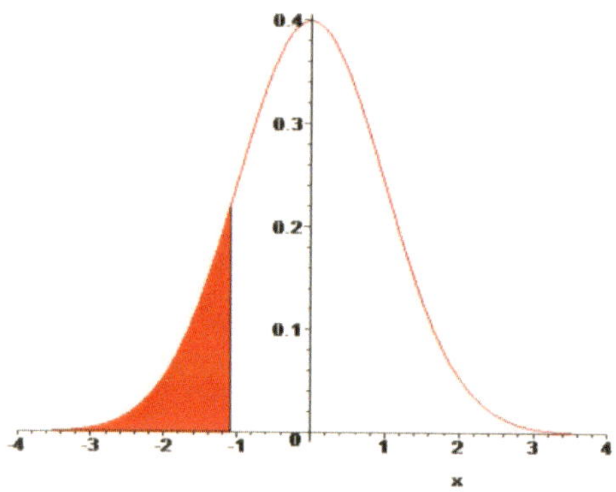

Mean and Variance

Expected value (mean) of $X: \mu = \int_{-\infty}^{\infty} xf(x)dx$

Variance of X: $\sigma^2 = \int_{-\infty}^{\infty} (x-\mu)^2 f(x)dx = \int_{-\infty}^{\infty} x^2 f(x)dx - \mu^2$

Note that the mean and variance may not be well defined for distributions with broad tails. The mode is the value of x where $f(x)$ is maximum (which may not be unique). The median is given by the value of x where

$$\int_{-\infty}^{x} f(x')dx' = \frac{1}{2}.$$

Uniform Distribution

The continuous random variable X has the Uniform distribution between θ_1 and θ_2 with $\theta_1 < \theta_2$ if

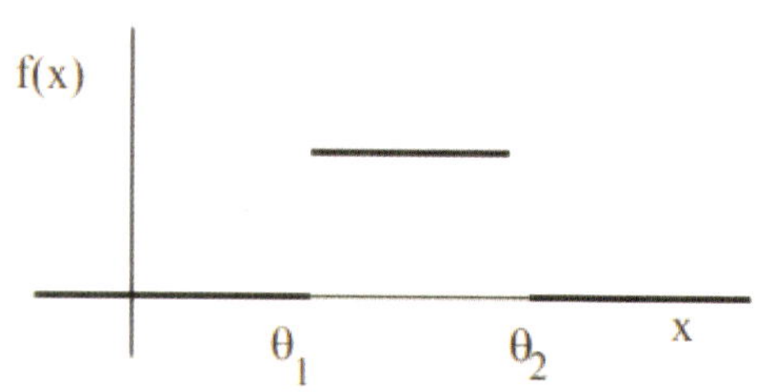

$$f(x)=\begin{cases}\dfrac{1}{\theta_2-\theta_1} & \theta_1 \le x \le \theta_2 \\ 0 & \text{ohterwise}\end{cases}$$

$X \sim U(\theta_1, \theta_2)$, for short.

Roughly speaking, $X \sim U(\theta_1, \theta_2)$, if X can only take values between θ_1 and θ_2 , and any value X of within these values is as likely as any other value.

Mean and variance: for $U(\theta_1, \theta_2)$, $\mu = \dfrac{(\theta_1+\theta_2)}{2}$ and $\sigma^2 = \dfrac{(\theta_2-\theta_1)^2}{12}$

Proof:

Let y be the distance from the mid-point, $y = x-(\theta_2-\theta_1)/2$, and the width be $w=\theta_2-\theta_1$. Then since means add

$$\mu = \langle x\rangle = \frac{\theta_2+\theta_1}{2} + \langle y\rangle = \frac{\theta_2+\theta_1}{2} + \int_{-\frac{w}{2}}^{\frac{w}{2}} y\frac{1}{w}dy = \frac{\theta_1+\theta_2}{2}.$$

Unsurprisingly the mean is the midpoint.

$$\sigma^2 = \int_{-\infty}^{\infty}(x-\mu)^2 f(x)dx = \int_{-\frac{w}{2}}^{\frac{w}{2}} y^2\frac{1}{w}dy = \frac{1}{w}\left[\frac{y^3}{3}\right]_{-\frac{w}{2}}^{\frac{w}{2}} = \frac{1}{3w}\left(\frac{w^3}{8}+\frac{w^3}{8}\right) = \frac{w^2}{12}$$

$$= \frac{(\theta_2-\theta_1)^2}{12}$$

Occurrence of the Uniform Distribution

1) Waiting times from random arrival time until a regular event

2) Engineering tolerances: e.g. if a diameter is quoted "±0.1mm", it sometimes assumed (probably incorrectly) that the error has a U(-0.1, 0.1) distribution.

3) Simulation: programming languages often have a standard routine for simulating the U(0, 1) distribution. This can be used to simulate other probability distributions.

Example: Disk wait times

In a hard disk drive, the disk rotates at 7200rpm. The wait time is defined as the time between the read/write head moving into position and the beginning of the required information appearing under the head.

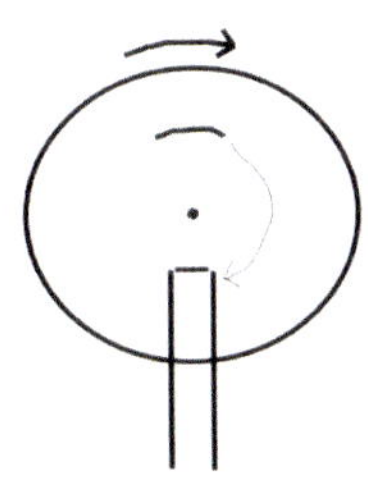

(a) Find the distribution of the wait time.

(b) Find the mean and standard deviation of the wait time.

(c) Booting a computer requires that 2000 pieces of information are read from random positions. What is the total expected contribution of the wait time to the boot time, and rms deviation?

Solution:

Rotation time = 8.33ms. Wait time can be anything between 0 and 8.33ms and each time in this range is as likely as any other time. Therefore, distribution of the wait time is U(0, 8.33ms) (i. . θ_1 = 0 and θ_2 = 8.33ms)

$$\mu = \frac{0+8.33}{2}\text{ ms} = 4.2\text{ ms}$$

$$\sigma^2 = \frac{(8.33-0)^2}{12}\text{ ms} = 5.8\text{ ms}^2; \Rightarrow \sigma = 2.4\text{ ms}$$

For 2000 reads the mean time is 2000 x 4.2ms = 8.3s. The variance is 2000 x 5.7ms² = 0.012s², so σ =0.11s.

Exponential Distribution

The continuous random variable y has the Exponential distribution, parameter υ if:

$$f(y) = \begin{cases} \upsilon e^{-\upsilon y}, & y > 0 \\ 0, & y < 0 \end{cases}$$

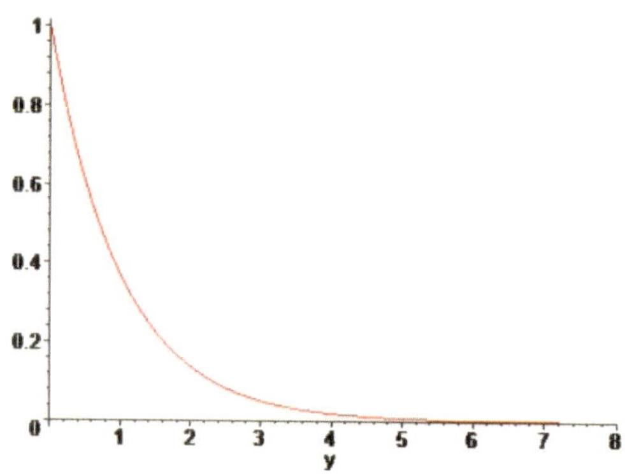

Relation to Poisson distribution: If a Poisson process has constant rate υ, the mean after a time is $\lambda = \upsilon t$. The probability of no-occurrences in this time is

$$P(k=0) = e^{-\lambda} = e^{-\upsilon t}.$$

If $f(t)$ is the pdf for the first occurrence, then the probability of no occurrences is also given by

$P(k=0) = 1 - P$ (first occurrence had happened by t) $= 1 - \int_0^t f(t)dt$

So equating the two ways of calculating the probability we have

$$e^{-vt} = 1 - \int_0^t f(t)dt$$

Now we can differentiate with respect to t giving

$$-ve^{-vt} = -\frac{d}{dt}\int_0^t f(t)dt = -f(t)$$

Hence $f(t) = ve^{-vt}$:. the time until the first occurrence (and between subsequent occurrences) has the Exponential distribution, parameter v.

Occurrence

1) Time until the failure of a part.

2) Times between randomly happening events.

Mean and Variance

$$\mu = \int_{-\infty}^{\infty} y\, f(y)dy = \int_0^{\infty} yve^{-vy}dy = [-ye^{-vy}]_0^{\infty} + \int_0^{\infty} e^{-vy}dy = \left[-\frac{e^{-vy}}{v}\right]_0^{\infty} = \frac{1}{v}$$

$$\sigma^2 = \int_{-\infty}^{\infty} y^2\, f(y)dy - \mu^2 = \int_{-\infty}^{\infty} y^2 ve^{-vy}dy - \frac{1}{v^2}$$

$$= [-y^2e^{-vy}]_0^{\infty} + 2\int_0^{\infty} y\, e^{-vy}dy - \frac{1}{v^2} = 0 + 2\frac{\mu}{v} - \frac{1}{v^2} = \frac{1}{v^2}$$

Example: Reliability

The time till failure of an electronic component has an Exponential distribution and it is known that 10% of components have failed by 1000 hours.

a. What is the probability that a component is still working after 5000 hours?

b. Find the mean and standard deviation of the time till failure.

Solution

(a) Let Y = time till failure in hours; $f(y) = ve^{-vy}$

$$P(Y \le 1000) = \int_0^{1000} ve^{-vy} = [-e^{-vy}]_0^{1000} = 1 - e^{-1000v} = 0.1$$

$$\Rightarrow e^{-1000v} = 0.9 \Rightarrow -1000v = \ln 0.9 = -0.10536$$

$$\Rightarrow V \approx \sim 1.05 \times 10^{-4}$$

$$P(Y > 5000) = \int_{5000}^{\infty} ve^{-vy}dy = [-e^{-vy}]_{5000}^{\infty} = e^{-5000v} \approx 0.59$$

(b) Mean= $1/\upsilon$ = 9491 hours.

Standard deviation $= \sqrt{\text{Variance}} = \sqrt{\frac{1}{\upsilon^2}} = 1/\upsilon = 9491$ hours.

Normal Distribution

The continuous random variable X has the Normal distribution if the pdf is:

$$f(x) = \frac{1}{\sqrt{2\pi\sigma^2}} e^{\frac{-(x-\mu)^2}{2\sigma^2}} \quad (-\infty < x < \infty)$$

The parameter μ is the mean and the variance is σ^2. The distribution is also sometimes called a Gaussian distribution.

The pdf is symmetric about μ. X lies between μ - 1.96σ and μ + 1.96σ with probability 0.95 i.e. X lies within 2 standard deviations of the mean approximately 95% of the time.

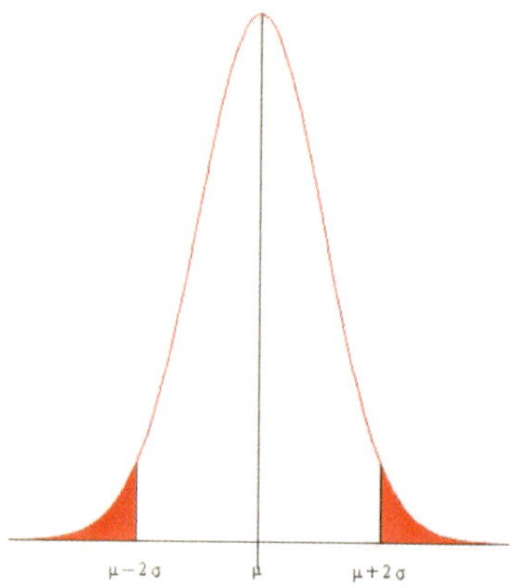

Normalization

$f(x)$ cannot be integrated analytically for general ranges, but the full range can be integrated as follows. Define

$$I = \int_{-\infty}^{\infty} dx\, e^{\frac{-(x-\mu)^2}{2\sigma^2}} = \int_{-\infty}^{\infty} dx\, e^{\frac{-x^2}{2\sigma^2}}$$

Then switching to polar co-ordinates we have

$$I^2 = \int_{-\infty}^{\infty} dx\, e^{\frac{-x^2}{2\sigma^2}} \int_{-\infty}^{\infty} dy\, e^{\frac{-y^2}{2\sigma^2}} = \int_{-\infty}^{\infty}\int_{-\infty}^{\infty} dxdy\, e^{\frac{x^2+y^2}{2\sigma^2}} = \int_0^{\infty}\int_0^{2\pi} rdrd\theta\, e^{\frac{r^2}{2\sigma^2}} = 2\pi \int_0^{\infty} rdre^{\frac{r^2}{2\sigma^2}}$$

$$= 2\pi \left[-\sigma^2 e^{\frac{r^2}{2\sigma^2}} \right]_0^{\infty} = 2\pi\sigma^2$$

Hence $I = \sqrt{2\pi\sigma^2}$ and the normal distribution integrates to one.

Mean and Variance

The mean is μ because the distribution is symmetric about μ (or you can check explicitly by integrating by parts). The variance can be also be checked by integrating by parts:

$$\int_{-\infty}^{\infty}(x-\mu)^2\frac{e^{\frac{-(x-\mu)^2}{2\sigma^2}}}{\sqrt{2\pi\sigma^2}}dx=\int_{-\infty}^{\infty}y^2\frac{e^{\frac{-y^2}{2\sigma^2}}}{\sqrt{2\pi\sigma^2}}dy=\int_{-\infty}^{\infty}y\times y\frac{e^{\frac{-y^2}{2\sigma^2}}}{\sqrt{2\pi\sigma^2}}dy$$

$$=\left[-y\sigma^2\frac{e^{\frac{-y^2}{2\sigma^2}}}{\sqrt{2\pi\sigma^2}}\right]_{-\infty}^{\infty}+\int_{-\infty}^{\infty}\sigma^2\frac{e^{\frac{-y^2}{2\sigma^2}}}{\sqrt{2\pi\sigma^2}}dy=\sigma^2\int_{-\infty}^{\infty}f(y)dy=\sigma^2$$

Occurrence of the Normal distribution

- Quite a few variables, e.g. human height, measurement errors, detector noise. (Bell-shaped histogram).
- Sample means and totals, Central Limit Theorem.
- Approximation to several other distributions.

Change of Variable

The probability for X in a range dx around x is for a distribution $f(x)$ is given by $f(x)\,dx$. The probability should be the same if it is written in terms of another variable $y = y(x)$. Hence

$$f(x)\,dx=f(y)dy\Rightarrow f(y)=f(x)\frac{dx}{dy}.$$

Standard Normal Distribution

There is no simple formula for $\int_a^b f(x)dx$, so numerical integration (or tables) must be used. The following result means that it is only necessary to have tables for one value of μ and σ^2.

If $X \sim N(\mu, \sigma^2)$, then $Z = \dfrac{X-\mu}{\sigma} \sim N(0,1)$

This follows when changing variables since $\dfrac{dz}{dx} = \dfrac{1}{\sigma}$ hence

$$f(z) = f(x)\frac{dz}{dx} = \frac{1}{\sqrt{2\pi\sigma^2}} e^{\frac{-(x-\mu)^2}{2\sigma^2}} \times \sigma = \frac{1}{\sqrt{q\pi}} e^{\frac{-z^2}{2}}.$$

Z is the standardised value of X; N(0, 1) is the standard Normal distribution. The Normal tables give values of $Q = P(Z \le z)$, also called $\Phi(z)$, for z between 0 and 3.59.

$$Q(z) = \Phi(z) = P(Z \le z) = \int_{-\infty}^{z} f(x)dx$$

Example: Using standard Normal tables

If $Z \sim N(0,1)$:

(a) $P(Z \le 1.0)\,\Phi(1.0) = 0.8413$

(b) $P(Z \le = -1.0) = P(Z \ge 1.0)$ (by symmetry) $= 1 - P(Z < 1.0)$

= 1 - 0.8413 = 0.1587

(c) $P(Z > -0.5) = P(Z \le 0.5) = \Phi(0.5)$

= 0.6915.

(d) $P(0.5 < Z < 1.5) = P(Z < 1.5) - P(Z < 0.5)$

$= \Phi(1.5) - \Phi(0.5)$

= 0.9332 - 0.6915

= 0.2417.

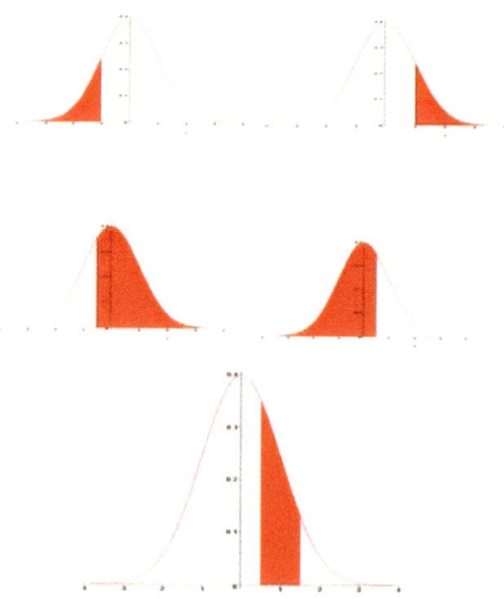

(e) $P(Z < 1.356)$ - Using interpolation:

$$P(Z < 1.356) = P(Z < 1.35) + \frac{1.356 - 1.35}{1.36 - 1.35}(P(Z < 1.36) - P(Z < 1.35)) = 0.9125$$

(f) $0.8 = P(Z \le c) = \Phi(c)$

Using tables "in reverse", $c \approx 0.842$.

(g) Finding a range of values within which *Z* lies with probability 0.95:

The answer is not unique; but suppose we want an interval which is symmetric about zero i.e. between -*d* and *d*.

Tail area = 0.025

$$\therefore P(Z \le d) = \Phi(d) = 0.975$$

Using the tables "in reverse", d = 1.96.

$\therefore$ range is -1.96 to 1.96.

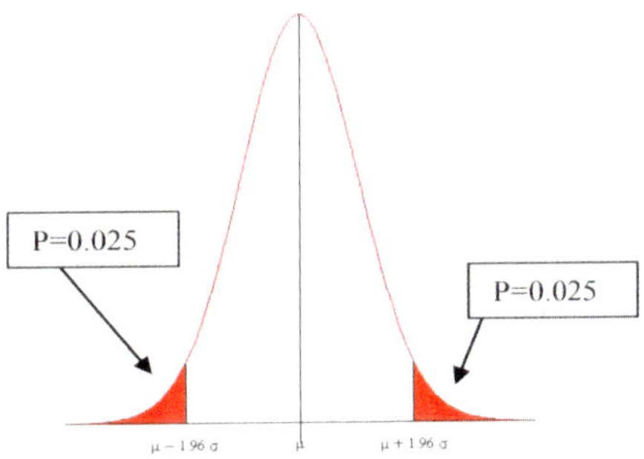

Example: Manufacturing variability

The outside diameter, *X* mm, of a copper pipe is $N(15.00, 0.02^2)$ and the fittings for joining the pipe have inside diameter *Y* mm, where $Y \sim N(15.07, 0.022^2)$.

(i) Find the probability that *X* exceeds 14.99 mm.

(ii) Within what range will *X* lie with probability 0.95?

(iii) Find the probability that a randomly chosen pipe fits into a randomly chosen fitting (i.e. $X < Y$).

Solution

(i) $X \sim N(15.0, 0.02^2)$

$$P(X > 14.99) = P\left(Z > \frac{14.99 - 15.0}{0.02}\right) = P(Z > -0.5) = P(Z < 0.5) \approx 0.6915$$

(ii) From previous example $Z = \dfrac{X - 15.0}{0.02}$ lies in (-1.96, 1.96) with probability 0.95. i.e.

$$P(-1.96 < \frac{X - 15.0}{0.02} < 1.96) = 0.95$$

$$\Rightarrow P(15 - 0.02 \times 1.96 < X < 15 + 0.02 \times 1.96) = 0.95$$

$$\Rightarrow P(14.96 < X < 15.04) = 0.95$$

i.e. the required range is 14.96mm to 15.04mm.

(iii) For $X < Y$ we want $P(Y - X > 0)$. To answer this we need to know the distribution of $Y - X$.

Distribution of the Sum of Normal Variates

Remember than means and variances of independent random variables just add. So if $X_1, X_2, \ldots, X_n$ are independent and each have a normal distribution $X_i \sim N(\mu_i, \sigma_i^2)$, we can easily calculate the mean and variance of the sum. A special property of the Normal distribution is that the distribution of the sum of Normal variates is also a Normal distribution. So if $c_1, c_2, \ldots, c_n$ are constants then:

$$c_1 X_1 + c_2 X_2 + \ldots + c_n X_n \sim N(c_1\mu_1 + \ldots + c_n\mu_n, c_1^2\sigma_1^2 + c_2^2\sigma_2^2 + \ldots + c_n^2\sigma_n^2)$$

Proof that the distribution of the sum is Normal is beyond scope. Useful special cases for two variables are

$$X_1 + X_2 \sim N(\mu_1 + \mu_2, \sigma_1^2 + \sigma_2^2)$$
$$X_1 - X_2 \sim N(\mu_1 + \mu_2, \sigma_1^2 + \sigma_2^2)$$

If all the X's have the same distribution i.e. $\mu_1 = \mu_2 = \ldots = \mu_n = \mu$, say and $\sigma_1^2 = \sigma_2^2 = \ldots = \sigma_n^2 = \sigma^2$, say, then:

(i) All $c_i = 1$: $X_1 + X_2 + \ldots + X_n \sim N(n\mu, n\sigma^2)$

(ii) All $c_i = 1/n$: $\bar{X} = \dfrac{X_1 + X_2 + \ldots + X_n}{n} \sim N(\mu, \sigma^2/n)$

The last result tells you that if you average n identical independent noisy measurements, the error decreases by $1/\sqrt{n}$. (variance goes down as $1/n$.

Example: Manufacturing variability

Find the probability that a randomly chosen pipe fits into a randomly chosen fitting (i.e. $X < Y$).

Using the above results

$$Y - X \sim N(\mu_Y - \mu_X, \sigma_Y^2 + \sigma_X^2) = N(15.07 - 15, 0.02^2 + 0.022^2)$$

$$= N(0.07, 0.000884)$$

Hence

$$P(Y - X > 0) = P\left(Z > \frac{0 - 0.07}{\sqrt{0.0.000884}}\right) = P(Z > -2.35) = P(Z < 2.35) \approx 0.991$$

Example: Detector noise

A detector on a satellite can measure $T+g$, the temperature T of a source with a random noise g, where $g \sim N(0, 1K^2)$. How many detectors with independent noise would you need to measure T to an rms error of 0.1K?

Solution: We can estimate the temperature from n detectors by calculating the mean from each. The variance of the mean will be $1K^2 /n$ where n is the number of detectors. An rms error of 0.1K corresponds to a variance of 0.01 K^2, hence we need n=100 detectors.

Expected Value

In probability theory, the expected value of a random variable, intuitively, is the long-run average value of repetitions of the experiment it represents. For example, the expected value in rolling a six-sided die is 3.5, because the average of all the numbers that come up in an extremely large number of rolls is close to 3.5. Less roughly, the law of large numbers states that the arithmetic mean of the values almost surely converges to the expected value as the number of repetitions approaches infinity. The expected value is also known as the expectation, mathematical expectation, EV, average, mean value, mean, or first moment.

More practically, the expected value of a discrete random variable is the probability-weighted average of all possible values. In other words, each possible value the random variable can assume is multiplied by its probability of occurring, and the resulting products are summed to produce the expected value. The same principle applies to an absolutely continuous random variable, except that an integral of the variable with respect to its probability density replaces the sum. The formal definition subsumes both of these and also works for distributions which are neither discrete nor absolutely continuous; the expected value of a random variable is the integral of the random variable with respect to its probability measure.

The expected value does not exist for random variables having some distributions with large "tails", such as the Cauchy distribution. For random variables such as these, the long-tails of the distribution prevent the sum or integral from converging.

The expected value is a key aspect of how one characterizes a probability distribution; it is one type of location parameter. By contrast, the variance is a measure of dispersion of the possible values of the random variable around the expected value. The variance itself is defined in terms of two expectations: it is the expected value of the squared deviation of the variable's value from the variable's expected value.

The expected value plays important roles in a variety of contexts. In regression analysis, one desires a formula in terms of observed data that will give a "good" estimate of the parameter giving the effect of some explanatory variable upon a dependent variable. The formula will give different estimates using different samples of data, so the estimate it gives is itself a random variable. A formula is typically considered good in this context if it is an unbiased estimator— that is if the expected value of the estimate (the average value it would give over an arbitrarily large number of separate samples) can be shown to equal the true value of the desired parameter.

In decision theory, and in particular in choice under uncertainty, an agent is described as making an optimal choice in the context of incomplete information. For risk neutral agents, the choice involves using the expected values of uncertain quantities, while for risk averse agents it involves maximizing the expected value of some objective function such as a von Neumann–Morgenstern utility function. One example of using expected value in reaching optimal decisions is the Gordon–Loeb model of information security investment. According to the model, one can conclude that the amount a firm spends to protect information should generally be only a small fraction of the expected loss (i.e., the expected value of the loss resulting from a cyber or information security breach).

Finite Case

Let X be a random variable with a finite number of finite outcomes x_1, x_2, ..., x_k occurring with probabilities p_1, p_2, ..., p_k, respectively. The expectation of X is defined as

$$\mathrm{E}[X] = x_1 p_1 + x_2 p_2 + \cdots + x_k p_k.$$

Since all probabilities p_i add up to 1 ($p_1 + p_2 + \ldots + p_k = 1$), the expected value is the weighted average, with p_i's being the weights.

If all outcomes x_i are equiprobable (that is, $p_1 = p_2 = \ldots = p_k$), then the weighted average turns into the simple average. This is intuitive: the expected value of a random variable is the average of all values it can take; thus the expected value is what one expects to happen *on average*. If the outcomes x_i are not equiprobable, then the simple average must be replaced with the weighted average, which takes into account the fact that some outcomes are more likely than the others. The intuition however remains the same: the expected value of X is what one expects to happen *on average*.

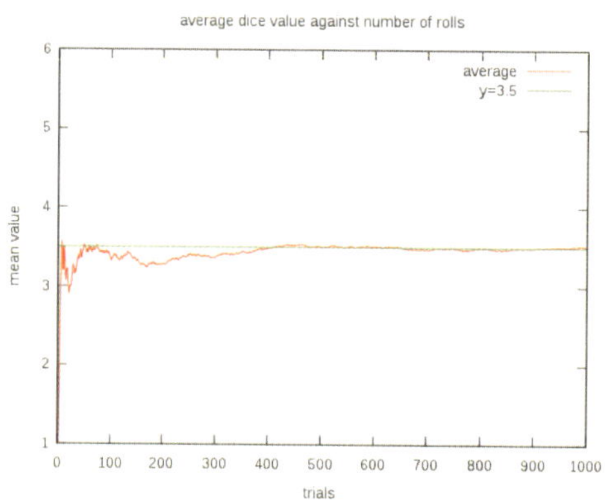

The convergence of sequence averages of rolls of a die to the expected value of 3.5 as the number of rolls (trials) grows.

Examples

- Let X represent the outcome of a roll of a fair six-sided die. More specifically, will be the number of pips showing on the top face of the die after the toss. The possible values for X are 1, 2, 3, 4, 5, and 6, all equally likely (each having the probability of 1/6). The expectation of X is:

$$\mathrm{E}[X]=1\cdot\frac{1}{6}+2\cdot\frac{1}{6}+3\cdot\frac{1}{6}+4\cdot\frac{1}{6}+5\cdot\frac{1}{6}+6\cdot\frac{1}{6}=3.5.$$

If one rolls the die n times and computes the average (arithmetic mean) of the results, then as n grows, the average will almost surely converge to the expected value, a fact known as the strong law of large numbers. One example sequence of ten rolls of the die is 2, 3, 1, 2, 5, 6, 2, 2, 2, 6, which has the average of 3.1, with the distance of 0.4 from the expected value of 3.5. The convergence is relatively slow: the probability that the average falls within the range 3.5 ± 0.1 is 21.6% for ten rolls, 46.1% for a hundred rolls and 93.7% for a thousand rolls. Figure for an illustration of the averages of longer sequences of rolls of the die and how they converge to the expected value of 3.5. More generally, the rate of convergence can be roughly quantified by e.g. Chebyshev's inequality and the Berry–Esseen theorem.

- The roulette game consists of a small ball and a wheel with 38 numbered pockets around the edge. As the wheel is spun, the ball bounces around randomly until it settles down in one of the pockets. Suppose random variable X represents the (monetary) outcome of a \$1 bet on a single number ("straight up" bet). If the bet wins (which happens with probability 1/38 in American roulette), the payoff is \$35; otherwise the player loses the bet. The expected profit from such a bet will be

$$\mathrm{E}[\text{gain from \$1 bet}]=-\$1\cdot\frac{37}{38}+\$35\cdot\frac{1}{38}=-\$0.0526.$$

That is, the bet of \$1 stands to lose \$0.0526, so its expected value is -\$0.0526.

Countably Infinite Case

Let X be a random variable with a countable set of finite outcomes x_1, x_2, ..., occurring with probabilities p_1, p_2, ..., respectively, such that the infinite sum $\sum_{i=1}^{\infty}|x_i|p_i$ converges. The expected value of X is defined as the series

$$\mathrm{E}[X]=\sum_{i=1}^{\infty}x_i p_i.$$

Remark 1. Observe that $\left|\mathrm{E}[X]\right|\leq\sum_{i=1}^{\infty}|x_i|p_i<\infty.$.

Remark 2. Due to absolute convergence, the expected value does not depend on the order in which the outcomes are presented. By contrast, a conditionally convergent series can be made to converge or diverge arbitrarily, via the Riemann rearrangement theorem.

Example

- Suppose $x_i=i$ and $p_i=\frac{k}{i2^i}$, for $i=1,2,3,\ldots$, where $k=\frac{1}{\ln 2}$ (with $\ln$ being the natural logarithm) is the scale factor such that the probabilities sum to 1. Then

$$\mathrm{E}[X]=1\left(\frac{k}{2}\right)+2\left(\frac{k}{8}\right)+3\left(\frac{k}{24}\right)+\ldots=\frac{k}{2}+\frac{k}{4}+\frac{k}{8}+\ldots=k.$$

Since this series converges absolutely, the expected value of X is k.

- For an example that is not absolutely convergent, suppose random variable X takes values 1, −2, 3, −4,..., with respective probabilities $\frac{c}{1^2},\frac{c}{2^2},\frac{c}{3^2},\frac{c}{4^2}$, ..., where $c=\frac{6}{\pi^2}$ is a normalizing constant that ensures the probabilities sum up to one. Then the infinite sum

  $$\sum_{i=1}^{\infty} x_i\,p_i = c\left(1-\frac{1}{2}+\frac{1}{3}-\frac{1}{4}+\cdots\right)$$

  converges and its sum is equal to $\frac{6\ln 2}{\pi^2}\approx 0.421383$. However it would be incorrect to claim that the expected value of X is equal to this number—in fact $\mathrm{E}[X]$ does not exist (finite or infinite), as this series does not converge absolutely (see Alternating harmonic series).

- An example that diverges arises in the context of the St. Petersburg paradox. Let $x_i=2^i$ and $p_i=\frac{1}{2^i}$ for $i=1,2,3,\ldots$. The expected value calculation gives

  $$\sum_{i=1}^{\infty} x_i\,p_i = 2\cdot\frac{1}{2}+4\cdot\frac{1}{4}+8\cdot\frac{1}{8}+16\cdot\frac{1}{16}+\cdots=1+1+1+1+\cdots.$$

  Since this does not converge but instead keeps growing, the expected value is infinite.

Absolutely Continuous Case

If X is a random variable whose cumulative distribution function admits a density $f(x)$, then the expected value is defined as the following Lebesgue integral:

$$\mathrm{E}[X]=\int_{\mathbb{R}} xf(x)dx.$$

Remark: From computational perspective, the integral in the definition of $\mathrm{E}[X]$ may often be treated as an improper Riemann integral $\int_{-\infty}^{+\infty} xf(x)dx$. Specifically, if the function $xf(x)$ is Riemann-integrable on every finite interval $[a,b]$, and

$$\min\left((-1)\cdot(\mathrm{R})\int_{-\infty}^{0} xf(x)dx,(\mathrm{R})\int_{0}^{+\infty} xf(x)dx\right)<\infty,$$

then the values (whether finite or infinite) of both integrals agree.

General Case

In general, if X is a random variable defined on a probability space $(\Omega,\Sigma,\mathrm{P})$, then the expected value of X, denoted by $\mathrm{E}[X]$, $\langle X\rangle$, or $\bar{X}$,, is defined as the Lebesgue integral

$$\mathrm{E}[X] = \int_{\Omega} X(\omega) d\,\mathrm{P}(\omega).$$

Remark 1: If $X_+(\omega) = \max(X(\omega), 0)$ and $X_-(\omega) = -\min(X(\omega), 0)$, then $X = X_+ - X_-$. The functions X_+ and X_- can be shown to be measurable (hence, random variables), and, by definition of Lebesgue integral,

$$\begin{aligned}\mathrm{E}[X] &= \int_{\Omega} X(\omega) d\,\mathrm{P}(\omega) \\ &= \int_{\Omega} X_+(\omega) d\,\mathrm{P}(\omega) - \int_{\Omega} X_-(\omega) d\,\mathrm{P}(\omega) \\ &= \mathrm{E}[X_+] - \mathrm{E}[X_-],\end{aligned}$$

where $\mathrm{E}[X_+]$ and $\mathrm{E}[X_-]$ are non-negative and possibly infinite.

The following scenarios are possible:

- $\mathrm{E}[X]$ is finite, i.e. $\max(\mathrm{E}[X_+], \mathrm{E}[X_-]) < \infty$;
- $\mathrm{E}[X]$ is infinite, i.e. $\max(\mathrm{E}[X_+], \mathrm{E}[X_-]) = \infty$ and $\min(\mathrm{E}[X_+], \mathrm{E}[X_-]) < \infty$;
- $\mathrm{E}[X]$ is neither finite nor infinite, i.e. $\mathrm{E}[X_+] = \mathrm{E}[X_-] = \infty$.

Remark 2: If $F_X(x) = \mathrm{P}(X \le x)$ is the cumulative distribution function of X, then

$$[X] = \int_{-\infty}^{+\infty} x\,dF_X(x),$$

where the integral is interpreted in the sense of Lebesgue–Stieltjes.

Remark 3: An example of a distribution for which there is no expected value is Cauchy distribution.

Remark 4: For multidimensional random variables, their expected value is defined per component, i.e.

$$\mathrm{E}[(X_1, \ldots, X_n)] = (\mathrm{E}[X_1], \ldots, \mathrm{E}[X_n])$$

and, for a random matrix X with elements $_{ij}$

$$(\mathrm{E}[X])_{ij} = \mathrm{E}[X_{ij}].$$

Basic Properties

The properties below replicate or follow immediately from those of Lebesgue integral.

$$\mathrm{E}(\mathbf{1}_A) = \mathrm{P}(A)$$

If A is an event, then $\mathrm{E}(1_A) = \mathrm{P}(A)$, where 1_A is the indicator function of the set A.

Proof: By definition of Lebesgue integral of the simple function $1_A = 1_A(\omega)$,

$$\mathrm{E}(1_A) = 1 \cdot \mathrm{P}(A) + 0 \cdot \mathrm{P}(\Omega \setminus A) = \mathrm{P}(A).$$

If $X = Y$ (a.s.) then $\mathrm{E}(X) = \mathrm{E}(Y)$

The statement follows from the definition of Lebesgue integral if we notice that $X_+ = Y_+$ (a.s.), $X_- = Y_-$ (a.s.), and that changing a simple random variable on a set of probability zero does not alter the expected value.

Expected Value of a Constant

If X is a random variable, and $X = c$ (a.s.), where $c \in [-\infty, +\infty]$, then $\mathrm{E}[X] = c$. In particular, for an arbitrary random variable X, $\mathrm{E}[\mathrm{E}[X]] = \mathrm{E}[X]$..

Proof:

Let C be a constant random variable, i.e. $C \equiv c$. It follows from the definition of Lebesgue integral that $\mathrm{E}[C] = c$. It also follows that $X = C$ (a.s.). By the previous property,

$$\mathrm{E}[X] = \mathrm{E}[C] = c.$$

Linearity

The expected value operator (or expectation operator) $\mathrm{E}[\cdot]$ is linear in the sense that

$$\mathrm{E}[X+Y] = \mathrm{E}[X] + \mathrm{E}[Y],$$
$$\mathrm{E}[aX] = a\mathrm{E}[X],$$

where X and Y are (arbitrary) random variables, and a is a scalar.

More rigorously, let X and Y be random variables whose expected values are defined (different from $\infty - \infty$).

- If $\mathrm{E}[X] + \mathrm{E}[Y]$ is also defined (i.e. differs from $\infty - \infty$), then

  $\mathrm{E}[X+Y] = \mathrm{E}[X] + \mathrm{E}[Y].$

- Let $\mathrm{E}[X]$ be finite, and $a \in \mathbb{R}$ be a finite scalar. Then $\mathrm{E}[aX] = a\mathrm{E}[X].$

1. We prove additivity in several steps.

1a. If X and Y are simple and non-negative, taking intersections where necessary, one can re-write X and Y in the form

$$X = \sum_{i=1}^{n} x_i \cdot \mathbf{1}_{A_i}$$

and

$$= \sum_{i=1}^{n} y_i \cdot \mathbf{1}_{A_i},$$

for some measurable pairwise-disjoint sets $\{A_i\}_{i=1}^{n}$ partitioning Ω,, and $1_{A_i} = 1_{A_i}(\omega)$ being the indicator function of the set A_i. By a straightforward check, the additivity follows.

1b. Assuming that X and Y are arbitrary and non-negative, recall that every non-negative measurable function is a pointwise limit of a pointwise non-decreasing sequence of simple non-negative ones. Let $\{X_n\}$ and $\{Y_n\}$ be such sequences converging to X and Y respectively. We see that $\{X_n + Y_n\}$ pointwise non-decreases, and $X_n + Y_n \to X + Y$ pointwise. By monotone convergence theorem and case 1a,

$$\begin{aligned}
\mathrm{E}[X+Y] &= \mathrm{E}[\lim_n (X_n + Y_n)] \\
&= \lim_n \mathrm{E}[X_n + Y_n] \\
&= \lim_n (\mathrm{E}[X_n] + \mathrm{E}[Y_n]) \\
&= \lim_n \mathrm{E}[X_n] + \lim_n \mathrm{E}[Y_n] \\
&= \mathrm{E}[\lim_n X_n] + \mathrm{E}[\lim_n Y_n] \\
&= \mathrm{E}[X] + \mathrm{E}[Y].
\end{aligned}$$

(The reader can verify that using the monotone convergence theorem this way does *not* lead to circular logic).

1c. In the general case, if $Z = X + Y$, then $Z_+ + X_- + Y_- = Z_- + X_+ + Y_+$, and

$$\mathrm{E}[Z_+ + X_- + Y_-] = \mathrm{E}[Z_- + X_+ + Y_+].$$

Splitting up,

$$\mathrm{E}[Z_+] + \mathrm{E}[X_-] + \mathrm{E}[Y_-] = \mathrm{E}[Z_-] + \mathrm{E}[X_+] + \mathrm{E}[Y_+],$$

which is equivalent to,

$$\mathrm{E}[Z_+] - \mathrm{E}[Z_-] = \mathrm{E}[X_+] + \mathrm{E}[Y_+] - \mathrm{E}[X_-] - \mathrm{E}[Y_-],$$

and finally,

$$\mathrm{E}[Z] = \mathrm{E}[X] + \mathrm{E}[Y].$$

2. To prove homogeneity, we first assume that the scalar $\mathrm{E}[\quad]$ above is non-negative. The finiteness of X implies that $a \cdot X$ is finite (a.s.). Therefore, $a \cdot X$ is also finite (a.s.), which guarantees that $\mathrm{E}[aX]$ is finite. The equality, thus, is a straightforward check based on the definition of Lebesgue integral.

If $a < 0$, then we first prove that $\mathrm{E}[-X] = -\mathrm{E}[X]$ by observing that $(-X)_+ = X_-$ and vice versa. $\mathrm{E}(X)$ is finite if and only if is

The following statements regarding a random variable X are equivalent:

- $\mathrm{E}(X)$ is finite.
- Both $\mathrm{E}(X_+)$ and $\mathrm{E}(X_-)$ are finite.
- $\mathrm{E}|X|$ is finite.

Sketch of proof: Indeed, $|X| = X_+ + X_-$. By linearity, $\mathrm{E}|X| = \mathrm{E}(X_+) + \mathrm{E}(X_-)$. The above equivalency relies on the definition of Lebesgue integral and measurability of X.

Remark: For the reasons above, the expressions "X is integrable" and "the expected value of X is finite" are used interchangeably when speaking of a random variable.

If $X \geq 0$ (a.s.) then $\mathrm{E}(X) \geq 0$

Proof:

Denote

$$\mathrm{SF} = \{s : \Omega \to \mathbb{R} \mid s \text{ is a simple random variable, and } 0 \leq s \leq X_+\}.$$

If $s \in \mathrm{SF}$, then $\mathrm{E}[s] \in [0, +\infty)$, and hence, by definition of Lebesgue integral,

$$\mathrm{E}[X_+] = \sup_{s \in \mathrm{SF}} \mathrm{E}[s] \geq 0.$$

On the other hand, $X_- = 0$ (a.s.), so, through a similar argument, $\mathrm{E}[X_-] = 0$, and therefore $\mathrm{E}[X] = \mathrm{E}[X_+] - \mathrm{E}[X_-] = \mathrm{E}[X_+] \geq 0$.

Monotonicity

If $X \leq Y$ (a.s.), and both $\mathrm{E}[X]$ and $\mathrm{E}[Y]$ exist, then $\mathrm{E}[X] \leq \mathrm{E}[Y]$..

Remark: $\mathrm{E}[X]$ and $\mathrm{E}[Y]$ *exist* in the sense that $\min(\mathrm{E}[X_+], \mathrm{E}[X_-]) < \infty$ and $\min(\mathrm{E}[Y_+], \mathrm{E}[Y_-]) < \infty$.

Proof: follows from the linearity and the previous property if we set $Z = Y - X$ and notice that $Z \geq 0$ (a.s.).

If $|X| \leq Y$ (a.s.) and $\mathrm{E}(Y)$ is finite then so is $\mathrm{E}(X)$

Let X and Y be random variables such that $|X| \leq Y$ (a.s.) and $\mathrm{E}(Y) < \infty$. Then $\mathrm{E}(X) \neq \pm\infty$.

Proof: Due to non-negativity of $|X|$, $\mathrm{E}|X|$ exists, finite or infinite. By monotonicity, $\mathrm{E}|X| \leq \mathrm{E}(Y) < \infty$, so $\mathrm{E}|X|$ is finite which, as we saw earlier, is equivalent to $\mathrm{E}(X)$ being finite.

If $\mathrm{E}|X^\beta| < \infty$ and $0 < \alpha < \beta$ then $\mathrm{E}|X^\alpha| < \infty$

The proposition below will be used to prove the extremal property of $\mathrm{E}[X]$ later on.

Proposition. If X is a random variable, then so is X^α, for every $\alpha > 0$. If, in addition, $\mathrm{E}|X^\beta| < \infty$ and $0 < \alpha < \beta$, , then $\mathrm{E}|X^\alpha| < \infty$.

Proof:

To see why the first statement holds, observe that X^α is a composition of X with $x \mapsto x^\alpha$. As a composition of two measurable functions, X^α is measurable.

To prove the second statement, define

$$Y(\omega) = \max(|X(\omega)|^{\beta}, 1).$$

The reader can verify that Y is a random variable and $|X|^{\alpha} \leq Y$. By non-negativity,

$$\begin{aligned} E[Y] &= \int_{\{\omega \| X(\omega)|^{\beta} \leq 1\}} Y\,dP + \int_{\{\omega \| X(\omega)|^{\beta} > 1\}} Y\,dP \\ &= P\left(|X(\omega)|^{\beta} \leq 1\right) + \int_{\{\omega \| X(\omega)|^{\beta} > 1\}} |X|^{\beta}\,dP \\ &\leq 1 + E|X^{\beta}| < \infty. \end{aligned}$$

By monotonicity,

$$E|X^{\alpha}| \leq E[Y] \leq 1 + E|X^{\beta}| < \infty.$$

Counterexample for infinite measure

The requirement that $P(\Omega) < \infty$ is essential. By way of counterexample, consider the measurable space

$$([1,+\infty), \mathcal{B}_{\mathbb{R}[1,+\infty)}, \lambda),$$

where $\mathcal{B}_{\mathbb{R}[1,+\infty)}$ is the Borel σ-algebra on the interval $[1,+\infty)$, and λ is the linear Lebesgue measure. The reader can prove that $\int_{[1,+\infty)} \frac{1}{x}dx = \infty$, even though $\int_{[1,+\infty)} \frac{1}{x^2}dx = 1$. (Sketch of proof: $\int_S \frac{1}{x}dx$ and $\int_S \frac{1}{x^2}dx$ define a measure μ on $[1,+\infty) = \cup_{n=1}^{\infty}[1,n]$. Use "continuity from below" w.r. to and reduce to Riemann integral on each finite subinterval $[1,n]$).

Extremal Property

Recall, as we proved early on, that if X is a random variable, then so is X^2.

Proposition (extremal property of $E[X]$)). Let X be a random variable, and $E[X^2] < \infty$. Then $E[X]$ and Var[ ] are finite, and $E[X]$ is the best least squares approximation for X among constants. Specifically,

- For every $c \in \mathbb{R}$, $E[X-c]^2 \geq \mathrm{Var}[X]$;
- Equality holds if and only if $c = E[X]$.

($\mathrm{Var}[X]$ denotes the variance of X).

In intuitive terms, the extremal property says that if one is asked to predict the outcome of a *trial* of a random variable X, then $E[X]$, in some practically useful sense, is one's best bet *if no advance information about the outcome is available*. If, on the other hand, one does have some advance knowledge regarding the outcome, then — again, in some practically useful sense — one's bet may be improved upon by using conditional expectations $E[X \mid \mathcal{F}]$ (of which $E[X]$ is a special case) rather than $E[X]$.

Proof of proposition: By the above properties, both $\mathrm{E}[X]$ and $\mathrm{Var}[X]=\mathrm{E}[X^2]-\mathrm{E}^2[X]$ are finite, and

$$\begin{aligned}\mathrm{E}[X-c]^2 &= \mathrm{E}[X^2-2cX+c^2]\\ &= \mathrm{E}[X^2]-2c\,\mathrm{E}[X]+c^2\\ &= (c-\mathrm{E}[X])^2+\mathrm{E}[X^2]-\mathrm{E}^2[X]\\ &= (c-\mathrm{E}[X])^2+\mathrm{Var}[X],\end{aligned}$$

whence the extremal property follows.

Non-degeneracy

If $\mathrm{E}|X|=0$, then $X=0$ (a.s.).

Proof:

For every positive constant $r\in\mathbb{R}_{>0}$, $\mathrm{P}(|X|\geq r)=0$. Indeed,

$$r\cdot 1_{|X|\geq r}\leq |X|\cdot 1_{|X|\geq r}\leq |X|,$$

where $1_{|X|\geq r}=1_{|X|\geq r}(\omega)$ is the indicator function of the set $\{\omega\in\Omega \| X(\omega)|\geq r\}$.. By a property above, the finiteness of $\mathrm{E}|X|$ guarantees that the expected values $\mathrm{E}(r\cdot 1_{|X|\geq r})$ and $\mathrm{E}(|X|\cdot 1_{|X|\geq r})$ are also finite. By monotonicity,

$$r\cdot \mathrm{P}(|X|\geq r)=\mathrm{E}(r\cdot 1_{|X|\geq r})\leq \mathrm{E}(|X|\cdot 1_{|X|\geq r})\leq \mathrm{E}|X|=0.$$

For some integer $n>0$, set $r=\frac{1}{n}$. Define $S_n=\{\omega\in\Omega \| X(\omega)|\geq \frac{1}{n}\}$, and

$$S=\{\omega\in\Omega \| X(\omega)|>0\}.$$

The chain of sets

$$S_1\subseteq\ldots\subseteq S_n\subseteq S_{n+1}\subseteq\ldots\subseteq S$$

monotonically non-decreases, and $\mathrm{S}=\cup_{n=1}^{\infty}\mathrm{S}_n$. By "continuity from below", $\mathrm{P}(S)=\lim_n \mathrm{P}(S_n)$. Applying this formula, obtain

$$\mathrm{P}(X\neq 0)=\mathrm{P}(|X|>0)=\lim_n \mathrm{P}\left(|X|\geq\frac{1}{n}\right)=\lim_n 0=0,$$

as required. If $\mathrm{E}(X)<+\infty$ then $X<+\infty$ (a.s.)

Proof:

Since $\mathrm{E}[X]$ is defined (i.e. $\min(\mathrm{E}[X_+],\mathrm{E}[X_-])<\infty$), and $\mathrm{E}[X]=\mathrm{E}[X_+]-\mathrm{E}[X_-]$, we know that $\mathrm{E}[X_+]$ is finite, and we want to show that $X_+<+\infty$ (a.s.). We will show that $\mathrm{P}(\Omega_\infty)=0$, where

$$\Omega_\infty=\{\omega\in\Omega\,|\,X_+(\omega)=+\infty\}.$$

If $\Omega_\infty=\varnothing$, then $\mathrm{P}(\Omega_\infty)=0$, and the proof is complete. Assuming that $\Omega_\infty\neq\varnothing$, define

$$\mathrm{SF} = \{s \mid s \text{ is a simple random variable s.t. } 0 \le s \le X_+\}.$$

Given that $\mathrm{SF} \neq \varnothing$, pick $f \in \mathrm{SF}$. For every $n > \sup_\Omega f$, define

$$f_n(\omega) = \begin{cases} n & \text{if } \omega \in \Omega_\infty \\ f(\omega) & \text{if } \omega \notin \Omega_\infty. \end{cases}$$

Clearly, $\in \mathrm{SF}$, and

$$\mathrm{E}[f_n] = n \cdot \mathrm{P}(\Omega_\infty) + h,$$

for some constant $h \ge 0$ independent from n. (One can easily see that, in fact, $h = \mathrm{E}[f \cdot 1_{\Omega \setminus \Omega_\infty}]$, but this is of no interest to us here).

Suppose that $\mathrm{P}(\Omega_\infty) > 0$. The sequence $\{\mathrm{E}[f_n]\}$ strictly increases, so, by definition of Lebesgue integral,

$$\mathrm{E}[X_+] = \sup_{s \in \mathrm{SF}} \mathrm{E}[s] \ge \sup_{n > \sup_\Omega f} \mathrm{E}[f_n] = +\infty \cdot \mathrm{P}(\Omega_\infty) + h = +\infty,$$

in contradiction with an earlier conclusion that $\mathrm{E}[X_+]$ is finite. Corollary: if $\mathrm{E}(X) > -\infty$ then $X > -\infty$ (a.s.)

Corollary: if $\mathrm{E}|X| < \infty$ then $X \neq \pm\infty$ (a.s.)

$$\mathrm{E}(X)| \le \mathrm{E}|X|$$

For an arbitrary random variable X, $|\mathrm{E}(X)| \le \mathrm{E}|X|$.

Proof By definition of Lebesgue integral,

$$\begin{aligned} |\mathrm{E}(X)| &= \left|\mathrm{E}(X_+) - \mathrm{E}(X_-)\right| \le \left|\mathrm{E}(X_+)\right| + \left|\mathrm{E}(X_-)\right| \\ &= \mathrm{E}(X_+) + \mathrm{E}(X_-) = \mathrm{E}(X_+ + X_-) \\ &= \mathrm{E}|X|. \end{aligned}$$

Note that this result can also be proved based on Jensen's inequality.

Non-multiplicativity

In general, the expected value operator is not multiplicative, i.e. $\mathrm{E}[XY]$ is not necessarily equal to $\mathrm{E}[X] \cdot \mathrm{E}[Y]$. Indeed, let X assume the values of 1 and -1 with probability 0.5 each. Then

$$\mathrm{E}^2[X] = \left(\frac{1}{2} \cdot (-1) + \frac{1}{2} \cdot 1\right)^2 = 0,$$

and

$$\mathrm{E}[X^2] = \frac{1}{2} \cdot (-1)^2 + \frac{1}{2} \cdot 1^2 = 1,$$

so $E[X^2] \neq E^2[X]$.

The amount by which the multiplicativity fails is called the covariance:

$$\mathrm{Cov}(X,Y) = E[XY] - E[X]E[Y].$$

If, however, the random variables $X \in (\Omega_1, \mathcal{F}_1, P_1)$ and $Y \in (\Omega_2, \mathcal{F}_2, P_2)$ are independent, then $E[XY] = E[X]E[Y]$, and $\mathrm{Cov}(X,Y) = 0$.

Counterexample: $E[X_i] \not\to E[X]$ despite $X_i \to X$ point wise

Let $([0,1], \mathcal{B}_{[0,1]}, P)$ be the probability space, where $\mathcal{B}_{[0,1]}$ is the Borel σ-algebra on $[0,1]$ and P the linear Lebesgue measure. For $i \geq 1$, define a sequence of random variables

$$X_i = i \cdot 1_{\left[0,\frac{1}{i}\right]}$$

and a random variable

$$X = \begin{cases} +\infty & \text{if } x = 0 \\ 0 & \text{otherwise.} \end{cases}$$

on $[0,1]$, with 1_S being the indicator function of the set $S \subseteq [0,1]$. .

For every $x \in [0,1]$, as $i \to +\infty$, $X_i(x) \to X(x)$, and

$$E[X_i] = i \cdot P\left(\left[0, \frac{1}{i}\right]\right) = i \cdot \frac{1}{i} = 1,$$

so $\lim_{i\to\infty} E[X_i] = 1$. On the other hand, $P(\{0\}) = 0$, and hence $E[X] = 0$.

Countable Non-additivity

In general, the expected value operator is not σ-additive, i.e.

$$E\left[\sum_{i=0}^{\infty} X_i\right] \neq \sum_{i=0}^{\infty} E[X_i].$$

By way of counterexample, let $([0,1], \mathcal{B}_{[0,1]}, P)$ be the probability space, where $\mathcal{B}_{[0,1]}$ is the Borel σ-algebra on $[0,1]$ and P the linear Lebesgue measure. Define a sequence of random variables $X_i = (i+1) \cdot 1_{\left[0,\frac{1}{i+1}\right]} - i \cdot 1_{\left[0,\frac{1}{i}\right]}$ on $[0,1]$, with 1_S being the indicator function of the set

$S \subseteq [0,1]$. For the pointwise sums, we have

$$\sum_{i=0}^{n} X_i = (n+1) \cdot 1_{\left[0,\frac{1}{n+1}\right]},$$

$$\sum_{i=0}^{\infty} X_i(x) = \begin{cases} +\infty & \text{if } x = 0 \\ 0 & \text{otherwise.} \end{cases}$$

By finite additivity,

$$\sum_{i=0}^{\infty} \mathrm{E}[X_i] = \lim_{n\to\infty} \sum_{i=0}^{n} \mathrm{E}[X_i] = \lim_{n\to\infty} \mathrm{E}\left[\sum_{i=0}^{n} X_i\right] = 1.$$

On the other hand, $\mathrm{P}(\{0\}) = 0$, and hence

$$\mathrm{E}\left[\sum_{i=0}^{\infty} X_i\right] = 0 \neq 1 = \sum_{i=0}^{\infty} \mathrm{E}[X_i].$$

Countable Additivity for Non-negative Random Variables

Let $\{X_i\}_{i=0}^{\infty}$ be non-negative random variables. It follows from monotone convergence theorem that

$$\mathrm{E}\left[\sum_{i=0}^{\infty} X_i\right] = \sum_{i=0}^{\infty} \mathrm{E}[X_i].$$

$\mathrm{E}(XY) = \mathrm{E}(X)\mathrm{E}(Y)$ for independent X and Y

Let X and Y be independent random variables with finite expectations $\mathrm{E}[X]$ and $\mathrm{E}[Y]$. Then $\mathrm{E}[XY] = \mathrm{E}[X]\mathrm{E}[Y]$.

Proof:

1. The case of non-negative $\mathbb{Q}$-valued random variables.

Given a positive integer n, let the random variables $X:\Omega_1 \to \mathbb{R}$ and $Y:\Omega_2 \to \mathbb{R}$ assume their values in the set

$$\left\{\frac{m}{n} \middle| m = 0,1,2,3,\ldots\right\} \subset \mathbb{Q}_{\geq 0}.$$

Then $X = \sum_{m\geq 0} \frac{m}{n} \cdot 1_{X_{mn}}$, $Y = \sum_{m\geq 0} \frac{m}{n} \cdot 1_{Y_{mn}}$, and

$$XY = \sum_{m_1\geq 0} \sum_{m_2\geq 0} \frac{m_1}{n} 1_{X_{m_1 n}} \cdot \frac{m_2}{n} 1_{Y_{m_2 n}},$$

$$= \frac{1}{n^2} \sum_{i\geq 0} i \cdot \sum_{m_1 \cdot m_2 = i} 1_{X_{m_1 n}} \times Y_{m_2 n},$$

or equivalently,

$$XY(\omega) = \frac{i}{n^2} \Leftrightarrow \omega \in \bigsqcup_{m_1 m_2 = i} (X_{m_1 n} \times Y_{m_2 n}),$$

where 1_S is the indicator function of the set S,

$$X_{mn} = \left\{\omega \in \Omega_1 \middle| X(\omega) = \frac{m}{n}\right\},$$

$$Y_{mn} = \left\{\omega \in \Omega_2 \middle| Y(\omega) = \frac{m}{n}\right\},$$

and $\sqcup$ denotes disjoint union. By definition of expected value,

$$\begin{aligned}\mathrm{E}[XY] &= \int_{\Omega_1 \times \Omega_2} XY d\mathrm{P} \\ &= \frac{1}{n^2}\sum_{i \geq 0} i \cdot \sum_{m_1 \cdot m_2 = i} \mathrm{P}(X \in X_{m_1 n}, Y \in Y_{m_2 n})\end{aligned}$$

Due to independence,

$$\mathrm{P}(X \in X_{m_1 n}, Y \in Y_{m_2 n}) = \mathrm{P}(X \in X_{m_1 n}) \cdot \mathrm{P}(Y \in Y_{m_2 n}),$$

whence

$$\begin{aligned}\mathrm{E}[XY] &= \frac{1}{n^2}\sum_{i \geq 0} i \cdot \sum_{m_1 \cdot m_2 = i} \mathrm{P}(X \in X_{m_1 n}) P(Y \in Y_{m_2 n}) \\ &= \sum_{m_1 \geq 0}\sum_{m_2 \geq 0} \frac{m_1}{n} P(X \in X_{m_1 n}) \cdot \frac{m_2}{n} P(Y \in Y_{m_2 n}) \\ &= \left(\sum_{m_2 \geq 0} \frac{m_1}{n} P(X \in X_{m_1 n})\right) \cdot \left(\sum_{m_2 \geq 0} \frac{m_2}{n} P(Y \in Y_{m2n})\right) \\ &= E[X]E[Y].\end{aligned}$$

2. The Case of Non-negative random Variables

Let X and Y be (arbitrary) non-negative random variable. Define

$$X_n(\omega) = \begin{cases} \dfrac{m}{n} & \text{if } \dfrac{m}{n} \leq X(\omega) < \dfrac{m+1}{n}, \\ 0 & \text{if } X(\omega) = +\infty, \end{cases}$$

for an arbitrary $\omega \in \Omega_1$. Note that $X_n : \Omega_1 \to \mathbb{R}$ is a random variable and

$$\mathrm{Range}(X_n) \subseteq \left\{\frac{m}{n} \middle| m = 0, 1, 2, 3, \ldots\right\} \subset \mathbb{Q}_{\geq 0}.$$

As we saw previously, the finiteness of $\mathrm{E}[X]$ implies that X is finite almost sure, and consequently, $|X_n - X| \leq \frac{1}{n}$ (a.s.) on Ω_1. This, in turn, implies that $\mathrm{E}|X_n - X| \leq \frac{1}{n}$.

Let the random variable Y_n be defined the same way but with respect to Y. We have

$$
\begin{aligned}
&\left|\mathrm{E}[XY]-\mathrm{E}[X]\mathrm{E}[Y]\right|= \\
&=\left|\mathrm{E}[XY]-\mathrm{E}[X_nY]+\mathrm{E}[X_nY]-\mathrm{E}[X]\mathrm{E}[Y]\right| \\
&=\left|\mathrm{E}[(X-X_n)Y]+\mathrm{E}[X_nY]-\mathrm{E}[X]\mathrm{E}[Y]\right| \\
&\le\frac{1}{n}\mathrm{E}|Y|+\left|\mathrm{E}[X_nY]-\mathrm{E}[X]\mathrm{E}[Y]\right| \\
&=\frac{1}{n}\mathrm{E}|Y|+\left|\mathrm{E}[X_nY]-\mathrm{E}[X_nY_n]+\mathrm{E}[X_nY_n]-\mathrm{E}[X]\mathrm{E}[Y]\right| \\
&\le\frac{1}{n}\mathrm{E}|Y|+\frac{1}{n}\mathrm{E}|X_n|+\left|\mathrm{E}[X_nY_n]-\mathrm{E}[X]\mathrm{E}[Y]\right| \\
&=\frac{1}{n}\mathrm{E}|Y|+\frac{1}{n}\mathrm{E}|X_n-X+X|+\left|\mathrm{E}[X_nY_n]-\mathrm{E}[X]\mathrm{E}[Y]\right| \\
&\le\frac{1}{n}\mathrm{E}|Y|+\frac{\mathrm{E}|X_n-X|+\mathrm{E}|X|}{n}+\left|\mathrm{E}[X_nY_n]-\mathrm{E}[X]\mathrm{E}[Y]\right| \\
&\le\frac{1}{n}\mathrm{E}|Y|+\frac{1}{n^2}+\frac{\mathrm{E}|X|}{n}+\left|\mathrm{E}[X_nY_n]-\mathrm{E}[X]\mathrm{E}[Y]\right|.
\end{aligned}
$$

X_n and Y_n were shown to satisfy $\mathrm{E}[X_nY_n]=\mathrm{E}[X_n]\mathrm{E}[Y_n]$. Therefore,

$$
\begin{aligned}
&\left|\mathrm{E}[X_nY_n]-\mathrm{E}[X]\mathrm{E}[Y]\right|= \\
&=\left|\mathrm{E}[X_n]\mathrm{E}[Y_n]-\mathrm{E}[X]\mathrm{E}[Y]\right|= \\
&=\left|\mathrm{E}[X_n]\mathrm{E}[Y_n]-\mathrm{E}[X]\mathrm{E}[Y_n]+\mathrm{E}[X]\mathrm{E}[Y_n]-\mathrm{E}[X]\mathrm{E}[Y]\right| \\
&\le\left|\mathrm{E}[X_n]\mathrm{E}[Y_n]-\mathrm{E}[X]\mathrm{E}[Y_n]\right|+\left|\mathrm{E}[X]\mathrm{E}[Y_n]-\mathrm{E}[X]\mathrm{E}[Y]\right| \\
&\le\mathrm{E}|X_n-X|\cdot\mathrm{E}|Y_n|+\mathrm{E}|X|\cdot\mathrm{E}|Y_n-Y| \\
&\le\frac{\mathrm{E}|Y_n|+\mathrm{E}|X|}{n}=\frac{\mathrm{E}|Y_n-Y+Y|+\mathrm{E}|X|}{n} \\
&\le\frac{\mathrm{E}|Y_n-Y|+\mathrm{E}|Y|+\mathrm{E}|X|}{n} \\
&\le\frac{1}{n^2}+\frac{\mathrm{E}|Y|+\mathrm{E}|X|}{n}.
\end{aligned}
$$

It follows that, being independent from n, the constant value $\left|\mathrm{E}[XY]-\mathrm{E}[X]\mathrm{E}[Y]\right|$ can only be equal to 0.

3. The General Case

Let X and Y be arbitrary random variables. We have

$$
\begin{aligned}
[XY]&=\mathrm{E}[(X_+-X_-)(Y_+-Y_-)] \\
&=\mathrm{E}[X_+Y_+]-\mathrm{E}[X_+Y_-]-\mathrm{E}[X_-Y_+]+\mathrm{E}[X_-Y_-] \\
&=\mathrm{E}[X_+]\mathrm{E}[Y_+]-\mathrm{E}[X_+]\mathrm{E}[Y_-]-\mathrm{E}[X_-]\mathrm{E}[Y_+]+\mathrm{E}[X_-]\mathrm{E}[Y_-]
\end{aligned}
$$

$$= (\mathrm{E}[X_+] - \mathrm{E}[X_-])(\mathrm{E}[Y_+] - \mathrm{E}[Y_-])$$
$$= \mathrm{E}[X_+ - X_-]\mathrm{E}[Y_+ - Y_-]$$
$$= \mathrm{E}[X]\mathrm{E}[Y].$$

Inequalities

Cauchy–Bunyakovsky–Schwarz Inequality

The Cauchy–Bunyakovsky–Schwarz inequality states that

$$(\mathrm{E}[XY])^2 \leq \mathrm{E}[X^2] \cdot \mathrm{E}[Y^2].$$

Markov's Inequality

For a *nonnegative* random variable X and $a > 0$, the Markov's inequality states that

$$\mathrm{P}(X \geq a) \leq \frac{\mathrm{E}[X]}{a}.$$

Bienaymé-Chebyshev Inequality

Let X be an arbitrary random variable with finite expected value $\mathrm{E}[X]$ and finite variance $\mathrm{Var}[X] \neq 0$. The Bienaymé-Chebyshev inequality states that, for any real number $k > 0$,

$$\mathrm{P}\left(|X - \mathrm{E}[X]| \geq k\sqrt{\mathrm{Var}[X]}\right) \leq \frac{1}{k^2}.$$

Jensen's Inequality

Let $f : \mathbb{R} \to \mathbb{R}$ be a Borel convex function and X a random variable such that $\mathrm{E}|X| < \infty$. Jensen's inequality states that

$$f(\mathrm{E}(X)) \leq \mathrm{E}(f(X)).$$

Remark 1. The expected value $\mathrm{E}(f(X))$ is well-defined even if X is allowed to assume infinite values. Indeed, $\mathrm{E}|X| < \infty$ implies that $X \neq \pm\infty$ (a.s.), so the random variable $f(X(\omega))$ is defined almost sure, and therefore there is enough information to compute $\mathrm{E}(f(X))$.

Remark 2. Jensen's inequality implies that $|\mathrm{E}(X)| \leq \mathrm{E}|X|$ since the absolute value function is convex.

Lyapunov's Inequality

Let $0 < s < t$. Lyapunov's inequality states that

$$|X|^s)^{1/s} \leq \left(\mathrm{E}|X|^t\right)^{1/t}.$$

Proof: Applying Jensen's inequality to $|X|^s$ and $g(x)=|x|^{t/s}$, obtain $\left|E|X^s|\right|^{t/s} \leq E|X^s|^{t/s} = E|X|^t$. Taking the t th root of each side completes the Proof:

Corollary.

$$E|X| \leq \left(E|X|^2\right)^{1/2} \leq \ldots \leq \left(E|X|^n\right)^{1/n} \leq$$

Hölder's Inequality

Let p and q satisfy $1 \leq p \leq \infty$, $1 \leq q \leq \infty$, and $1/p + 1/q = 1$. The Hölder's inequality states that

$$E|XY| \leq (E|X|^p)^{1/p}(E|Y|^q)^{1/q}.$$

Minkowski Inequality

Let p be an integer satisfying $1 \leq p \leq \infty$. Let, in addition, $E|X|^p < \infty$ and $E|Y|^p < \infty$. Then, according to the Minkowski inequality, $E|X+Y|^p < \infty$ and

$$\left(E|X+Y|^p\right)^{1/p} \leq \left(E|X|^p\right)^{1/p} + \left(E|Y|^p\right)^{1/p}.$$

Taking Limits Under the E Sign

Monotone Convergence Theorem

Let the sequence of random variables $\{X_n\}$ and the random variables X and Y be defined on the same probability space (Ω, Σ, P). Suppose that

- All the expected values $E[X_n]$, $E[X]$, and $E[Y]$ are defined (differ from $\infty - \infty$);
- $E[Y] > -\infty$;
- For every n,

  $-\infty \leq Y \leq X_n \leq X_{n+1} \leq +\infty$ (a.s.);
- X is the pointwise limit of $\{X_n\}$ (a.s.), i.e. $X(\omega) = \lim_n X_n(\omega)$ (a.s.).

The monotone convergence theorem states that

$$\lim_n E[X_n] = E[X].$$

Proof:

Observe that, by monotonicity, the sequence $\{E[X_n]\}$ monotonically non-decreases, and $E[Y] \leq E[X_n] \leq E[X]$.

If $E[Y] = +\infty$, then $E[Y] = E[X_n] = E[X]$, and we are done.

If $E[Y] < +\infty$, then, following the assumption that $E[Y] > -\infty$, we conclude that $E[Y]$ is finite which, in turn, implies, as we saw previously, that Y is finite (a.s.).

Denote $Z_n = X_n - Y$ and $Z = X - Y$. The finiteness of Y (a.s.) implies that the differences $Z_n = X_n - Y$ and $Z = X - Y$ are defined (do not have the form $\infty - \infty$)) everywhere outside of a null set. On that null set, Z_n and Z may be defined arbitrarily (e.g. as zero or in any other way, as long as measurability is preserved) without affecting this Proof: As a difference of two random variables, Z_n and Z are also random variables.

It follows from the definition that $Z_n \geq 0$ (a.s.), $Z \geq 0$ (a.s.), the sequence $\{Z_n\}$ pointwise non-decreases (a.s.), and $Z_n \to Z$ pointwise (a.s.).

By (the general version of) monotone convergence theorem,

$$\begin{aligned}(\lim_n \mathrm{E}[X_n]) - \mathrm{E}[Y] &= \lim_n (\mathrm{E}[X_n] - \mathrm{E}[Y]) \\ &= \lim_n \mathrm{E}[X_n - Y] \\ &= \lim_n \mathrm{E}[Z_n] \\ &= \mathrm{E}[Z] \\ &= \mathrm{E}[X - Y] \\ &= \mathrm{E}[X] - \mathrm{E}[Y],\end{aligned}$$

whence the assertion follows.

Fatou's Lemma

Let the sequence of random variables $\{X_n\}$ and the random variable Y be defined on the same probability space $(\Omega, \Sigma, \mathrm{P})$. Suppose that

- All the expected values $\mathrm{E}[X_n]$, $\mathrm{E}[\liminf_n X_n]$, and $\mathrm{E}[Y]$ are defined (differ from $\infty - \infty$);
- $\mathrm{E}[Y] > -\infty$;
- $-\infty \leq Y \leq X_n \leq +\infty$ (a.s.), for every n.

Fatou's lemma states that

$$\mathrm{E}[\liminf_n X_n] \leq \liminf_n \mathrm{E}[X_n].$$

$\liminf_n X_n$ is a random variable, for every n, by the properties of limit inferior.

Proof:

If $\mathrm{E}[Y] = +\infty$, then, by monotonicity, $\mathrm{E}[Y] = \mathrm{E}[X_n] = +\infty$, so $\liminf_n \mathrm{E}[X_n] = +\infty$, and the assertion follows.

If $\mathrm{E}[Y] < +\infty$, then, following the assumption that $\mathrm{E}[Y] > -\infty$, we conclude that $\mathrm{E}[Y]$ is finite which, in turn, implies, as we saw previously, that Y is finite (a.s.).

Denote $Z_n = X_n - Y$. Then $Z_n \geq 0$ (a.s.). The finiteness of Y (a.s.) implies that Z_n is defined (does not have the form $\infty - \infty$) everywhere outside of a null set. On that null set Z_n may be defined arbitrarily (e.g. as zero or in any other way, as long as measurability is preserved) without affecting this Proof: As a difference of two random variables, Z_n is a random variable.

By (the general version of) Fatou's lemma,

$$\begin{aligned}\mathrm{E}[\liminf_n X_n]-\mathrm{E}[Y] &= \mathrm{E}[\liminf_n (X_n - Y)] \\ &= \mathrm{E}[\liminf_n Z_n] \\ &\leq \liminf_n \mathrm{E}[Z_n] \\ &= \liminf_n \mathrm{E}[X_n - Y] \\ &= \liminf_n (\mathrm{E}[X_n]-\mathrm{E}[Y]) \\ &= (\liminf_n \mathrm{E}[X_n])-\mathrm{E}[Y],\end{aligned}$$

whence the assertion follows. Corollary. Let

- $X_n \to X$ pointwise (a.s.);
- $\mathrm{E}[X_n] \leq C$, for some constant C (independent from n);
- $\mathrm{E}[Y] > -\infty$;
- $-\infty \leq Y \leq X_n \leq +\infty$ (a.s.), for every n.

Then $\mathrm{E}[X] \leq C$.

Proof is by observing that $X = \liminf_n X_n$ (a.s.) and applying Fatou's lemma.

Dominated Convergence Theorem

Let $\{X_n\}_n$ be a sequence of random variables. If $X_n \to X$ pointwise (a.s.), $|X_n| \leq Y \leq +\infty$ (a.s.), and $\mathrm{E}[Y] < \infty$. Then, according to the dominated convergence theorem,

- The function X is measurable (hence a random variable);
- $\mathrm{E}|X| < \infty$;
- All the expected values $\mathrm{E}[X_n]$ and $\mathrm{E}[X]$ are defined (do not have the form $\infty - \infty$);
- $\lim_n \mathrm{E}[X_n] = \mathrm{E}[X]$ (both sides may be infinite);
- $\lim_n \mathrm{E}|X_n - X| = 0$.

Relationship with Characteristic Function

The probability density function f_X of a scalar random variable X is related to its characteristic function φ_X by the inversion formula:

$$f_X(x) = \frac{1}{2\pi}\int_{\mathbb{R}} e^{-itx}\varphi_X(t)dt$$

For the expected value of $g(X)$ (where $g:\mathbb{R}\to\mathbb{R}$ is a Borel function), we can use this inversion formula to obtain

$$\mathrm{E}[g(X)] = \frac{1}{2\pi}\int_{\mathbb{R}} g(x)\left[\int_{\mathbb{R}} e^{-itx}\varphi_X(t)dt\right]dx.$$

If $\mathrm{E}[g(X)]$ is finite, changing the order of integration, we get, in accordance with Fubini-Tonelli theorem,

$$\mathrm{E}[g(X)] = \frac{1}{2\pi}\int_{\mathbb{R}} G(t)\varphi_X(t)dt,$$

where

$$G(t) = \int_{\mathbb{R}} g(x)e^{-itx}\,dx$$

is the Fourier transform of $g(x)$. The expression for $\mathrm{E}[g(X)]$ also follows directly from Plancherel theorem.

Uses and Applications

It is possible to construct an expected value equal to the probability of an event by taking the expectation of an indicator function that is one if the event has occurred and zero otherwise. This relationship can be used to translate properties of expected values into properties of probabilities, e.g. using the law of large numbers to justify estimating probabilities by frequencies.

The expected values of the powers of X are called the moments of X; the moments about the mean of X are expected values of powers of $X - \mathrm{E}[X]$. The moments of some random variables can be used to specify their distributions, via their moment generating functions.

To empirically estimate the expected value of a random variable, one repeatedly measures observations of the variable and computes the arithmetic mean of the results. If the expected value exists, this procedure estimates the true expected value in an unbiased manner and has the property of minimizing the sum of the squares of the residuals (the sum of the squared differences between the observations and the estimate). The law of large numbers demonstrates (under fairly mild conditions) that, as the size of the sample gets larger, the variance of this estimate gets smaller.

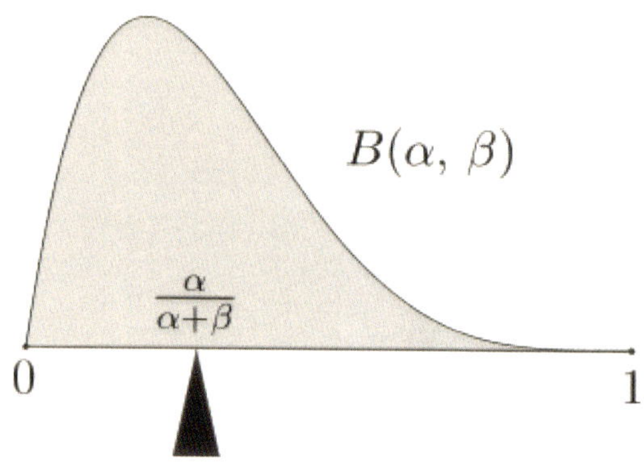

The mass of probability distribution is balanced at the expected value, here a Beta(α,β) distribution with expected value α/(α+β).

This property is often exploited in a wide variety of applications, including general problems of statistical estimation and machine learning, to estimate (probabilistic) quantities of interest via Monte Carlo methods, since most quantities of interest can be written in terms of expectation, e.g. $P(X \in \mathcal{A}) = E[1_{\mathcal{A}}]$, where $1_{\mathcal{A}}$ is the indicator function of the set $\mathcal{A}$.

In classical mechanics, the center of mass is an analogous concept to expectation. For example, suppose X is a discrete random variable with values x_i and corresponding probabilities p_i. Now consider a weightless rod on which are placed weights, at locations x_i along the rod and having masses p_i (whose sum is one). The point at which the rod balances is E[X].

Expected values can also be used to compute the variance, by means of the computational formula for the variance

$$\mathrm{Var}(X) = E[X^2] - (E[X])^2.$$

A very important application of the expectation value is in the field of quantum mechanics. The expectation value of a quantum mechanical operator $\hat{A}$ operating on a quantum state vector $|\psi\rangle$ is written as $\langle \hat{A} \rangle = \langle \psi | A | \psi \rangle$. The uncertainty in $\hat{A}$ can be calculated using the formula $(\Delta A)^2 = \langle \hat{A}^2 \rangle - \langle \hat{A} \rangle^2$.

The law of the Unconscious Statistician

The expected value of a measurable function of X, $g(X)$, given that X has a probability density function $f(x)$, is given by the inner product of f and g:

$$E[g(X)] = \int_{\mathbb{R}} g(x) f(x) dx.$$

This formula also holds in multidimensional case, when g is a function of several random variables, and f is their joint density.

Formula for Non-negative Random Variables

Finite and Countably Infinite Case

For a non-negative integer-valued random variable $X : \Omega \to \{0,1,2,3,\ldots\} \cup \{+\infty\}$,

$$E[X] = \sum_{i=1}^{\infty} P(X \geq i).$$

Proof

If $P(X = +\infty) > 0$, then $E[X] = +\infty$. On the other hand,

$$P(X \geq i) \geq P(X = +\infty) > 0,$$

so the series on the right diverges to $+\infty$, and the equality holds.

If $P(X = +\infty) = 0$, then

$$\sum_{i=1}^{\infty} P(X \geq i) = \sum_{i=1}^{\infty}\sum_{j=i}^{\infty} P(X = j).$$

Let

$$M = \begin{bmatrix} P(X=1) & P(X=2) & P(X=3) & \ldots & P(X=n) & \ldots \\ & P(X=2) & P(X=3) & \ldots & P(X=n) & \ldots \\ & & P(X=3) & \ldots & P(X=n) & \ldots \\ \ldots & \ldots & \ldots & \ldots & \ldots & \ldots \\ & & & & P(X=n) & \cdots \\ \ldots & \ldots & \ldots & \ldots & \ldots & \ldots \end{bmatrix}$$

be an infinite upper triangular matrix. The double series $\sum_{i=1}^{\infty}\sum_{j=i}^{\infty} P(X = j)$ is the sum of M's elements if summation is done row by row. Since every summand is non-negative, the series either converges absolutely or diverges to $+\infty$. In both cases changing summation order does not affect the sum. Changing summation order, from row-by-row to column-by-column, gives us

$$\begin{aligned} \sum_{i=1}^{\infty}\sum_{j=i}^{\infty} P(X = j) &= \sum_{j=1}^{\infty}\sum_{i=1}^{j} P(X = j) \\ &= \sum_{j=1}^{\infty} j P(X = j) \\ &= E[X]. \end{aligned}$$

Example

In a coin tossing experiment, let the probability of heads be p. Including the final attempt, how many tosses can we expect until the first head?

Solution: If N is the random variable indicating the numbers of coin tosses before and including the first head, then, for $i \geq 1$,

$$\begin{aligned} P(N \geq i) &= 1 - P(N \leq i-1) \\ &= 1 - \sum_{j=0}^{i-1} P(N = j) \\ &= 1 - \sum_{j=1}^{i-1} (1-p)^{j-1} p \\ &= 1 - \frac{1-(1-p)^{i-1}}{p} \cdot p \\ &= (1-p)^{i-1}, \end{aligned}$$

where we took into account the geometric series summation formula. We now compute

$$\begin{aligned}E[N] &= \sum_{i=1}^{\infty} P(N \geq i) \\ &= \sum_{i=1}^{\infty} (1-p)^{i-1} \\ &= \frac{1}{p}.\end{aligned}$$

General Case

If $X : \Omega \to [0, +\infty]$ is a non-negative random variable, then

$$E[X] = \int_{[0,+\infty)} P(X \geq x) dx = \int_{[0,+\infty)} P(X > x) dx$$

and

$$E[X] = (R)\int_0^{\infty} P(X \geq x) dx = (R)\int_0^{\infty} P(X > x) dx,$$

where $(R)\int_0^{\infty}$ denotes improper Riemann integral.

Proof

1. For every $\omega \in \Omega$, ,

$$X(\omega) = \int_{(0,\, X(\omega))} dx = \int_{(0,\, +\infty)} 1_{(0,\, X(\omega))}(x) dx = \int_{(0,\, +\infty)} 1_{(0,\, X(\omega)]}(x) dx,$$

Where $1_{(0,X(\omega))}$ and $1_{(0,X(\omega)]}$ are the indicator functions of $(0, X(\omega))$ and $(0, X(\omega)]$, respectively. Substituting this into the definition of $E[X]$, obtain

$$\begin{aligned}E[X] &= \int_{\Omega} X dP \\ &= \int_{\Omega} \int_{[0,+\infty)} 1_{(0,X(\omega)]}(x) dx dP(\omega)\end{aligned}$$

Since $X(\omega) \geq 0$ and $1_{(0,X(\omega)]}(x) \geq 0$, this integral (finite or infinite) meets the requirements of Tonelli's theorem. Changing the order of integration gives us

$$\begin{aligned}&\int_{[0,+\infty)} \int_{\Omega} 1_{(0,X(\omega)]}(x) dP(\omega) dx \\ &= \int_{[0,+\infty)} P(X \geq x) dx.\end{aligned}$$

2a. The function $y(x) = P(X \geq x)$ is Riemann-integrable on each finite interval $[a,b]$. Indeed, since $y(x)$ is non-increasing, the set D of its discontinuities is countable. Due to countable additivity, D is a null set with respect to the linear Lebesgue measure. Furthermore, $0 \le y(x) \le 1$, for all $x \in [-\infty, +\infty]$. Using the Lebesgue criterion, Riemann integrability of $y(x)$ follows. We also conclude that

$$\int_{[a,b]} \mathrm{P}(X \ge x)dx = (\mathrm{R})\int_a^b \mathrm{P}(X \ge x)dx.$$

2b. By "continuity from below",

$$\begin{aligned}\int_{[0,+\infty)} \mathrm{P}(X \ge x)dx &= \lim_{t\to+\infty} \int_{[0,t]} \mathrm{P}(X \ge x)dx \\ &= \lim_{t\to+\infty} (\mathrm{R})\int_0^t \mathrm{P}(X \ge x)dx \\ &= (\mathrm{R})\int_0^\infty \mathrm{P}(X \ge x)dx.\end{aligned}$$

The case of $\mathrm{P}(X > x)$ is similar.

Formula for Non-positive Random Variables

If $X : \Omega \to [-\infty, 0]$ is a non-positive random variable, then

$$\mathrm{E}[X] = -\int_{(-\infty,0]} \mathrm{P}(X \le x)dx = -\int_{(-\infty,0]} \mathrm{P}(X < x)dx$$

and

$$\mathrm{E}[X] = -(\mathrm{R})\int_{-\infty}^0 \mathrm{P}(X \le x)dx = -(\mathrm{R})\int_{-\infty}^0 \mathrm{P}(X < x)dx,$$

where $(\mathrm{R})\int_{-\infty}^0$ denotes improper Riemann integral.

This formula follows from that for the non-negative case applied to $-X$.

If, in addition, X is integer-valued, i.e. $X : \Omega \to \{\ldots, -3, -2, -1, 0\} \cup \{-\infty\}$, then

$$\mathrm{E}[X] = -\sum_{i=-1}^{-\infty} \mathrm{P}(X \le i).$$

General Case

If X can be both positive and negative, then $\mathrm{E}[X] = \mathrm{E}[X_+] - \mathrm{E}[X_-]$, and the above results may be applied to X_+ and X_- separately.

Independent Random Variables

Two random variables are independent if they convey no information about each other and, as a consequence, receiving information about one of the two does not change our assessment of the probability distribution of the other.

That two events A and B are independent if and only if

$$P(A \cap B) = P(A)P(B)$$

This definition is extended to random variables as follows.

Definition Two random variables X and Y are said to be independent if and only if

$$P(\{X \in A\} \cap \{Y \in B\}) = P(\{X \in A\} P\{Y \in B\})$$

for any couple of events $\{X \in A\}$ and $\{Y \in B\}$, where $A \subset \mathbb{R}$ and $B \subseteq \mathbb{R}$.

In other words, two random variables are independent if and only if the events related to those random variables are independent events.

The independence between two random variables is also called statistical independence.

Independence Criterion

Checking the independence of all possible couples of events related to two random variables can be very difficult. This is the reason why the above explanation is seldom used to verify whether two random variables are independent. The following criterion is more often used instead.

Proposition Two random variables X and Y are independent if and only if

$$F_{XY}(x,y) = F_X(x)F_Y(y) \quad \forall x,y \in \mathbb{R}$$

where $F_{XY}(x,y)$ is their joint distribution function and $F_X(x)$ and $F_Y(y)$ are their marginal distribution functions.

Proof

Example Let X and Y be two random variables with marginal distribution functions

$$F_X(x) = \begin{cases} 0 & \text{if } x < 0 \\ 1 - \exp(-x) & \text{if } x \geq 0 \end{cases}$$

$$F_Y(y) = \begin{cases} 0 & \text{if } y < 0 \\ 1 - \exp(-y) & \text{if } y \geq 0 \end{cases}$$

and joint distribution function

$$F_{X,Y}(x,y) = \begin{cases} 0 & \text{if } x < 0 \text{ or } y < 0 \\ 1 - \exp(-x) - \exp(-y) + \exp(-x-y) & \text{if } x \geq 0 \text{ and } y \geq 0 \end{cases}$$

X and Y are independent if and only if

$$F_{XY}(x,y) = F_X(x)F_Y(y)$$

which is straightforward to verify. When $x<0$ or $y<0$, then

$$F_X(x)F_Y(y) = 0 = F_{X,Y}(x,y)$$

When $x \geq 0$ and $y \geq 0$, then:

$$\begin{aligned} F_X(x)F_Y(y) &= [1-\exp(-x)][1-\exp(-y)] \\ &= 1-\exp(-x)-\exp(-y)+\exp(-x)\exp(-y) \\ &= 1-\exp(-x)-\exp(-y)+\exp(-x-y) \\ &= F_{X,Y}(x,y) \end{aligned}$$

Independence between Discrete Random Variables

When the two variables, taken together, form a discrete random vector, independence can also be verified using the following proposition:

Proposition Two random variables *X* and *Y*, forming a discrete random vector, are independent if and only if

$$p_{XY}(x,y) = p_X(x)p_y(y) \quad \forall x,y \in \mathbb{R}$$

where $p_{XY}(x,y) = p_X(x)p_Y(y)\ \forall x,y \in \mathbb{R}$ is their joint probability mass function and $p_X(x)$ and $p_Y(y)$ are their marginal probability mass functions.

The following example illustrates how this criterion can be used.

Example Let $[X\ Y]$ be a discrete random vector with support

$$R_{XY} = \{[1\ 1],[2\ 0],[0\ 0]\}$$

Let its joint probability mass function be

$$p_{X,Y}(x,y) = \begin{cases} \frac{1}{3} & \text{if } [x\ \ y]=[1\ \ 1] \\ \frac{1}{3} & \text{if } [x\ \ y]=[2\ \ 0] \\ \frac{1}{3} & \text{if } [x\ \ y]=[0\ \ 0] \\ 0 & \text{otherwise} \end{cases}$$

In order to verify whether *X* and *Y* are independent, we first need to derive the marginal probability mass functions of *X* and *Y*. The support of *X* is

$$R_X = \{0, 1, 2\}$$

and the support of Y is

$$R_Y = \{0, 1\}$$

We need to compute the probability of each element of the support X of :

$$p_X(0) = \sum_{y c R_Y} p_{XY}(0, y) = p_{XY}(0, 0) + p_{XY}(0, 1) = \frac{1}{3} + 0 = \frac{1}{3}$$

$$p_X(1) = \sum_{y c R_Y} p_{XY}(1, y) = p_{XY}(1, 0) + p_{XY}(1, 1) = 0 + \frac{1}{3} = \frac{1}{3}$$

$$p_X(2) = \sum_{y c R_Y} p_{XY}(2, y) = p_{XY}(2, 0) + p_{XY}(2, 1) = \frac{1}{3} + 0 = \frac{1}{3}$$

Thus, the probability mass function of X is

$$p_X(x) \sum_{y c R_Y} p_{XY}(x, y) = \begin{cases} \frac{1}{3} & if\ x = 0 \\ \frac{1}{3} & if\ x = 1 \\ \frac{1}{3} & if\ x = 2 \\ 0 & otherwise \end{cases}$$

We need to compute the probability of each element of the support of Y:

$$p_Y(0) = \sum_{y c R_Y} p_{XY}(x, 0) = p_{XY}(0, 0) + p_{XY}(1, 0) + p_{XY}(2, 0) = \frac{1}{3} + 0 + \frac{1}{3} = \frac{2}{3}$$

$$p_Y(1) = \sum_{y c R_Y} p_{XY}(x, 1) = p_{XY}(0, 1) + p_{XY}(1, 1) + p_{XY}(2, 1) = 0 + \frac{1}{3} + 0 = \frac{1}{3}$$

Thus, the probability mass function of Y is

$$p_X(x) \sum_{y c R_Y} p_{XY}(x, y) = \begin{cases} \frac{2}{3} & if\ y = 0 \\ \frac{1}{3} & if\ y = 0 \\ 0 & otherwise \end{cases}$$

The product of the marginal probability mass functions is

$$p_X(x)p_Y(y) = \begin{cases} \frac{1}{3}\frac{2}{3} = \frac{2}{9} & if [x\ y] = [0\ 0] \\ \frac{1}{3}\frac{2}{3} = \frac{2}{9} & if [x\ y] = [1\ 0] \\ \frac{1}{3}\frac{2}{3} = \frac{2}{9} & if [x\ y] = [2\ 0] \\ \frac{1}{3}\frac{1}{3} = \frac{1}{9} & if [x\ y] = [0\ 1] \\ \frac{1}{3}\frac{1}{3} = \frac{1}{9} & if [x\ y] = [1\ 1] \\ \frac{1}{3}\frac{1}{3} = \frac{1}{9} & if [x\ y] = [2\ 1] \\ 0 & otherwise \end{cases}$$

which is obviously different from $p_{XY}(x, y)$. Therefore, *X* and *Y* are not independent.

Independence between Absolutely Continuous Random Variables

When the two variables, taken together, form an absolutely continuous random vector, independence can also be verified by means of the following proposition.

Proposition Two random variables *X* and *Y*, forming an absolutely continuous random vector, are independent if and only if

$$f_{XY}(x, y) = f_X(x) f_Y(y) \quad \forall x, y \in \mathbb{R}$$

Where $f_{XY}(x, y)$ is their joint probability density function and $f_X(x)$ and $f_Y(y)$ their marginal probability density functions.

The following example illustrates how this criterion can be used.

Example Let the joint probability density function of *X* and *Y* be

$$f_{X,Y}(x, y) = \begin{cases} 0 \ \ if \ x \notin [0,1] \ or \ y \notin [0,1] \\ 1 \ \ if \ x \in [0,1] \ or \ y \in [0,1] \end{cases}$$

Its marginals are

$$\begin{aligned} f_X(x) &= \int_{-\infty}^{\infty} f_{X,Y}(x, y)dy \\ &= \int_{-\infty}^{0} f_{X,Y}(x, y)dy + = \int_{0}^{1} f_{X,Y}(x, y)dy + = \int_{1}^{\infty} f_{X,Y}(x, y)dy \\ &= 0 + \int_{0}^{1} f_{X,Y}(x, y)dy + 0 \\ &= \begin{cases} 0 \ \ if \ x \notin [0,1] \\ 1 \ \ if \ x \in [0,1] \end{cases} \end{aligned}$$

and

$$f_Y(y) = \int_{-\infty}^{\infty} f_{X,Y}(x,y)dx$$
$$= \int_{-\infty}^{0} f_{X,Y}(x,y)dx + = \int_{0}^{1} f_{X,Y}(x,y)dx + = \int_{1}^{\infty} f_{X,Y}(x,y)dx$$
$$= 0 + \int_{0}^{1} f_{X,Y}(x,y)dx + 0$$
$$= \begin{cases} 0 \ if \ x \notin [0,1] \\ 1 \ if \ x \in [0,1] \end{cases}$$

Verifying that $f_{XY}(x,y) = f_X(x) f_Y(y)$ is straightforward. When $x \notin [0,1]$ or $y \notin [0,1]$, then. $f_X(x) f_Y(y) = 0 = f_{X,Y}(x,y)$. When $x \in [0,1]$ and $y \in [0,1]$, then

$$f_X(x) f_Y(y) = 1 \cdot 1 = 1 = f_{X,Y}(x,y)$$

Mutually Independent Random Variables

The explanation of mutually independent random variables extends the mutually independent events to random variables.

We can say that n random variables $X_1,...,X_n$ are mutually independent (or jointly independent) if and only if

$$p\left(\bigcap_{j=1}^{k} \{X_{i_j} \in A_j\}\right) = \prod_{j=1}^{k} P(\{X_{i_j} \in A_j\})$$

for any sub-collection of k random variables $X_{i1},....,X_{ik}$ (where $k \leq n$) and for any collection of events $\{X_{i1} \in A_1\},...,\{X_{ik} \in A_k\}$, where $A_1,...,A_k \subset \mathbb{R}$

In other words, n random variables are mutually independent if the events related to those random variables are mutually independent events.

Denote by X a random vector whose components are $X_1,...,X_n$. The above condition for mutual independence can be replaced:

- In general, by a condition on the joint distribution function of X:

$$F_X(x_1,...,x_n) = \prod_{j=1}^{n} F_{X_j}(x_j)$$

- For discrete random variables, by a condition on the joint probability mass function of X:

$$P_X(x_1,...,x_n) = \prod_{j=1}^{n} P_{X_j}(x_j)$$

- For absolutely continuous random variables, by a condition on the joint probability density function of X :

$$f_X(x_1,...,x_n)=\prod_{j-1}^{n} f_{X_j}(x_j)$$

Mutual independence via expectations

It can be proved that n random variables $X_1,....,X_n$ are mutually independent if and only if

$$E\left[\prod_{j-1}^{n} g_j(X_j)\right]=\prod_{j-1}^{n} \mathrm{E}\left[g_j(X_j)\right]$$

for any n functions $g_1,...,g_n$ such that the above expected values exist and are well-defined.

Independence and zero covariance

If two random variables X_1 and X_2 are independent, then their covariance is zero:

$$\mathrm{Cov}[X_1,X_2]=0$$

Proof:

The converse is not true: two random variables that have zero covariance are not necessarily independent.

Independent random vectors

The above notions are easily generalized to the case in which x and y are two random vectors, having dimensions $K_X \times 1$ and $K_Y \times 1$ respectively. Denote their joint distribution functions by $F_X(X)$ and $F_Y(Y)$ and the joint distribution function of x and y together by

$$F_{XY}(X,Y)=F_{X_1,X_2,...,X_{kx},Y_1,Y_2,...,Y_{ky}}\left(x_1,x_2,...,x_{kx},\mathrm{y}_1,\mathrm{y}_2,...,\mathrm{y}_{k_y}\right)$$

Also, if the two vectors are discrete or absolutely continuous replace F with p or f to denote the corresponding probability mass or density functions.

Two random vectors X and Y are independent if and only if one of the following equivalent conditions is satisfied:

Condition 1:

$$P(\{X\in A\}\cap\{Y\in B\})=P(\{X\in A\}P\{X\in B\})$$

for any couple of events $\{X\in A\}$ and $\{Y\in B\}$, where $A\subseteq\mathbb{R}^{K_x}$ and $B\subseteq\mathbb{R}^{K_x}$

Condition 2:

$$F_{XY}(x,\mathrm{y})=F_X(x)F_Y(\mathrm{y})$$

for any $x\in\mathbb{R}^{K_x}$ and $\mathrm{y}\in\mathbb{R}^{K_y}$ (replace F with p or f when the distributions are discrete or absolutely continuous)

Condition 3:

$$E\left[g_1(X)g_2(Y)\right]=E\left[g_1(X)\right]E\left[g_2(Y)\right]$$

for any functions $g_1:\mathbb{R}^{K_x}\to\mathbb{R}$ and $g_2:\mathbb{R}^{K_y}\to\mathbb{R}$ such that the above expected values exist and are well-defined.

Mutually Independent Random Vectors

Also the definition of mutual independence extends in a straightforward manner to random vectors.

We say that n random vectors $x_1,\ldots,x_n$ are mutually independent (or jointly independent) if and only if

$$P\left(\bigcap_{j-1}^{k}\left\{X_{ij}\in A_j\right\}\right)=\prod_{j-1}^{k}P\left(\left\{X_{ij}\in A_j\right\}\right)$$

for any sub-collection of k random vectors $x_{i1},\ldots,x_{ik}$ (where $k\le n$) and for any collection of events $\{X_{1i}\in A_1\},\ldots,\{X_{ik}\in A_k\}$

All the equivalent conditions for the joint independence of a set of random variables apply with obvious modifications also to random vectors.

Solved exercises

Below you can find some exercises with explained solutions.

Exercise 1

Consider two random variables x and y having marginal distribution functions

$$F_{X(x)}=\begin{cases}0 & \text{if } x<1\\ \frac{1}{2} & \text{if } 1\le x<2\\ 1 & \text{if } x\ge 2\end{cases}$$

$$F_Y(y)=\begin{cases}0 & \text{if } y<0\\ 1-\frac{1}{2}\exp(-y)-\frac{1}{2}\exp(-2y) & \text{if } y\ge 0\end{cases}$$

If x and y are independent, what is their joint distribution function?

Solution

For X and Y to be independent, their joint distribution function must be equal to the product of their marginal distribution functions:

$$F_{X,Y}(x,y)=\begin{cases}0 & if\, x<1 or\, y<0\\ \frac{1}{2}-\frac{1}{4}\exp(-y)-\frac{1}{4}\exp(-2y) & if\, 1\leq x<2 and\, y\geq 0\\ 1-\frac{1}{2}\exp(-y)-\frac{1}{2}\exp(-2y) & if\, x\geq 2 and\, y\geq 0\end{cases}$$

Exercise 2

Let $[X\ Y]$ be a discrete random vector with support:

$$R_{XY}=\{[0\ 0],[0\ 1],[1\ 0],[1\ 1]\}$$

Let its joint probability density function be

$$P_{X,Y}(x,y)=\begin{cases}\frac{1}{4}\text{if}[x\ y]=[0\ \ 0]\\ \frac{1}{4}\text{if}[x\ y]=[0\ \ 1]\\ \frac{1}{4}\text{if}[x\ y]=[1\ \ 0]\\ \frac{1}{4}\text{if}[x\ y]=[1\ \ 1]\\ 0\ \text{otherwise}\end{cases}$$

Are X and Y independent?

Solution

In order to verify whether X and Y are independent, we first need to derive the marginal probability mass functions of X and Y. The support of X is

$$R_X=\{0,1\}$$

and the support of Y is

$$R_Y=\{0,1\}$$

We need to compute the probability of each element of the support of X:

$$\text{px}(0)=\sum_{yeRy}\text{pxy}(0,y)=\text{pxy}(0,0)+\text{pxy}(0,1)=\frac{1}{2}$$

$$\text{px}(1)=\sum_{yeRy}\text{pxy}(1,y)=\text{pxy}(1,0)+\text{pxy}(1,1)=\frac{1}{2}$$

Thus, the probability mass function of X is

$$p_X(x) = \sum_{yeRy} P_{XY}(x,y) = \begin{cases} \frac{1}{2} & \text{if } x = 0 \\ \frac{1}{2} & \text{if } x = 1 \\ 0 & \textit{otherwise} \end{cases}$$

We need to compute the probability of each element of the support of Y :

$$p_Y(0) = \sum_{xeRx} p_{XY}(x,0) = p_{XY}(0,0) + p_{XY}(1,0) = \frac{1}{2}$$

$$p_Y(1) = \sum_{xeRx} p_{XY}(x,1) = p_{XY}(0,1) + p_{XY}(1,1) = \frac{1}{2}$$

Thus, the probability mass function of y is

$$p_Y(y) = \sum_{xeRx} p_{XY}(x,y) = \begin{cases} \frac{1}{2} & \text{if } y = 0 \\ \frac{1}{2} & \text{if } y = 0 \\ 0 & \text{otherwise} \end{cases}$$

The product of the marginal probability mass functions is

$$p_X(x)p_Y(y) = \begin{cases} \frac{1}{2}\frac{1}{2} = \frac{1}{4} & \text{if} [x \ y] = [0 \ 0] \\ \frac{1}{2}\frac{1}{2} = \frac{1}{4} & \text{if} [x \ y] = [1 \ 0] \\ \frac{1}{2}\frac{1}{2} = \frac{1}{4} & \text{if} [x \ y] = [0 \ 1] \\ \frac{1}{2}\frac{1}{2} = \frac{1}{4} & \text{if} [x \ y] = [1 \ 1] \\ 0 & \text{otherwise} \end{cases}$$

Which is equal to $p_{XY}(x,y)$. Therefore, X and Y are independent.

Exercise 3

Let [X Y] be an absolutely continuous random vector with support

$$R_{XY} = [0,\infty] \times (2,3)$$

and its joint probability density function be

$$f_{XY}(x,y) = \begin{cases} \exp(-x) if\ x \in [0,\infty) and\ y \in [2,3] \\ 0 \quad \text{otherwise} \end{cases}$$

Are X and Y independent?

The support of Y is

$$Ry = [2,3]$$

When $y \notin [2,3]$, the marginal probability density function of Y is 0, while, when $y \notin [2,3]$, the marginal probability density function of Y is

$$fx(x) f_Y(y) = \exp(-x) \cdot 1 = \exp(-x) = fx,y(x,y)$$

Thus, summing up, the marginal probability density function of Y is

$$f_y(y) = \begin{Bmatrix} 1 \text{ if } y \in [2,3] \\ 0\ otherwise \end{Bmatrix}$$

The support of X is

$$Rx = [0,\infty)$$

When $x \notin [0,\infty)$, the marginal probability density function of X is 0, while, when $x \notin [0,\infty)$, the marginal probability density function of X is

$$\begin{aligned} f_x(x) &= \int_{-\infty}^{\infty} f_{xy}(x,y)dy \\ &= \int_2^3 \exp(-x)dy \\ &= \exp(-x) \int_2^3 dy \\ &= \exp(-x) \end{aligned}$$

Thus, the marginal probability density function of X is

$$f_X(x) = \begin{cases} \exp(-x)\ ifx \in [0,\infty) \\ 0 \quad \text{otherwise} \end{cases}$$

Verifying that $f_{xy}(x,y) = f_X(x) f_y(y)$ is straightforward. When $x \in [0,\infty)$ or $y \notin [2,3]$, then $f_x(x) f_y(y) = 0 = f_{x,y}(x,y)$. When $x \in [0,\infty)$ and $y \in [2,3)$, then

$$f_x(x) f_Y(y) = \exp(-x) \cdot 1 = \exp(-x) = f_{x,y}(x,y)$$

Thus, X and Y are independent.

Conditional Probability

A conditional probability is a probability that a certain event will occur given some knowledge about the outcome or some other event.

Suppose that a fair die is thrown and the score noted. Let A be the event, the score is 'even'. Then

$$A = \{2, 4, 6\}$$

$$\therefore \quad P(A) = \frac{3}{6} = \frac{1}{2}.$$

Now suppose we are told that the score is greater than 3. With this additional information what will be P(A)?

Let B be the event, 'the score is greater than 3'. Then B is {4, 5 and 6}. When we say that B has occurred, the event 'the score is less than or equal to 3' is no longer possible. Hence the sample space has changed from 6 to 3 points only. Out of these three points 4, 5 and 6; 4 and 6 are even scores.

Thus, given that B has occurred, P (A) must be $\frac{2}{3}$

Let us denote the probability of A given that B has already occurred by P (A|B).

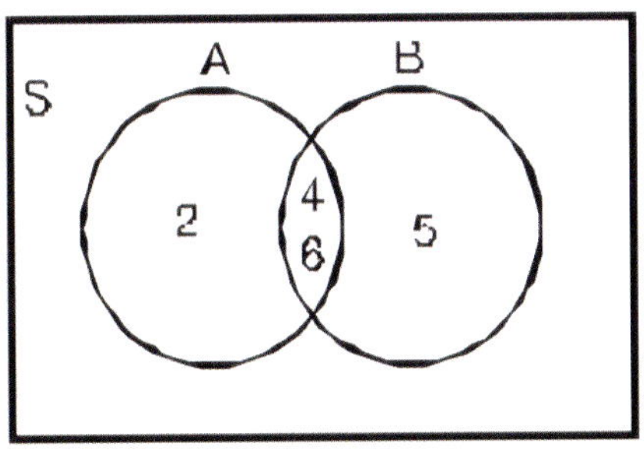

Again, consider the experiment of drawing a single card from a deck of 52 cards. We are interested in the event A consisting of the outcome that a black ace is drawn.

Since we may assume that there are 52 equally likely possible outcomes and there are two black aces in the deck, so we have

$$P(A) = \frac{2}{52}$$

However, suppose a card is drawn and we are informed that it is a spade. How should this information be used to reappraise the likelihood of the event A?

Clearly, since the event B "A spade has been drawn" has occurred, the event "not spade" is no longer possible. Hence, the sample space has changed from 52 playing cards to 13 spade cards. The number of black aces that can be drawn has now been reduced to 1.

Therefore, we must compute the probability of event A relative to the new sample space B

The event A is " a black ace is drawn'. We have computed the probability of the event A knowing that B has occurred. This means that we are computing a probability relative to a new sample space B. That is, B is treated as the universal set. We should consider only that part of A which is included in B.

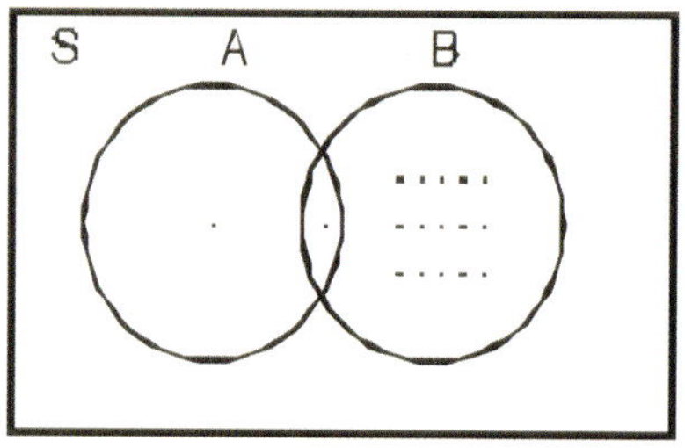

Hence, we consider $A \cap B$.

Thus, the probability of A. given B, is the ratio of the number of entries in $A \cap B$ to the number of entries in B. Since $n(A \cap B) = 1$ and $n(B) = 13$,

then $$P(A|B) = \frac{n(A \cap B)}{n(B)} = \frac{1}{13}$$

Notice that $n(A \cap B) = 1 \Rightarrow P(A \cap B) = \frac{1}{52}$

$$n(B) = 13 \Rightarrow P(B) = \frac{13}{52}$$

$$\therefore \quad P(A|B) = \frac{1}{13} = \frac{\frac{1}{52}}{\frac{13}{52}} = \frac{P(A \cap B)}{P(B)}$$

This leads to the definition of conditional probability as given below:

Let A an B be two events defined on a sample space S. Let P(B) > 0, then the conditional probability of A, provided B has already occurred, is denoted by P(A|B) and mathematically written as :

$$P(A|B) = \frac{P(A \cap B)}{P(B)}, \quad P(B) > 0$$

Similary, $P(B|A) = \frac{P(A \cap B)}{P(A)}, P(A) > 0$

The symbol P(A | B) is usually read as "the probability of A given B"

Example: Consider all families "with two children (not twins). Assume that all the elements of the sample space {BB, BG, GB, GG} are equally likely. (Here, for instance, BG denotes the birth sequence "boy girls"). Let A be the event {BB} and B be the event that 'at least one boy'. Calculate P (A | B).

Solution:

$$\text{Here} \quad A = \{BB\}$$

$$B = \{BB, BG, GB\}$$

$$A \cap B = \{BB\}$$

$$\therefore \quad P(A \cap B) = \frac{1}{4}$$

$$P(B) = \frac{1}{4} + \frac{1}{4} + \frac{1}{4} = \frac{3}{4}$$

$$\text{Hence,} \quad P(A|B) = \frac{P(A \cap B)}{P(B)}$$

$$\frac{\frac{1}{4}}{\frac{3}{4}} = \frac{1}{3}.$$

Example: Assume that a certain school contains equal number of female and male students. 5 % of the male population is football players. Find the probability that a randomly selected student is a football player male.

Solution: Let M = Male

F = Football player

We wish to calculate $P(M \cap F)$ From the given data,

$P(M) = \frac{1}{2}$ ($\because$ School contains equal number of male and female students)

$$P(F|M) = 0.05$$

But from definition of conditional probability, we have

$$P(F|M) = \frac{P(M \cap F)}{P(M)}$$

$$\Rightarrow \quad P(M \cap F) = P(M) \times P(F|M)$$

$$= \frac{1}{2} \times 0.05 = 0.025$$

Example: If A and B are two events, such that P (A) = 0.8,

$$P(B)=0.6, \quad P(A\cap B)=0.5, \text{ find the value of}$$
$$(i)\,P(A\cap B)\ (ii)\,P(B|A)\ \ (iii)\ P(A|B).$$

Solution:

$$(i)\,P(A\cup B)=P(A)+P(B)-P(A\cap B)$$
$$=0.8+0.6-0.5=0.9$$

$$(ii)\qquad P(B|A)=\frac{P(A\cap B)}{P(A)}=\frac{0.5}{0.8}=\frac{5}{8}$$

$$(iii)\qquad P(A|B)=\frac{P(A\cap B)}{P(B)}$$
$$=\frac{0.5}{0.6}=\frac{5}{6}$$

Example: Find the chance of drawing 2 white balls in succession from a bag containing 5 red and 7 white balls, the balls drawn not being replaced.

Solution: Let A be the event that ball drawn is white in the first draw. B be the event that ball drawn is white in the second draw.

$$\therefore \qquad P(A\cap B)=P(A)P(B|A)$$

$$\text{Here, } P(A)=\frac{7}{12}, P(B|A)=\frac{6}{11}$$

$$\therefore \quad P(A\cap B)=\frac{7}{12}\times\frac{6}{11}=\frac{7}{22}$$

Example: A coin is tossed until a head appears or until it has been tossed three times. Given that head does not occur on the first toss, what is the probability that coin is tossed three times?

Solution: Here, it is given that head does not occur on the first toss. That is, we may get the head on the second toss or on the third toss or even no head.

Let B be the event, “no heads on first toss”.

Then B = {TH, TTH, TTT}

These events are mutually exclusive

$$P(B)=P(TH)+P(TTH)+P(TTT)$$

Now $P(TH)=\frac{1}{4}$ ($\because$ This event has the sample space of four outcomes)

And $P(TTH) = P(TTT) = \frac{1}{8}$ ($\because$ This event has the sample space of eight outcomes)

Putting these values in $P(B) = P(TH) + P(TTH) + P(TTT)$, we get

$$P(B) = \frac{1}{4} + \frac{1}{8} + \frac{1}{8} + = \frac{4}{8} + = \frac{1}{2}$$

Let A be the event "coin is tossed three times"

Then $A = \{TTH, TTT\}$

$\therefore$ We have to find P (A | B).

$$P(A|B) = \frac{P(A \cap B)}{P(B)}$$

Here, $A \cap B = A$

$$\therefore \quad P(A|B) = \frac{\frac{1}{4}}{\frac{1}{2}} = \frac{1}{2}$$

Theorems on Multiplication Law of Probability and Conditional Probability

Theorem 1: For two events A and B,

$$P(A \cap B) = P(A).P(B|A),$$

$$\text{and } P(A \cap B) = P(B)P(A|B),$$

where $P(B|A)$ represents the conditional probability of occurrence of B, when the event A has already occurred and $P(A|B)$ is the conditional probability of happening of A, given that B has already happened.

Proof : Let n (S) denote the total number of equally likely cases, n (A) denote the cases favourable to the event A, n (B) denote the cases favourable to B and $n(A \cap B)$ denote the cases favourable to both A and B.

$$\therefore \quad P(A) = \frac{n(A)}{n(S)},$$

$$P(B) = \frac{n(B)}{n(S)}$$

$$P(A \cap B) = \frac{n(A \cap B)}{n(S)}$$

For the conditional event $A|B$, the favourable outcomes must be one of the sample points of B, i.e., for the event A|B, the sample space is B and out of the n (B) sample points, n $(A\cap B)$ pertain to the occurrence of the event A, Hence,

$$P(A|B)=\frac{n(A\cap B)}{n(B)}$$

Rewriting $P(A\cap B)=\frac{n(A\cap B)}{n(S)}$, we get $P(A\cap B)=\frac{n(B)}{n(S)}.\frac{n(A\cap B)}{n(B)}=P(B).P(A|B)$

Similarly, we can prove

$$P(A\cap B)=P(A).P(B|A)$$

Note: If A and B are independent events, then

$$P(A|B)=P(A) \text{and } P(B|A)=P(B)$$
$$P(A\cap B)=P(A).P(B)$$

Theorem 2: Two events A and B of the sample space S are independent, if and only if

$$P(A\cap B)=P(A).P(B)$$

Proof : If A and B are independent events,

Then $P(A|B)=P(A)$

We know that $P(A|B)=\frac{P(A\cap B)}{P(B)}$

$$\Rightarrow P(A\cap B)=P(A)P(B)$$

Hence, if A and B are independent events, then the probability of 'A and B' is equal to the product of the probability of A and probability of B.

Conversely, if $P(A\cap B)=P(A)P(B)$,then

$$P(A|B)=\frac{P(A\cap B)}{P(B)} \text{gives}$$
$$P(A|B)=\frac{P(A)P(B)}{P(B)}=P(A)$$

That is, A and B are independent events.

Probability Distributions

A probability distribution is a table or an equation that links each outcome of a statistical experiment with its probability of occurrence.

An example will make clear the relationship between random variables and probability distributions. Suppose you flip a coin two times. This simple statistical experiment can have four possible outcomes: HH, HT, TH, and TT. Now, let the variable X represent the number of Heads that result from this experiment. The variable X can take on the values 0, 1, or 2. In this example, X is a random variable; because its value is determined by the outcome of a statistical experiment.

Cumulative Probability Distributions

A cumulative probability refers to the probability that the value of a random variable falls within a specified range.

Let us return to the coin flip experiment. If we flip a coin two times, we might ask: What is the probability that the coin flips would result in one or fewer heads? The answer would be a cumulative probability. It would be the probability that the coin flip experiment results in zero heads plus the probability that the experiment results in one head.

$$P(X <= 1) = P(X = 0) + P(X = 1) = 0.25 + 0.50 = 0.75$$

Like a probability distribution, a cumulative probability distribution can be represented by a table or an equation. In the table below, the cumulative probability refers to the probability than the random variable X is less than or equal to x.

Number of heads: x	Probability: P(X <= x)	Cumulative Probability: P(X < x)
0	0.25	0.25
1	0.50	0.75
2	0.25	1.00

Uniform Probability Distribution

The simplest probability distribution occurs when all of the values of a random variable occur with equal probability. This probability distribution is called the uniform distribution.

Uniform Distribution: Suppose the random variable X can assume k different values. Suppose also that the $P(X = x_k)$ is constant. Then,

$$P(X = x_k) = 1/k$$

Example: Suppose a die is tossed. What is the probability that the die will land on 5?

Solution: When a die is tossed, there are 6 possible outcomes represented by: S = {1, 2, 3, 4, 5 and 6}. Each possible outcome is a random variable (X), and each outcome is equally likely to occur. Thus, we have a uniform distribution. Therefore, the P(X = 5) = 1/6.

Example: Suppose we repeat the dice tossing experiment described in Example 1. This time, we ask what is the probability that the die will land on a number that is smaller than 5?

Solution: When a die is tossed, there are 6 possible outcomes represented by: S = { 1, 2, 3, 4, 5, 6 }. Each possible outcome is equally likely to occur. Thus, we have a uniform distribution.

This problem involves a cumulative probability. The probability that the die will land on a number smaller than 5 is equal to:

$P(X < 5) = P(X = 1) + P(X = 2) + P(X = 3) + P(X = 4)$

$P(X < 5) = 1/6 + 1/6 + 1/6 + 1/6 = 2/3$

The Binomial Distribution

A distribution where only two outcomes are possible, such as success or failure, gain or loss, win or lose and where the probability of success and failure is same for all the trials is called a Binomial Distribution.

The outcomes need not be equally likely. Remember the example of a fight between me and Undertaker? So, if the probability of success in an experiment is 0.2 then the probability of failure can be easily computed as q = 1 – 0.2 = 0.8.

Each trial is independent since the outcome of the previous toss doesn't determine or affect the outcome of the current toss. An experiment with only two possible outcomes repeated n number of times is called binomial. The parameters of a binomial distribution are n and p where n is the total number of trials and p is the probability of success in each trial.

On the basis of the above explanation, the properties of a Binomial Distribution are

1. Each trial is independent.
2. There are only two possible outcomes in a trial- either a success or a failure.
3. A total number of n identical trials are conducted.
4. The probability of success and failure is same for all trials. (Trials are identical.)

The mathematical representation of binomial distribution is given by:

$$P(x) = \frac{n!}{(n-x)!x!} p^x q^{n-x}$$

A binomial distribution graph where the probability of success does not equal the probability of failure looks like

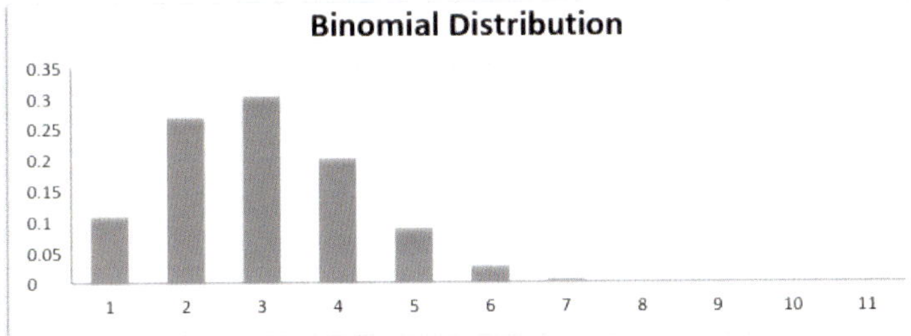

Now, when probability of success = probability of failure, in such a situation the graph of binomial distribution looks like

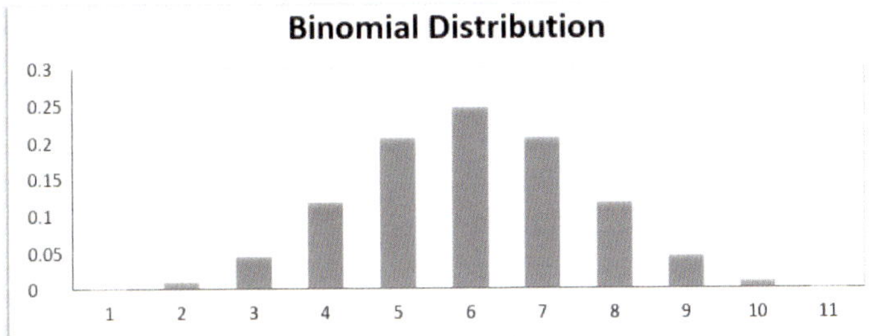

The mean and variance of a binomial distribution are given by:

Mean -> $\mu = n*p$

Variance -> $Var(X) = n*p*q$

Cumulative Distribution Function

The cumulative distribution function can be expressed as:

$$F(k;n,p) = \Pr(X \le k) = \sum_{i=0}^{\lfloor k \rfloor} \binom{n}{i} p^i (1-p)^{n-i}$$

where $\lfloor k \rfloor$ is the "floor" under k, i.e. the greatest integer less than or equal to k.

It can also be represented in terms of the regularized incomplete beta function, as follows:

$$\begin{aligned} F(k;n,p) &= \Pr(X \le k) \\ &= I_{1-p}(n-k, k+1) \\ &= (n-k)\binom{n}{k}\int_0^{1-p} t^{n-k-1}(1-t)^k \, dt. \end{aligned}$$

Some closed-form bounds for the cumulative distribution function are given below.

Example

Suppose a biased coin comes up heads with probability 0.3 when tossed. What is the probability of achieving 0, 1,..., 6 heads after six tosses?

$$\Pr(0 \text{ heads}) = f(0) = \Pr(X=0) = \binom{6}{0} 0.3^{0}(1-0.3)^{6-0} = 0.117649$$

$$\Pr(1 \text{ heads}) = f(1) = \Pr(X=1) = \binom{6}{1} 0.3^{1}(1-0.3)^{6-1} = 0.302526$$

$$\Pr(2 \text{ heads}) = f(2) = \Pr(X=2) = \binom{6}{2} 0.3^{2}(1-0.3)^{6-2} = 0.324135$$

$$\Pr(3 \text{ heads}) = f(3) = \Pr(X=3) = \binom{6}{3} 0.3^{3}(1-0.3)^{6-3} = 0.18522$$

$$\Pr(4 \text{ heads}) = f(4) = \Pr(X=4) = \binom{6}{4} 0.3^{4}(1-0.3)^{6-4} = 0.059535$$

$$\Pr(5 \text{ heads}) = f(5) = \Pr(X=5) = \binom{6}{5} 0.3^{5}(1-0.3)^{6-5} = 0.010206$$

$$\Pr(6 \text{ heads}) = f(6) = \Pr(X=6) = \binom{6}{6} 0.3^{6}(1-0.3)^{6-6} = 0.000729$$ [2]

Mean

If $X \sim B(n, p)$, that is, X is a binomially distributed random variable, n being the total number of experiments and p the probability of each experiment yielding a successful result, then the expected value of X is:

$$\mathrm{E}[X] = np.$$

For example, if $n = 100$, and $p = 1/4$, then the average number of successful results will be 25.

Proof: We calculate the mean, μ, directly calculated from its definition

$$\mu = \sum_{i=0}^{n} x_i p_i,$$

and the binomial theorem:

$$\begin{aligned}
\mu &= \sum_{k=0}^{n} k \binom{n}{k} p^{k}(1-p)^{n-k} \\
&= np \sum_{k=0}^{n} k \frac{(n-1)!}{(n-k)!k!} p^{k-1}(1-p)^{(n-1)-(k-1)} \\
&= np \sum_{k=1}^{n} \frac{(n-1)!}{((n-1)-(k-1))!(k-1)!} p^{k-1}(1-p)^{(n-1)-(k-1)}
\end{aligned}$$

$$= np\sum_{k=1}^{n}\binom{n-1}{k-1}p^{k-1}(1-p)^{(n-1)-(k-1)}$$

$$= np\sum_{\ell=0}^{n-1}\binom{n-1}{\ell}p^{\ell}(1-p)^{(n-1)-\ell} \qquad \text{with } \ell := k-1$$

$$= np\sum_{\ell=0}^{m}\binom{m}{\ell}p^{\ell}(1-p)^{m-\ell} \qquad \text{with } m := n-1$$

$$= np(p+(1-p))^{m}$$

$$= np$$

It is also possible to deduce the mean from the equation $X = X_1 + \cdots + X_n$ whereby all X_i are Bernoulli distributed random variables with $E[X_i] = p$ ($X_i = 1$ if the ith experiment succeeds and $X_i = 0$ otherwise). We get: $E[X] = E[X_1 + \cdots + X_n] = E[X_1] + \cdots + E[X_n] = \underbrace{p + \cdots + p}_{n \text{ times}} = np$

Variance

The variance is:

$$\text{Var}(X) = np(1-p).$$

Proof: Let $X = X_1 + \cdots + X_n$ where all X_i are independently Bernoulli distributed random variables. Since $\text{Var}(X_i) = p(1-p)$, we get:

$$\text{Var}(X) = \text{Var}(X_1 + \cdots + X_n) = \text{Var}(X_1) + \cdots + \text{Var}(X_n) = n\,\text{Var}(X_1) = np(1-p).$$

Mode

Usually the mode of a binomial $B(n, p)$ distribution is equal to $\lfloor (n+1)p \rfloor$, where $\lfloor . \rfloor$ is the floor function. However, when $(n + 1)p$ is an integer and p is neither 0 nor 1, then the distribution has two modes: $(n + 1)p$ and $(n + 1)p - 1$. When p is equal to 0 or 1, the mode will be 0 and n correspondingly. These cases can be summarized as follows:

$$\text{mode} = \begin{cases} \lfloor (n+1)p \rfloor & \text{if } (n+1)p \text{ is 0 or a noninteger,} \\ (n+1)p \text{ and } (n+1)p-1 & \text{if } (n+1)p \in \{1,\ldots,n\}, \\ n & \text{if } (n+1)p = n+1. \end{cases}$$

Proof: Let

$$f(k) = \binom{n}{k}p^{k}q^{n-k}$$

For $p = 0$ only $f(0)$ has a nonzero value with $f(0) = 1$. For $p = 1$ we find $f(n) = 1$ and $f(k) = 0$ for $k \neq n$. This proves that the mode is 0 for $p = 0$ and n for $p = 1$.

Let $0 < p < 1$. We find

$$\frac{f(k+1)}{f(k)} = \frac{(n-k)p}{(k+1)(1-p)}.$$

From this follows

$$\begin{aligned} k > (n+1)p - 1 &\Rightarrow f(k+1) < f(k) \\ k = (n+1)p - 1 &\Rightarrow f(k+1) = f(k) \\ k < (n+1)p - 1 &\Rightarrow f(k+1) > f(k) \end{aligned}$$

So when $(n+1)p-1$ is an integer, then $(n+1)p-1$ and $(n+1)p$ is a mode. In the case that $(n+1)p-1 \notin \mathbb{Z}$, then only $\lfloor (n+1)p-1 \rfloor + 1 = \lfloor (n+1)p \rfloor$ is a mode.

Median

In general, there is no single formula to find the median for a binomial distribution, and it may even be non-unique. However several special results have been established:

- If np is an integer, then the mean, median, and mode coincide and equal np.
- Any median m must lie within the interval $\lfloor np \rfloor \le m \le \lceil np \rceil$.
- A median m cannot lie too far away from the mean: $|m - np| \le \min\{\ln 2, \max\{p, 1-p\}\}$.
- The median is unique and equal to $m = \text{round}(np)$ in cases when either $p \le 1 - \ln 2$ or $p \ge \ln 2$ or $|m - np| \le \min\{p, 1-p\}$ (except for the case when $p = ½$ and n is odd).
- When $p = 1/2$ and n is odd, any number m in the interval $½(n-1) \le m \le ½(n+1)$ is a median of the binomial distribution. If $p = 1/2$ and n is even, then $m = n/2$ is the unique median.

Covariance between two Binomials

If two binomially distributed random variables X and Y are observed together, estimating their covariance can be useful. The covariance is

$$\text{Cov}(X,Y) = \text{E}(XY) - \mu_X \mu_Y.$$

In the case $n = 1$ (the case of Bernoulli trials) XY is non-zero only when both X and Y are one, and μ_X and μ_Y are equal to the two probabilities. Defining p_B as the probability of both happening at the same time, this gives

$$\text{Cov}(X,Y) = p_B - p_X p_Y,$$

and for n independent pairwise trials

$$\text{Cov}(X,Y)_n = n(p_B - p_X p_Y).$$

If X and Y are the same variable, this reduces to the variance formula given above.

Related Distributions

Sums of Binomials

If $X \sim B(n, p)$ and $Y \sim B(m, p)$ are independent binomial variables with the same probability p, then $X + Y$ is again a binomial variable; its distribution is $Z=X+Y \sim B(n+m, p)$:

$$P(Z=k)=\sum_{i=0}^{k}\left[\binom{n}{i}p^{i}(1-p)^{n-i}\right]\left[\binom{m}{k-i}p^{k-i}(1-p)^{m-k+i}\right]$$

$$=\binom{n+m}{k}p^{k}(1-p)^{n+m-k}$$

However, if X and Y do not have the same probability p, then the variance of the sum will be smaller than the variance of a binomial variable distributed as $B(n+m,\bar{p})$.

Ratio of two Binomial Distributions

This result was first derived by Katz.

Let p_1 and p_2 be the probabilities of success in the binomial distributions B(X,n) and B(Y,m) respectively. Let T = (X/n)/(Y/m).

Then log(T) is approximately normally distributed with mean $\log(p_1/p_2)$ and variance $((1/p_1) - 1)/n + ((1/p_2) - 1)/m$.

Conditional Binomials

If $X \sim B(n, p)$ and, conditional on X, $Y \sim B(X, q)$, then Y is a simple binomial variable with distribution $Y \sim B(n, pq)$.

For example, imagine throwing n balls to a basket U_X and taking the balls that hit and throwing them to another basket U_Y. If p is the probability to hit U_X then $X \sim B(n, p)$ is the number of balls that hit U_X. If q is the probability to hit U_Y then the number of balls that hit U_Y is $Y \sim B(X, q)$ and therefore $Y \sim B(n, pq)$.

Proof

Since $X \sim B(n,p)$ and $Y \sim B(X,q)$, by the law of total probability,

$$\Pr[Y=m]=\sum_{k=m}^{n}\Pr[Y=m\mid X=k]\Pr[X=k]$$

$$=\sum_{k=m}^{n}\binom{n}{k}\binom{k}{m}p^{k}q^{m}(1-p)^{n-k}(1-q)^{k-m},$$

Since $\binom{n}{k}\binom{k}{m}=\binom{n}{m}\binom{n-m}{k-m}$, the equation above can be expressed as

$$\Pr[Y=m]=\sum_{k=m}^{n}\binom{n}{m}\binom{n-m}{k-m}p^{k}q^{m}(1-p)^{n-k}(1-q)^{k-m}$$

Factoring $p^k = p^m p^{k-m}$ and pulling all the terms that don't depend on k out of the sum now yields

$$\Pr[Y=m] = \binom{n}{m} p^m q^m \left(\sum_{k=m}^{n} \binom{n-m}{k-m} p^{k-m} (1-p)^{n-k} (1-q)^{k-m} \right)$$

$$= \binom{n}{m} (pq)^m \left(\sum_{k=m}^{n} \binom{n-m}{k-m} \left(p(1-q)\right)^{k-m} (1-p)^{n-k} \right)$$

After substituting $i = k - m$ in the expression above, we get

$$\Pr[Y=m] = \binom{n}{m} (pq)^m \left(\sum_{i=0}^{n-m} \binom{n-m}{i} (p-pq)^i (1-p)^{n-m-i} \right)$$

Notice that the sum (in the parentheses) above equals $(p - pq + 1 - p)^{n-m}$ by the binomial theorem. Substituting this in finally yields

$$\Pr[Y=m] = \binom{n}{m} (pq)^m (p - pq + 1 - p)^{n-m}$$

$$= \binom{n}{m} (pq)^m (1-pq)^{n-m}$$

and thus $Y \sim B(n, pq)$ as desired.

Bernoulli Distribution

The Bernoulli distribution is a special case of the binomial distribution, where $n = 1$. Symbolically, $X \sim \mathrm{B}(1, p)$ has the same meaning as $X \sim \mathrm{B}(p)$. Conversely, any binomial distribution, $\mathrm{B}(n, p)$, is the distribution of the sum of n Bernoulli trials, $\mathrm{B}(p)$, each with the same probability p.

Poisson Binomial Distribution

The binomial distribution is a special case of the Poisson binomial distribution, or general binomial distribution, which is the distribution of a sum of n independent non-identical Bernoulli trials $\mathrm{B}(p_i)$.

Normal Approximation

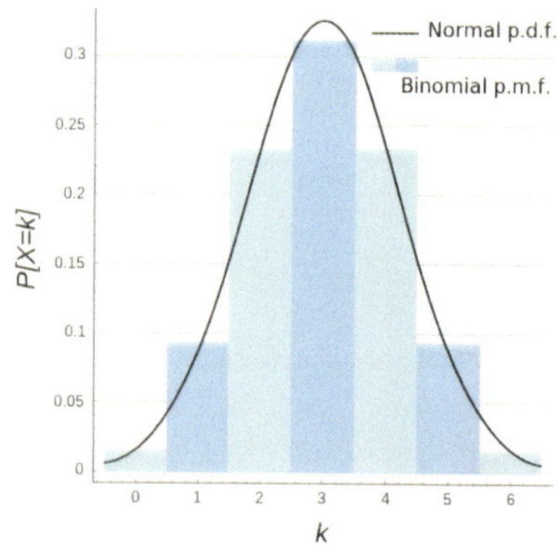

Binomial probability mass function and normal probability density function approximation for $n = 6$ and $p = 0.5$

If n is large enough, then the skew of the distribution is not too great. In this case a reasonable approximation to $B(n, p)$ is given by the normal distribution and this basic approximation can be improved in a simple way by using a suitable continuity correction.

$$\mathcal{N}(np, np(1-p)),$$

The basic approximation generally improves as n increases (at least 20) and is better when p is not near to 0 or 1. Various rules of thumb may be used to decide whether n is large enough, and p is far enough from the extremes of zero or one:

- One rule is that for $n > 5$ the normal approximation is adequate if the absolute value of the skewness is strictly less than 1/3; that is, if

$$\frac{|1-2p|}{\sqrt{np(1-p)}} = \frac{1}{\sqrt{n}}\left|\sqrt{\frac{1-p}{p}} - \sqrt{\frac{p}{1-p}}\right| < \frac{1}{3}.$$

- A stronger rule states that the normal approximation is appropriate only if everything within 3 standard deviations of its mean is within the range of possible values; that is, only if

$$\mu \pm 3\sigma = np \pm 3\sqrt{np(1-p)} \in (0, n).$$

This 3-standard-deviation rule is equivalent to the following conditions, which also imply the first rule above.

$$n > 9\frac{1-p}{p} \quad \text{and} \quad n > 9\frac{p}{1-p}.$$

Proof

The rule $np \pm 3\sqrt{np(1-p)} \in (0, n)$ is totally equivalent to request that

$$np - 3\sqrt{np(1-p)} > 0 \quad \text{and} \quad np + 3\sqrt{np(1-p)} < n.$$

Moving terms around yields:

$$np > 3\sqrt{np(1-p)} \quad \text{and} \quad n(1-p) > 3\sqrt{np(1-p)}.$$

Since $0 < p < 1$, we can apply the square power and divide by the respective factors np^2 and $n(1-p)^2$, to obtain the desired conditions:

$$n > 9\frac{1-p}{p} \quad \text{and} \quad n > 9\frac{p}{1-p}$$

Notice that these conditions automatically imply that $n > 9$. On the other hand, apply again the square root and divide by 3,

$$\frac{\sqrt{n}}{3} > \sqrt{\frac{1-p}{p}} > 0 \quad \text{and} \quad \frac{\sqrt{n}}{3} > \sqrt{\frac{p}{1-p}} > 0.$$

Subtracting the second set of inequalities from the first one yields:

$$\frac{\sqrt{n}}{3} > \sqrt{\frac{1-p}{p}} - \sqrt{\frac{p}{1-p}} > -\frac{\sqrt{n}}{3};$$

and so, the desired first rule is satisfied,

$$\left| \sqrt{\frac{1-p}{p}} - \sqrt{\frac{p}{1-p}} \right| < \frac{\sqrt{n}}{3}.$$

Another commonly used rule is that both values np and $n(1-p)$ must be greater than or equal to 5. However, the specific number varies from source to source, and depends on how good an approximation one wants. In particular, if one uses 9 instead of 5, the rule implies the results stated in the previous paragraphs.

Proof

Assume that both values np and $n(1-p)$ are greater than 9. Since $0 < p < 1$, we easily have that

$$np \geq 9 > 9(1-p) \quad \text{and} \quad n(1-p) \geq 9 > 9p.$$

We only have to divide now by the respective factors p and , to deduce the alternative form of the 3-standard-deviation rule:

$$n > 9\frac{1-p}{p} \quad \text{and} \quad n > 9\frac{p}{1-p}.$$

The following is an example of applying a continuity correction. Suppose one wishes to calculate $\Pr(X \leq 8)$ for a binomial random variable *X*. If *Y* has a distribution given by the normal approximation, then $\Pr(X \leq 8)$ is approximated by $\Pr(Y \leq 8.5)$. The addition of 0.5 is the continuity correction; the uncorrected normal approximation gives considerably less accurate results.

This approximation, known as de Moivre–Laplace theorem, is a huge time-saver when undertaking calculations by hand (exact calculations with large *n* are very onerous); historically, it was the first use of the normal distribution, introduced in Abraham de Moivre's book *The Doctrine of Chances* in 1738. Nowadays, it can be seen as a consequence of the central limit theorem since B(*n*, *p*) is a sum of *n* independent, identically distributed Bernoulli variables with parameter *p*. This fact is the basis of a hypothesis test, a "proportion z-test", for the value of p using *x/n*, the sample proportion and estimator of *p*, in a common test statistic.

For example, suppose one randomly samples *n* people out of a large population and ask them whether they agree with a certain statement. The proportion of people who agree will of course depend on the sample. If groups of *n* people were sampled repeatedly and truly randomly, the proportions would follow an approximate normal distribution with mean equal to the true proportion *p* of agreement in the population and with standard deviation

$$\sigma = \sqrt{\frac{p(1-p)}{n}}$$

Poisson Approximation

The binomial distribution converges towards the Poisson distribution as the number of trials goes to infinity while the product np remains fixed or at least p tends to zero. Therefore, the Poisson distribution with parameter $\lambda = np$ can be used as an approximation to $B(n, p)$ of the binomial distribution if n is sufficiently large and p is sufficiently small. According to two rules of thumb, this approximation is good if $n \geq 20$ and $p \leq 0.05$, or if $n \geq 100$ and $np \leq 10$.

Concerning the accuracy of Poisson approximation.

Limiting Distributions

- *Poisson limit theorem*: As n approaches ∞ and p approaches 0 with the product np held fixed, the Binomial(n, p) distribution approaches the Poisson distribution with expected value $\lambda = np$.
- *de Moivre–Laplace theorem*: As n approaches ∞ while p remains fixed, the distribution of approaches the normal distribution with expected value 0 and variance 1.

$$\frac{X - np}{\sqrt{np(1-p)}}$$

This result is sometimes loosely stated by saying that the distribution of X is asymptotically normal with expected value np and variance $np(1 - p)$. This result is a specific case of the central limit theorem.

Beta Distribution

Beta distributions provide a family of prior probability distributions for binomial distributions in Bayesian inference:

$$P(p;\alpha,\beta) = \frac{p^{\alpha-1}(1-p)^{\beta-1}}{B(\alpha,\beta)}.$$

Confidence Intervals

Even for quite large values of n, the actual distribution of the mean is significantly nonnormal. Because of this problem several methods to estimate confidence intervals have been proposed.

In the equations for confidence intervals below, the variables have the following meaning:

- n_1 is the number of successes out of n, the total number of trials
- $\hat{p} = \frac{n_1}{n}$ is the proportion of successes
- z is the $1-\frac{1}{2}\alpha$ quantile of a standard normal distribution (i.e., probit) corresponding to the target error rate α. For example, for a 95% confidence level the error α = 0.05, so $1-\frac{1}{2}\alpha = 0.975$ and z = 1.96.

Wald Method

$$\hat{p} \pm z \sqrt{\frac{\hat{p}(1-\hat{p})}{n}}.$$

A continuity correction of $0.5/n$ may be added.

Agresti–Coull Method

$$\tilde{p} \pm z \sqrt{\frac{\tilde{p}(1-\tilde{p})}{n+z^2}}.$$

Here the estimate of p is modified to

$$\tilde{p} = \frac{n_1 + \frac{1}{2} z^2}{n + z^2}$$

ArcSine Method

$$\sin^2\left(\arcsin\left(\sqrt{\hat{p}}\right) \pm \frac{z}{2\sqrt{n}}\right)$$

Wilson (Score) Method

The notation in the formula below differs from the previous formulas in two respects:

- Firstly, z_x has a slightly different interpretation in the formula below: it has its ordinary meaning of 'the xth quantile of the standard normal distribution', rather than being a shorthand for 'the $(1 - x)$-th quantile'.
- Secondly, this formula does not use a plus-minus to define the two bounds. Instead, one may use $z = z_{\alpha/2}$ to get the lower bound, or use $z = z_{1-\alpha/2}$ to get the upper bound. For example: for a 95% confidence level the error $\alpha = 0.05$, so one gets the lower bound by using $z = z_{\alpha/2} = z_{0.025} = -1.96$, and one gets the upper bound by using $z = z_{1-\alpha/2} = z_{0.975} = 1.96$.

$$\frac{\hat{p} + \frac{z^2}{2n} + z\sqrt{\frac{\hat{p}(1-\hat{p})}{n} + \frac{z^2}{4n^2}}}{1 + \frac{z^2}{n}}$$ [21]

Comparison

The exact (Clopper–Pearson) method is the most conservative.

The Wald method, although commonly recommended in textbooks, is the most biased.

Generating Binomial Random Variates

Methods for random number generation where the marginal distribution is a binomial distribution are well-established.

One way to generate random samples from a binomial distribution is to use an inversion algorithm. To do so, one must calculate the probability that P(X=k) for all values k from 0 through n. (These probabilities should sum to a value close to one, in order to encompass the entire sample space.) Then by using a pseudorandom number generator to generate samples uniformly between 0 and 1, one can transform the calculated samples U[0,1] into discrete numbers by using the probabilities calculated in step one.

Tail Bounds

For $k \le np$, upper bounds for the lower tail of the distribution function can be derived. Recall that $F(k;n,p) = \Pr(X \le k)$, the probability that there are at most k successes.

Hoeffding's inequality yields the bound

$$F(k;n,p) \le \exp\left(-2\frac{(np-k)^2}{n}\right),$$

and Chernoff's inequality can be used to derive the bound

$$F(k;n,p) \le \exp\left(-\frac{1}{2p}\frac{(np-k)^2}{n}\right).$$

Moreover, these bounds are reasonably tight when $p = 1/2$, since the following expression holds for all $k \ge 3n/8$

$$F(k;n,\tfrac{1}{2}) \le \frac{14}{15}\exp\left(-\frac{16(\frac{n}{2}-k)^2}{n}\right).$$

However, the bounds do not work well for extreme values of p. In particular, as $p \to 1$, value $F(k;n,p)$ goes to zero (for fixed k, n with $k < n$) while the upper bound above goes to a positive constant. In this case a better bound is given by

$$F(k;n,p) \le \exp\left(-nD\left(\frac{k}{n} \| p\right)\right) \qquad \text{if } 0 < \frac{k}{n} < p$$

where $D(a \,||\, p)$ is the relative entropy between an a-coin and a p-coin (i.e. between the Bernoulli(a) and Bernoulli(p) distribution):

$$D(a \,||\, p) = (a)\log\frac{a}{p} + (1-a)\log\frac{1-a}{1-p}.$$

Asymptotically, this bound is reasonably tight; see for details. An equivalent formulation of the bound is

$$\Pr(X \geq k) = F(n-k;n,1-p) \leq \exp\left(-nD\left(\frac{k}{n}\| \,p\right)\right) \qquad \text{if } p < \frac{k}{n} < 1$$

Both these bounds are derived directly from the Chernoff bound. It can also be shown that,

$$\Pr(X \geq k) = F(n-k;n,1-p) \geq \frac{1}{(n+1)^2} \exp\left(-nD\left(\frac{k}{n}\| \,p\right)\right) \qquad \text{if } p < \frac{k}{n} < 1$$

We can also change the $(n+1)^2$ in the denominator to $\sqrt{2n}$, by approximating the binomial coefficient with Stirling's formula.

The Poisson Distribution

In probability theory and statistics, the Poisson distribution, named after French mathematician Siméon Denis Poisson, is a discrete probability distribution that expresses the probability of a given number of events occurring in a fixed interval of time or space if these events occur with a known constant rate and independently of the time since the last event. The Poisson distribution can also be used for the number of events in other specified intervals such as distance, area or volume.

For instance, an individual keeping track of the amount of mail they receive each day may notice that they receive an average number of 4 letters per day. If receiving any particular piece of mail does not affect the arrival times of future pieces of mail, i.e., if pieces of mail from a wide range of sources arrive independently of one another, then a reasonable assumption is that the number of pieces of mail received in a day obeys a Poisson distribution. Other examples that may follow a Poisson include the number of phone calls received by a call center per hour and the number of decay events per second from a radioactive source.

Basics

The Poisson distribution is popular for modelling the *number of times an event occurs in an interval of time or space.*

Example

The Poisson distribution may be useful to model events such as

- The number of meteorites greater than 1 meter diameter that strike Earth in a year.
- The number of patients arriving in an emergency room between 10 and 11 pm.
- In a laser the number of photons hitting a detector in a particular time interval will follow Poisson distribution.

Assumptions: When is the Poisson Distribution an Appropriate Model?

The Poisson distribution is an appropriate model if the following assumptions are true.

- k is the number of times an event occurs in an interval and k can take values 0, 1, 2, ...
- The occurrence of one event does not affect the probability that a second event will occur. That is, events occur independently.
- The rate at which events occur is constant. The rate cannot be higher in some intervals and lower in other intervals.
- Two events cannot occur at exactly the same instant; instead, at each very small sub-interval exactly one event either occurs or does not occur.
- The actual probability distribution is given by a binomial distribution and the number of trials is sufficiently bigger than the number of successes one is asking about.

If these conditions are true, then k is a Poisson random variable, and the distribution of k is a Poisson distribution.

Probability of Events for a Poisson Distribution

An event can occur 0, 1, 2, ... times in an interval. The average number of events in an interval is designated λ (lambda). Lambda is the event rate, also called the rate parameter. The probability of observing k events in an interval is given by the equation

$$P(k \text{ events in interval}) = e^{-\lambda} \frac{\lambda^k}{k!}$$

where

- λ is the average number of events per interval
- e is the number 2.71828... (Euler's number) the base of the natural logarithms
- k takes values 0, 1, 2, ...
- $k! = k \times (k-1) \times (k-2) \times \ldots \times 2 \times 1$ is the factorial of k.

This equation is the probability mass function (PMF) for a Poisson distribution.

Notice that this equation can be adapted if, instead of the average number of events λ, we are given a time rate r for the events to happen. Then $\lambda = rt$ (with r in units of 1/time), and

$$P(k \text{ events in interval } t) = e^{-rt} \frac{(rt)^k}{k!}$$

Examples of Probability for Poisson Distributions

On a particular river, overflow floods occur once every 100 years on average. Calculate the probability of k = 0, 1, 2, 3, 4, 5, or 6 overflow floods in a 100-year interval, assuming the Poisson model is appropriate.

Because the average event rate is one overflow flood per 100 years, $\lambda = 1$

$$P(k \text{ overflow floods in 100 years}) = \frac{\lambda^k e^{-\lambda}}{k!} = \frac{1^k e^{-1}}{k!}$$

$$P(k = 0 \text{ overflow floods in 100 years}) = \frac{1^0 e^{-1}}{0!} = \frac{e^{-1}}{1} \approx 0.368$$

$$P(k = 1 \text{ overflow flood in 100 years}) = \frac{1^1 e^{-1}}{1!} = \frac{e^{-1}}{1} \approx 0.368$$

$$P(k = 2 \text{ overflow floods in 100 years}) = \frac{1^2 e^{-1}}{2!} = \frac{e^{-1}}{2} \approx 0.184$$

The table below gives the probability for 0 to 6 overflow floods in a 100-year period.

k	*P*(*k* overflow floods in 100 years)
0	0.368
1	0.368
2	0.184
3	0.061
4	0.015
5	0.003
6	0.0005

Ugarte and colleagues report that the average number of goals in a World Cup soccer match is approximately 2.5 and the Poisson model is appropriate.

Because the average event rate is 2.5 goals per match, $\lambda = 2.5$.

$$P(k \text{ goals in a match}) = \frac{2.5^k e^{-2.5}}{k!}$$

$$P(k = 0 \text{ goals in a match}) = \frac{2.5^0 e^{-2.5}}{0!} = \frac{e^{-2.5}}{1} \approx 0.082$$

$$P(k = 1 \text{ goal in a match}) = \frac{2.5^1 e^{-2.5}}{1!} = \frac{2.5e^{-2.5}}{1} \approx 0.205$$

$$P(k = 2 \text{ goals in a match}) = \frac{2.5^2 e^{-2.5}}{2!} = \frac{6.25e^{-2.5}}{2} \approx 0.257$$

The table below gives the probability for 0 to 7 goals in a match.

k	P(k goals in a World Cup soccer match)
0	0.082
1	0.205
2	0.257
3	0.213
4	0.133
5	0.067
6	0.028
7	0.010

Once in an Interval Events: The Special Case of λ = 1 and k = 0

Suppose that astronomers estimate that large meteorites (above a certain size) hit the earth on average once every 100 years ($\lambda = 1$ event per 100 years), and that the number of meteorite hits follows a Poisson distribution. What is the probability of $k = 0$ meteorite hits in the next 100 years?

$$P(k = 0 \text{ meteorites hit in next 100 years}) = \frac{1^0 e^{-1}}{0!} = \frac{1}{e} \approx 0.37$$

Under these assumptions, the probability that no large meteorites hit the earth in the next 100 years is roughly 0.37. The remaining 1 – 0.37 = 0.63 is the probability of 1, 2, 3, or more large meteorite hits in the next 100 years. In an example above, an overflow flood occurred once every 100 years ($\lambda = 1$). The probability of no overflow floods in 100 years was roughly 0.37, by the same calculation.

In general, if an event occurs on average once per interval ($\lambda = 1$), and the events follow a Poisson distribution, then P(0 events in next interval) = 0.37. In addition, P(exactly one event in next interval) = 0.37.

Examples that Violate the Poisson Assumptions

The number of students who arrive at the student union per minute will likely not follow a Poisson distribution, because the rate is not constant (low rate during class time, high rate between class times) and the arrivals of individual students are not independent (students tend to come in groups).

The number of magnitude 5 earthquakes per year in a country may not follow a Poisson distribution if one large earthquake increases the probability of aftershocks of similar magnitude.

Among patients admitted to the intensive care unit of a hospital, the number of days that the patients spend in the ICU is not Poisson distributed because the number of days cannot be zero. The distribution may be modeled using a Zero-truncated Poisson distribution.

Count distributions in which the number of intervals with zero events is higher than predicted by a Poisson model may be modeled using a Zero-inflated model.

Poisson Regression and Negative Binomial Regression

Poisson regression and negative binomial regression are useful for analyses where the dependent (response) variable is the count (0, 1, 2, ...) of the number of events or occurrences in an interval.

A discrete random variable X is said to have a Poisson distribution with parameter $\lambda > 0$, if, for $k = 0, 1, 2, \ldots$, the probability mass function of X is given by:

$$f(k;\lambda) = \Pr(X = k) = \frac{\lambda^k e^{-\lambda}}{k!},$$

where

- e is Euler's number ($e = 2.71828\ldots$)
- $k!$ is the factorial of k.

The positive real number λ is equal to the expected value of X and also to its variance

$$\lambda = \mathrm{E}(X) = \mathrm{Var}(X).$$

The Poisson distribution can be applied to systems with a large number of possible events, each of which is rare. How many such events will occur during a fixed time interval? Under the right circumstances, this is a random number with a Poisson distribution.

The conventional definition of the Poisson distribution contains two terms that can easily overflow on computers: λ^k and $k!$. The fraction of λ^k to $k!$ can also produce a rounding error that is very large compared to $e^{-\lambda}$, and therefore give an erroneous result. For numerical stability the Poisson probability mass function should therefore be evaluated as

$$f(k;\lambda) = \exp\{k \ln \lambda - \lambda - \ln \Gamma(k+1)\},$$

which is mathematically equivalent but numerically stable. The natural logarithm of the Gamma function can be obtained using the lgamma function in the C (programming language) standard library (C99 version), the gammaln function in MATLAB or SciPy, or the log_gamma function in Fortran 2008 and later.

Properties

Descriptive Statistics

- The expected value and variance of a Poisson-distributed random variable are both equal to λ.
- The coefficient of variation is $\lambda^{-1/2}$, while the index of dispersion is 1.
- The mean absolute deviation about the mean is

$$\mathrm{E}|X - \lambda| = 2\exp(-\lambda)\frac{\lambda^{\lfloor \lambda \rfloor + 1}}{\lfloor \lambda \rfloor !}.$$

- The mode of a Poisson-distributed random variable with non-integer λ is equal to , which is the largest integer less than or equal to λ. This is also written as floor(λ). When λ is a positive integer, the modes are λ and $\lambda - 1$.
- All of the cumulants of the Poisson distribution are equal to the expected value λ. The nth factorial moment of the Poisson distribution is λ^n.
- The expected value of a Poisson process is sometimes decomposed into the product of *intensity* and *exposure* (or more generally expressed as the integral of an "intensity function" over time or space, sometimes described as "exposure").

Median

Bounds for the median (v) of the distribution are known and are sharp:

$$\lambda - \ln 2 \le v < \lambda + \frac{1}{3}.$$

Higher Moments

- The higher moments m_k of the Poisson distribution about the origin are Touchard polynomials in λ:

$$m_k = \sum_{i=0}^{k} \lambda^i \begin{Bmatrix} k \\ i \end{Bmatrix},$$

where the {braces} denote Stirling numbers of the second kind. The coefficients of the polynomials have a combinatorial meaning. In fact, when the expected value of the Poisson distribution is 1, then Dobinski's formula says that the nth moment equals the number of partitions of a set of size n.

Sums of Poisson-distributed Random Variables

If $X_i \sim \text{Pois}(\lambda_i) i = 1,\ldots,n$ are independent, and $\lambda = \sum_{i=1}^{n} \lambda_i$, then $Y = \left(\sum_{i=1}^{n} X_i \right) \sim \text{Pois}(\lambda)$. . A converse is Raikov's theorem, which says that if the sum of two independent random variables is Poisson-distributed, then so are each of those two independent random variables.

Other Properties

- The Poisson distributions are infinitely divisible probability distributions.
- The directed Kullback–Leibler divergence of Pois(λ_0) from Pois(λ) is given by

$$D_{\text{KL}}(\lambda \mid \lambda_0) = \lambda_0 - \lambda + \lambda \log \frac{\lambda}{\lambda_0}.$$

- Bounds for the tail probabilities of a Poisson random variable $X \sim \text{Pois}(\lambda)$ can be derived using a Chernoff bound argument.

$$P(X \geq x) \leq \frac{e^{-\lambda}(e\lambda)^x}{x^x}, \text{ for } x > \lambda,$$

$$P(X \leq x) \leq \frac{e^{-\lambda}(e\lambda)^x}{x^x}, \text{ for } x < \lambda.$$

Poisson Races

Let $X \sim \text{Pois}(\lambda)$ and $Y \sim \text{Pois}(\mu)$ be independent random variables, with $\lambda < \mu$, then we have that

$$\frac{e^{-(\sqrt{\mu}-\sqrt{\lambda})^2}}{(\lambda+\mu)^2} - \frac{e^{-(\lambda+\mu)}}{2\sqrt{\lambda\mu}} - \frac{e^{-(\lambda+\mu)}}{4\lambda\mu} \leq P(X-Y \geq 0) \leq e^{-(\sqrt{\mu}-\sqrt{\lambda})^2}$$

The upper bound is proved using a standard Chernoff bound.

The lower bound can be proved by noting that $P(X-Y \geq 0 \mid X+Y=i)$ is the probability that $Z \geq \frac{i}{2}$, where $Z \sim \text{Bin}\left(i, \frac{\lambda}{\lambda+\mu}\right)$, which is bounded below by $\frac{1}{(i+1)^2} e^{\left(2D\left(0.5\| \frac{\lambda}{\lambda+\mu}\right)\right)}$, where D is relative entropy. Further noting that $X+Y \sim \text{Pois}(\lambda+\mu)$, and computing a lower bound on the unconditional probability gives the result. More details can be found in the appendix of.

Related Distributions

- If $X_1 \sim \text{Pois}(\lambda_1)$ and $X_2 \sim \text{Pois}(\lambda_2)$ are independent, then the difference $Y = X_1 - X_2$ follows a Skellam distribution.

- If $X_1 \sim \text{Pois}(\lambda_1)$ and $X_2 \sim \text{Pois}(\lambda_2)$ are independent, then the distribution of X_1 conditional on $X_1 + X_2$ is a binomial distribution.

  Specifically, given $X_1 + X_2 = k$, $X_1 \sim \text{Binom}(k, \lambda_1/(\lambda_1+\lambda_2))$.

  More generally, if $X_1, X_2, \ldots, X_n$ are independent Poisson random variables with parameters $\lambda_1, \lambda_2, \ldots, \lambda_n$ then

  given $\sum_{j=1}^{n} X_j = k$, $X_j = k$, $X_i \sim \text{Binom}\left(k, \frac{\lambda_i}{\sum_{j-1}^{n} \lambda_j}\right)$. In fact, $\{X_i\} \sim \text{Multinom}\left(k, \left\{\frac{\lambda_i}{\sum_{j=1}^{n} \lambda_j}\right\}\right)$.

- If $X \sim \text{Pois}(\lambda)$ and the distribution of Y, conditional on X = k, is a binomial distribution, $Y \mid (X=k) \sim \text{Binom}(k,p)$, then the distribution of Y follows a Poisson distribution $Y \sim \text{Pois}(\lambda \cdot p)$. In fact, if $\{Y_i\}$, conditional on X = k, follows a multinomial distribution, $\{Y_i\} \mid (X=k) \sim \text{Multinom}(k, p_i)$, then each Y_i. follows an independent Poisson distribution $Y_i \sim \text{Pois}(\lambda \cdot p_i), \rho(Y_i, Y_j) = 0$.
- The Poisson distribution can be derived as a limiting case to the binomial distribution as

the number of trials goes to infinity and the expected number of successes remains fixed — see law of rare events below. Therefore, it can be used as an approximation of the binomial distribution if n is sufficiently large and p is sufficiently small. There is a rule of thumb stating that the Poisson distribution is a good approximation of the binomial distribution if n is at least 20 and p is smaller than or equal to 0.05, and an excellent approximation if $n \geq 100$ and $np \leq 10$.

$$F_{\text{Binomial}}(k;n,p) \approx F_{\text{Poisson}}(k;\lambda = np)$$

- The Poisson distribution is a special case of the discrete compound Poisson distribution (or stuttering Poisson distribution) with only a parameter. The discrete compound Poisson distribution can be deduced from the limiting distribution of univariate multinomial distribution. It is also a special case of a compound Poisson distribution.
- For sufficiently large values of λ, (say λ>1000), the normal distribution with mean λ and variance λ (standard deviation $\sqrt{\lambda}$) is an excellent approximation to the Poisson distribution. If λ is greater than about 10, then the normal distribution is a good approximation if an appropriate continuity correction is performed, i.e., if $P(X \leq x)$, where x is a non-negative integer, is replaced by $P(X \leq x + 0.5)$.

$$F_{\text{Poisson}}(x;\lambda) \approx F_{\text{normal}}(x;\mu = \lambda, \sigma^2 = \lambda)$$

- Variance-stabilizing transformation: When a variable is Poisson distributed, its square root is approximately normally distributed with expected value of about $\sqrt{\lambda}$ and variance of about 1/4. Under this transformation, the convergence to normality (as λ increases) is far faster than the untransformed variable. Other, slightly more complicated, variance stabilizing transformations are available, one of which is Anscombe transform. See Data transformation (statistics) for more general uses of transformations.
- If for every $t > 0$ the number of arrivals in the time interval $[0, t]$ follows the Poisson distribution with mean λt, then the sequence of inter-arrival times are independent and identically distributed exponential random variables having mean $1/\lambda$.
- The cumulative distribution functions of the Poisson and chi-squared distributions are related in the following ways:

$$F_{\text{Poisson}}(k;\lambda) = 1 - F_{\chi^2}(2\lambda; 2(k+1)) \qquad \text{integer } k,$$

and

$$\Pr(X = k) = F_{\chi^2}(2\lambda; 2(k+1)) - F_{\chi^2}(2\lambda; 2k).$$

Occurrence

Applications of the Poisson distribution can be found in many fields related to counting:

- Telecommunication example: telephone calls arriving in a system.

- Astronomy example: photons arriving at a telescope.
- Chemistry example: the molar mass distribution of a living polymerization.
- Biology example: the number of mutations on a strand of DNA per unit length.
- Management example: customers arriving at a counter or call centre.
- Finance and insurance example: number of losses or claims occurring in a given period of time.
- Earthquake seismology example: an asymptotic Poisson model of seismic risk for large earthquakes.
- Radioactivity example: number of decays in a given time interval in a radioactive sample.

The Poisson distribution arises in connection with Poisson processes. It applies to various phenomena of discrete properties (that is, those that may happen 0, 1, 2, 3, ... times during a given period of time or in a given area) whenever the probability of the phenomenon happening is constant in time or space. Examples of events that may be modelled as a Poisson distribution include:

- The number of soldiers killed by horse-kicks each year in each corps in the Prussian cavalry. This example was made famous by a book of Ladislaus Josephovich Bortkiewicz .
- The number of yeast cells used when brewing Guinness beer. This example was made famous by William Sealy Gosset.
- The number of phone calls arriving at a call centre within a minute. This example was made famous by A.K. Erlang.
- Internet traffic.
- The number of goals in sports involving two competing teams.
- The number of deaths per year in a given age group.
- The number of jumps in a stock price in a given time interval.
- Under an assumption of homogeneity, the number of times a web server is accessed per minute.
- The number of mutations in a given stretch of DNA after a certain amount of radiation.
- The proportion of cells that will be infected at a given multiplicity of infection.
- The arrival of photons on a pixel circuit at a given illumination and over a given time period.
- The targeting of V-1 flying bombs on London during World War II investigated by R. D. Clarke in 1946.

Gallagher in 1976 showed that the counts of prime numbers in short intervals obey a Poisson distribution provided a certain version of an unproved conjecture of Hardy and Littlewood is true.

Law of Rare Events

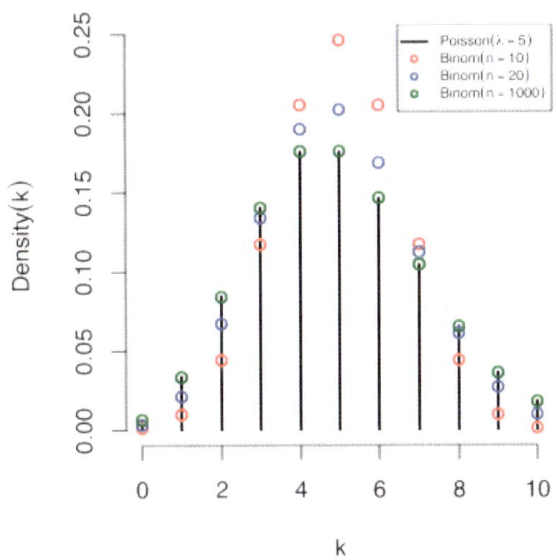

Comparison of the Poisson distribution (black lines) and the binomial distribution with $n = 10$ (red circles), $n = 20$ (blue circles), $n = 1000$ (green circles). All distributions have a mean of 5. The horizontal axis shows the number of events k. Notice that as n gets larger, the Poisson distribution becomes an increasingly better approximation for the binomial distribution with the same mean.

The rate of an event is related to the probability of an event occurring in some small subinterval (of time, space or otherwise). In the case of the Poisson distribution, one assumes that there exists a small enough subinterval for which the probability of an event occurring twice is "negligible". With this assumption one can derive the Poisson distribution from the Binomial one, given only the information of expected number of total events in the whole interval. Let this total number be λ. Divide the whole interval into λ subintervals $I_1,\ldots,I_n$ of equal size, such that $n > \lambda$ (since we are interested in only very small portions of the interval this assumption is meaningful). This means that the expected number of events in an interval I_i for each i is equal to . Now we assume that the occurrence of an event in the whole interval can be seen as a Bernoulli trial, where the ith trial corresponds to looking whether an event happens at the subinterval I_i with probability λ / n. The expected number of total events in n such trials would be λ, the expected number of total events in the whole interval. Hence for each subdivision of the interval we have approximated the occurrence of the event as a Bernoulli process of the form $\mathrm{B}(n, \lambda / n)$. As we have noted before we want to consider only very small subintervals. Therefore, we take the limit as n goes to infinity. In this case the binomial distribution converges to what is known as the Poisson distribution by the Poisson limit theorem.

In several of the above examples—such as, the number of mutations in a given sequence of DNA—the events being counted are actually the outcomes of discrete trials, and would more precisely be modelled using the binomial distribution, that is

$$X \sim \mathrm{B}(n, p).$$

In such cases n is very large and p is very small (and so the expectation np is of intermediate magnitude). Then the distribution may be approximated by the less cumbersome Poisson distribution.

$$X \sim \text{Pois}(np).$$

This approximation is sometimes known as the law of rare events, since each of the n individual Bernoulli events rarely occurs. The name may be misleading because the total count of success events in a Poisson process need not be rare if the parameter np is not small. For example, the number of telephone calls to a busy switchboard in one hour follows a Poisson distribution with the events appearing frequent to the operator, but they are rare from the point of view of the average member of the population who is very unlikely to make a call to that switchboard in that hour.

The word law is sometimes used as a synonym of probability distribution, and *convergence in law* means *convergence in distribution*. Accordingly, the Poisson distribution is sometimes called the law of small numbers because it is the probability distribution of the number of occurrences of an event that happens rarely but has very many opportunities to happen. *The Law of Small Numbers* is a book by Ladislaus Bortkiewicz (Bortkevitch) about the Poisson distribution, published in 1898.

Poisson Point Process

The Poisson distribution arises as the number of points of a Poisson point process located in some finite region. More specifically, if D is some region space, for example Euclidean space R^d, for which $|D|$, the area, volume or, more generally, the Lebesgue measure of the region is finite, and if $N(D)$ denotes the number of points in D, then

$$P(N(D)=k)=\frac{(\lambda\,|\,D\,|)^k e^{-\lambda|D|}}{k!}.$$

Other Applications in Science

In a Poisson process, the number of observed occurrences fluctuates about its mean λ with a standard deviation $\sigma_k = \sqrt{\lambda}$. These fluctuations are denoted as Poisson noise or (particularly in electronics) as *shot noise*.

The correlation of the mean and standard deviation in counting independent discrete occurrences is useful scientifically. By monitoring how the fluctuations vary with the mean signal, one can estimate the contribution of a single occurrence, *even if that contribution is too small to be detected directly*. For example, the charge e on an electron can be estimated by correlating the magnitude of an electric current with its shot noise. If N electrons pass a point in a given time t on the average, the mean current is $I = eN/t$; since the current fluctuations should be of the order $\sigma_I = e\sqrt{N}/t$ (i.e., the standard deviation of the Poisson process), the charge e can be estimated from the ratio $t\sigma_I^2/I$.

An everyday example is the graininess that appears as photographs are enlarged; the graininess is due to Poisson fluctuations in the number of reduced silver grains, not to the individual grains themselves. By correlating the graininess with the degree of enlargement, one can estimate the contribution of an individual grain (which is otherwise too small to be seen unaided). Many other molecular applications of Poisson noise have been developed, e.g., estimating the number density of receptor molecules in a cell membrane.

$$\Pr(N_t = k) = f(k; \lambda t) = \frac{e^{-\lambda t}(\lambda t)^k}{k!}.$$

In Causal Set theory the discrete elements of spacetime follow a Poisson distribution in the volume.

Generating Poisson-distributed Random Variables

A simple algorithm to generate random Poisson-distributed numbers (pseudo-random number sampling):

```
algorithm poisson random number (Knuth):
    init:
        Let L ← e^−λ, k ← 0 and p ← 1.
    do:
        k ← k + 1.
        Generate uniform random number u in [0,1] and let p ← p × u.
    while p > L.
    return k − 1.
```

The complexity is linear in the returned value k, which is λ on average. There are many other algorithms to improve this. Some are given in Ahrens & Dieter. Also, for large values of λ, there may be numerical stability issues because of the term $e^{-\lambda}$. This could be solved by a slight change to allow λ to be added into the calculation gradually:

```
algorithm poisson random number (Junhao, based on Knuth):
    init:
        Let λLeft ← λ, k ← 0 and p ← 1.
    do:
        k ← k + 1.
        Generate uniform random number u in (0,1) and let p ← p × u.
        while p < 1 and λLeft > 0:
            if λLeft > STEP:
                p ← p × e^STEP
                λLeft ← λLeft − STEP
            else:
                p ← p × e^λLeft
                λLeft ← 0
```

```
    while p > 1.
    return k - 1.
```

The choice of STEP depends on the threshold of overflow. For double precision floating point format, the threshold is near e^{700}, so 500 shall be a safe *STEP*.

Other solutions for large values of λ include rejection sampling and using Gaussian approximation.

Inverse transform sampling is simple and efficient for small values of λ, and requires only one uniform random number *u* per sample. Cumulative probabilities are examined in turn until one exceeds *u*.

```
algorithm Poisson generator based upon the inversion by sequential search:[38]
    init:
        Let x ← 0, p ← e^-λ, s ← p.
        Generate uniform random number u in [0,1].
    while u > s do:
        x ← x + 1.
        p ← p * λ / x.
        s ← s + p.
    return x.
```

"This algorithm, requires expected time proportional to λ as λ→∞. For large λ, round-off errors proliferate, which provides us with another reason for avoiding large values of λ."

Parameter Estimation

Maximum Likelihood

Given a sample of n measured values $k_i = 0, 1, 2, \ldots$, for $i = 1, \ldots, n$, we wish to estimate the value of the parameter λ of the Poisson population from which the sample was drawn. The maximum likelihood estimate is

$$\hat{\lambda}_{\mathrm{MLE}} = \frac{1}{n}\sum_{i=1}^{n} k_i.$$

Since each observation has expectation λ so does this sample mean. Therefore, the maximum likelihood estimate is an unbiased estimator of λ. It is also an efficient estimator, i.e. its estimation variance achieves the Cramér–Rao lower bound (CRLB). Hence it is minimum-variance unbiased. Also it can be proven that the sum (and hence the sample mean as it is a one-to-one function of the sum) is a complete and sufficient statistic for λ.

To prove sufficiency we may use the factorization theorem. Consider partitioning the probability mass function of the joint Poisson distribution for the sample into two parts: one that depends solely on the sample $\mathbf{x}$ (called $h(\mathbf{x})$) and one that depends on the parameter $\mathbf{x}$ and the sample $\mathbf{x}$ only through the function $T(\mathbf{x})$. Then $T(\mathbf{x})$ is a sufficient statistic for λ.

$$P(\mathbf{x}) = \prod_{i=1}^{n} \frac{\lambda^{x_i} e^{-\lambda}}{x_i!} = \frac{1}{\prod_{i=1}^{n} x_i!} \times \lambda^{\sum_{i=1}^{n} x_i} e^{-n\lambda}$$

Note that the first term, $h(\mathbf{x})$, depends only on $\mathbf{x}$. The second term, $g(T(\mathbf{x})|\lambda)$, depends on the sample only through $T(\mathbf{x}) = \sum_{i=1}^{n} x_i$. Thus, $T(\mathbf{x})$ is sufficient.

To find the parameter λ that maximizes the probability function for the Poisson population, we can use the logarithm of the likelihood function:

$$\begin{aligned}\ell(\lambda) &= \ln \prod_{i=1}^{n} f(k_i \mid \lambda) \\ &= \sum_{i=1}^{n} \ln\left(\frac{e^{-\lambda}\lambda^{k_i}}{k_i!}\right) \\ &= -n\lambda + \left(\sum_{i=1}^{n} k_i\right)\ln(\lambda) - \sum_{i=1}^{n} \ln(k_i!)\end{aligned}$$

We take the derivative of ℓ with respect to λ and compare it to zero:

$$\frac{\mathrm{d}}{\mathrm{d}\lambda}\ell(\lambda) = 0 \Leftrightarrow -n + \left(\sum_{i=1}^{n} k_i\right)\frac{1}{\lambda} = 0.$$

Solving for λ gives a stationary point.

$$\lambda = \frac{\sum_{i=1}^{n} k_i}{n}$$

So λ is the average of the k_i values. Obtaining the sign of the second derivative of L at the stationary point will determine what kind of extreme value λ is.

$$\frac{\partial^2 \ell}{\partial \lambda^2} = -\lambda^{-2}\sum_{i=1}^{n} k_i$$

Evaluating the second derivative *at the stationary point* gives:

$$\frac{\partial^2 \ell}{\partial \lambda^2} = -\frac{n^2}{\sum_{i=1}^{n} k_i}$$

which is the negative of n times the reciprocal of the average of the k_i. This expression is negative when the average is positive. If this is satisfied, then the stationary point maximizes the probability function.

For completeness, a family of distributions is said to be complete if and only if $E(g(T)) = 0$ implies that $P_\lambda(g(T)=0)=1$ for all λ. If the individual X_i are iid $\mathrm{Po}(\lambda)$, then $T(\mathbf{x})=\sum_{i=1}^{n}X_i \sim \mathrm{Po}(n\lambda)$. Knowing the distribution we want to investigate, it is easy to see that the statistic is complete.

$$E(g(T))=\sum_{t=0}^{\infty}g(t)\frac{(n\lambda)^t e^{-n\lambda}}{t!}=0$$

For this equality to hold $g(t)$ must be 0. This follows from the fact that none of the other terms will be 0 for all t in the sum and for all possible values of λ. Hence, $E(g(T))=0$ for all λ implies that $P_\lambda(g(T)=0)=1$, and the statistic has been shown to be complete.

Confidence Interval

The confidence interval for the mean of a Poisson distribution can be expressed using the relationship between the cumulative distribution functions of the Poisson and chi-squared distributions. The chi-squared distribution is itself closely related to the gamma distribution, and this leads to an alternative expression. Given an observation k from a Poisson distribution with mean μ, a confidence interval for μ with confidence level $1-\alpha$ is

$$\tfrac{1}{2}\chi^2(\alpha/2;2k)\le\mu\le\tfrac{1}{2}\chi^2(1-\alpha/2;2k+2),$$

or equivalently,

$$F^{-1}(\alpha/2;k,1)\le\mu\le F^{-1}(1-\alpha/2;k+1,1)$$

where $\chi^2(p;n)$ is the quantile function (corresponding to a lower tail area p) of the chi-squared distribution with n degrees of freedom and $F^{-1}(p;n,1)$ is the quantile function of a Gamma distribution with shape parameter n and scale parameter 1. This interval is 'exact' in the sense that its coverage probability is never less than the nominal $1-\alpha$.

When quantiles of the Gamma distribution are not available, an accurate approximation to this exact interval has been proposed (based on the Wilson–Hilferty transformation):

$$k\left(1-\frac{1}{9k}-\frac{z_{\alpha/2}}{3\sqrt{k}}\right)^3\le\mu\le(k+1)\left(1-\frac{1}{9(k+1)}+\frac{z_{\alpha/2}}{3\sqrt{k+1}}\right)^3$$

where $z_{\alpha/2}$ denotes the standard normal deviate with upper tail area $\alpha/2$.

For application of these formulae in the same context as above (given a sample of n measured values k_i each drawn from a Poisson distribution with mean λ), one would set

$$k=\sum_{i=1}^{n}k_i$$

calculate an interval for $\mu = n\lambda$, and then derive the interval for λ.

Bayesian Inference

In Bayesian inference, the conjugate prior for the rate parameter λ of the Poisson distribution is the gamma distribution. Let

$$\lambda \sim \text{Gamma}(\alpha, \beta)$$

denote that λ is distributed according to the gamma density g parameterized in terms of a shape parameter α and an inverse scale parameter β:

$$g(\lambda \mid \alpha, \beta) = \frac{\beta^{\alpha}}{\Gamma(\alpha)} \lambda^{\alpha-1} e^{-\beta\lambda} \qquad \text{for } \lambda > 0$$

Then, given the same sample of n measured values k_i as before, and a prior of Gamma(α, β), the posterior distribution is

$$\lambda \sim \text{Gamma}\left(\alpha + \sum_{i=1}^{n} k_i, \beta + n\right).$$

The posterior mean E[λ] approaches the maximum likelihood estimate $\hat{\lambda}_{\text{MLE}}$ in the limit as $\alpha \to 0, \beta \to 0,$ which follows immediately from the general expression of the mean of the gamma distribution.

The posterior predictive distribution for a single additional observation is a negative binomial distribution, sometimes called a Gamma–Poisson distribution.

Simultaneous Estimation of Multiple Poisson Means

Suppose $X_1, X_2, \ldots, X_p$ is a set of independent random variables from a set of p Poisson distributions, each with a parameter λ_i, $i = 1, \ldots, p$, and we would like to estimate these parameters. Then, Clevenson and Zidek show that under the normalized squared error loss $L(\lambda, \hat{\lambda}) = \sum_{i=1}^{p} \lambda_i^{-1} (\hat{\lambda}_i - \lambda_i)^2$, when $p > 1$, then, similar as in Stein's famous example for the Normal means, the MLE estimator $\hat{\lambda}_i = X_i$ is inadmissible.

In this case, a family of minimax estimators is given for any $0 < c \leq 2(p-1)$ and $b \geq (p - 2 + p^{-1})$ as

$$\hat{\lambda}_i = \left(1 - \frac{c}{b + \sum_{i=1}^{p} X_i}\right) X_i, \qquad i = 1, \ldots, p.$$

Bivariate Poisson Distribution

This distribution has been extended to the bivariate case. The generating function for this distribution is

$$g(u, v) = \exp[(\theta_1 - \theta_{12})(u-1) + (\theta_2 - \theta_{12})(v-1) + \theta_{12}(uv-1)]$$

with

$$\theta_1, \theta_2 > \theta_{12} > 0$$

The marginal distributions are Poisson(θ_1) and Poisson(θ_2) and the correlation coefficient is limited to the range

$$0 \le \rho \le \min\left\{\frac{\theta_1}{\theta_2}, \frac{\theta_2}{\theta_1}\right\}$$

A simple way to generate a bivariate Poisson distribution X_1, X_2 is to take three independent Poisson distributions Y_1, Y_2, Y_3 with means $\lambda_1, \lambda_2, \lambda_3$ and then set $X_1 = Y_1 + Y_3, X_2 = Y_2 + Y_3$. The probability function of the bivariate Poisson distribution is

$$\Pr(X_1 = k_1, X_2 = k_2)$$
$$= \exp\left(-\lambda_1 - \lambda_2 - \lambda_3\right)\frac{\lambda_1^{k_1}}{k_1!}\frac{\lambda_2^{k_2}}{k_2!}\sum_{k=0}^{\min(k_1,k_2)}\binom{k_1}{k}\binom{k_2}{k}k!\left(\frac{\lambda_3}{\lambda_1\lambda_2}\right)^k$$

Computer Software for the Poisson Distribution

Poisson Distribution Using R

The R function dpois(x, lambda) calculates the probability that there are x events in an interval, where the argument "lambda" is the average number of events per interval.

For example,

dpois(x=0, lambda=1) = 0.3678794

dpois(x=1, lambda=2.5) = 0.2052125

The following R code creates a graph of the Poisson distribution from x= 0 to 8, with lambda=2.5.

```
x=0:8

px = dpois(x, lambda=2.5)

plot(x, px, type="h", xlab="Number of events k", ylab="Probability of k events", ylim=c(0,0.5), pty="s", main="Poisson distribution \n Probability of events for lambda = 2.5")
```

Poisson Distribution using Excel

The Excel function POISSON(x, mean, cumulative) calculates the probability of x events where mean is lambda, the average number of events per interval. The argument cumulative specifies the cumulative distribution.

For example,

=POISSON(0, 1, FALSE) = 0.3678794

=POISSON(1, 2.5, FALSE) = 0.2052125

Poisson Distribution using Python (Scipy)

The function scipy.stats.distributions.poisson.pmf(x, poissonLambda) calculates the probability that there are x events in an interval, where the argument "poissonLambda" is the average number of events per interval.

Poisson Distribution using Mathematical

Mathematica supports the univariate Poisson distribution as Poisson Distribution[λ], and the bivariate Poisson distribution as Multivariate Poisson Distribution[θ_{12}, {$\theta_1 - \theta_{12}$, $\theta_2 - \theta_{12}$}], including computation of probabilities and expectation, sampling, parameter estimation and hypothesis testing.

The Normal Distribution

The normal distribution is the most important probability distribution in statistics because it fits many natural phenomena. For example, heights, blood pressure, measurement error, and IQ scores follow the normal distribution. It is also known as the Gaussian distribution and the bell curve.

The normal distribution is a probability function that describes how the values of a variable are distributed. It is a symmetric distribution where most of the observations cluster around the central peak and the probabilities for values further away from the mean taper off equally in both directions. Extreme values in both tails of the distribution are similarly unlikely.

Example of Normally Distributed Data: Heights

Height data are normally distributed. The distribution in this example fits real data that I collected from 14-year-old girls during a study.

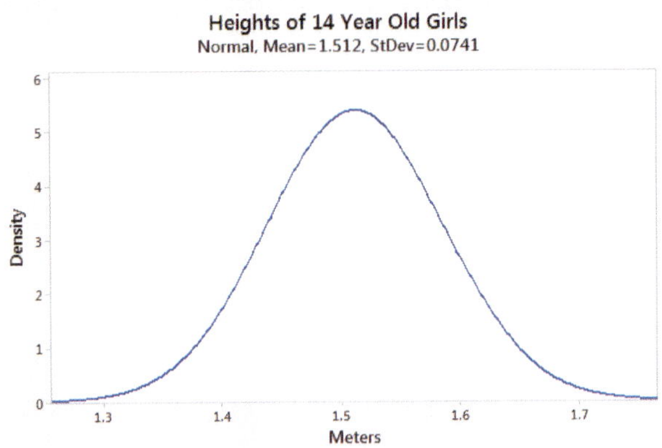

As you can see, the distribution of heights follows the typical pattern for all normal distributions. Most girls are close to the average (1.512 meters). Small differences between an individual's height and the mean occur more frequently than substantial deviations from the mean. The standard deviation is 0.0741m, which indicates the typical distance that individual girls tend to fall from mean height.

The distribution is symmetric. The number of girls shorter than average equals the number of girls taller than average. In both tails of the distribution, extremely short girls occur as infrequently as extremely tall girls.

Parameters of the Normal Distribution

As with any probability distribution, the parameters for the normal distribution define its shape and probabilities entirely. The normal distribution has two parameters, the mean and standard deviation. The normal distribution does not have just one form. Instead, the shape changes based on the parameter values, as shown in the graphs below.

Mean

The mean is the central tendency of the distribution. It defines the location of the peak for normal distributions. Most values cluster around the mean. On a graph, changing the mean shifts the entire curve left or right on the X-axis.

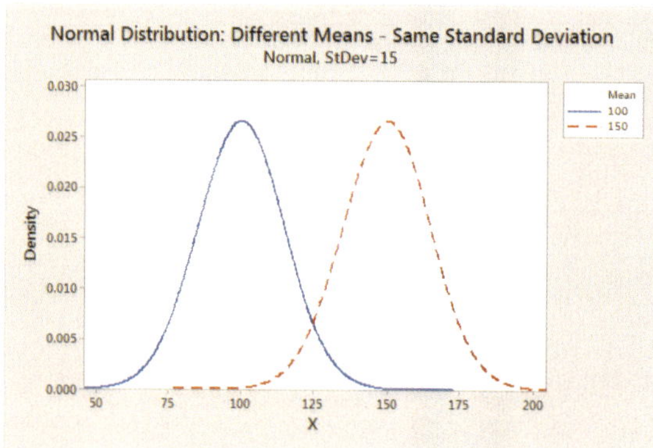

Standard Deviation

The standard deviation is a measure of variability. It defines the width of the normal distribution. The standard deviation determines how far away from the mean the values tend to fall. It represents the typical distance between the observations and the average.

On a graph, changing the standard deviation either tightens or spreads out the width of the distribution along the X-axis. Larger standard deviations produce distributions that are more spread out.

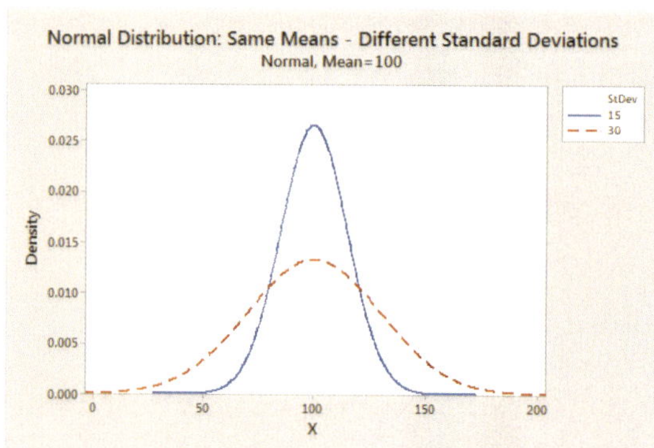

When you have narrow distributions, the probabilities are higher that values won't fall far from the mean. As you increase the spread of the distribution, the likelihood that observations will be further away from the mean also increases.

Population Parameters Versus Sample Estimates

The mean and standard deviation are parameter values that apply to entire populations. For the normal distribution, statisticians signify the parameters by using the Greek symbol μ (mu) for the population mean and σ (sigma) for the population standard deviation.

Unfortunately, population parameters are usually unknown because it's generally impossible to measure an entire population. However, you can use random samples to calculate estimates of these parameters. Statisticians represent sample estimates of these parameters using $\bar{x}$ for the sample mean and s for the sample standard deviation.

Common Properties for All Forms of the Normal Distribution

Despite the different shapes, all forms of the normal distribution have the following characteristic properties.

- They're all symmetric. The normal distribution cannot model skewed distributions.
- The mean, median, and mode are all equal.
- Half of the population is less than the mean and half is greater than the mean.
- The Empirical Rule allows you to determine the proportion of values that fall within certain distances from the mean.

While the normal distribution is essential in statistics, it is just one of many probability distributions, and it does not fit all populations.

The Empirical Rule for the Normal Distribution

When you have normally distributed data, the standard deviation becomes particularly valuable. You can use it to determine the proportion of the values that fall within a specified number of standard deviations from the mean. For example, in a normal distribution, 68% of the observations fall within (+/- 1) standard deviation from the mean. This property is part of the Empirical Rule, which describes the percentage of the data that fall within specific numbers of standard deviations from the mean for bell-shaped curves.

Mean +/- standard deviations	Percentage of data contained
1	68%
2	95%
3	99.7%

Let's look at a pizza delivery example. Assume that a pizza restaurant has a mean delivery time of 30 minutes and a standard deviation of 5 minutes. Using the Empirical Rule, we can determine that 68% of the delivery times are between 25-35 minutes (30 +/- 5), 95% are between 20-40 minutes (30 +/- 2*5), and 99.7% are between 15-45 minutes (30 +/-3*5). The chart below illustrates this property graphically.

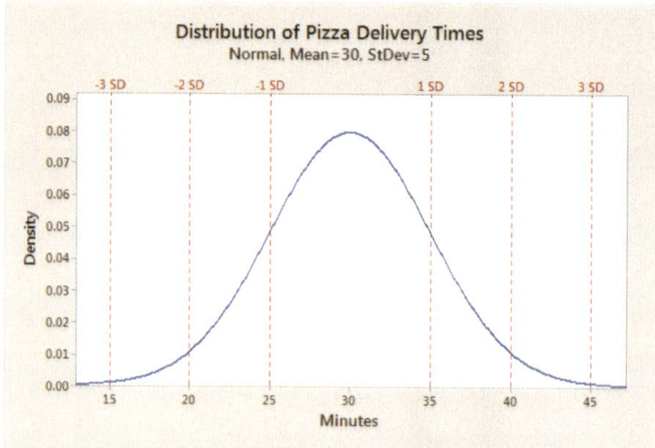

Standard Normal Distribution and Standard Scores

The normal distribution has many different shapes depending on the parameter values. However, the standard normal distribution is a special case of the normal distribution where the mean is zero and the standard deviation is 1. This distribution is also known as the Z-distribution.

A value on the standard normal distribution is known as a standard score or a Z-score. A standard score represents the number of standard deviations above or below the mean that a specific observation falls. For example, a standard score of 1.5 indicates that the observation is 1.5 standard deviations above the mean. On the other hand, a negative score represents a value below the average. The mean has a Z-score of 0.

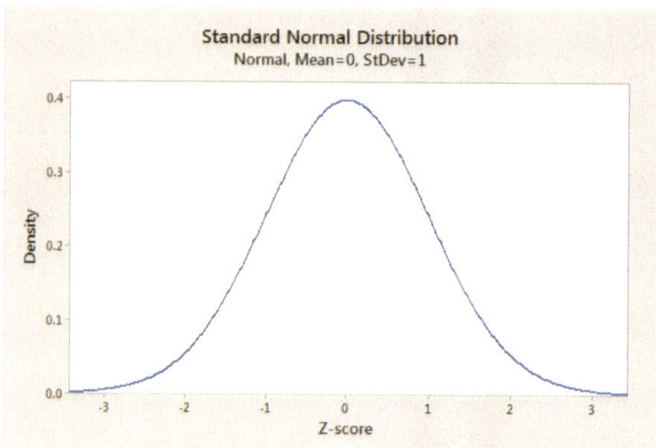

Suppose you weigh an apple and it weighs 110 grams. There's no way to tell from the weight alone how this apple compares to other apples. However, as you'll see, after you calculate its Z-score, you know where it falls relative to other apples.

Standardization: How to Calculate Z-scores

Standard scores are a great way to understand where a specific observation falls relative to the entire distribution. They also allow you to take observations drawn from normally distributed populations that have different means and standard deviations and place them on a standard scale. This standard scale enables you to compare observations that would otherwise be difficult.

This process is called standardization, and it allows you to compare observations and calculate probabilities across different populations. In other words, it permits you to compare apples to oranges.

To standardize your data, you need to convert the raw measurements into Z-scores.

To calculate the standard score for an observation, take the raw measurement, subtract the mean, and divide by the standard deviation. Mathematically, the formula for that process is the following:

$$Z = \frac{X - \mu}{\sigma}$$

X represents the raw value of the measurement of interest. Mu and sigma represent the parameters for the population from which the observation was drawn.

After you standardize your data, you can place them within the standard normal distribution. In this manner, standardization allows you to compare different types of observations based on where each observation falls within its own distribution.

Example of Using Standard Scores to Make an Apples to Oranges Comparison

Suppose we literally want to compare apples to oranges. Specifically, let's compare their weights. Imagine that we have an apple that weighs 110 grams and an orange that weighs 100 grams.

If we compare the raw values, it's easy to see that the apple weighs more than the orange. However, let's compare their standard scores. To do this, we'll need to know the properties of the weight distributions for apples and oranges. Assume that the weights of apples and oranges follow a normal distribution with the following parameter values:

	Apples	Oranges
Mean weight grams	100	140
Standard Deviation	15	25

Now we'll calculate the Z-scores:

- Apple = 110-100/15 = 0.667
- Orange = 100-140/25 = -1.6

The Z-score for the apple (0.667) is positive, which means that our apple weighs more than the average apple. It's not an extreme value by any means, but it is above average for apples. On the other hand, the orange has fairly negative Z-score (-1.6). It's pretty far below the mean weight for oranges.

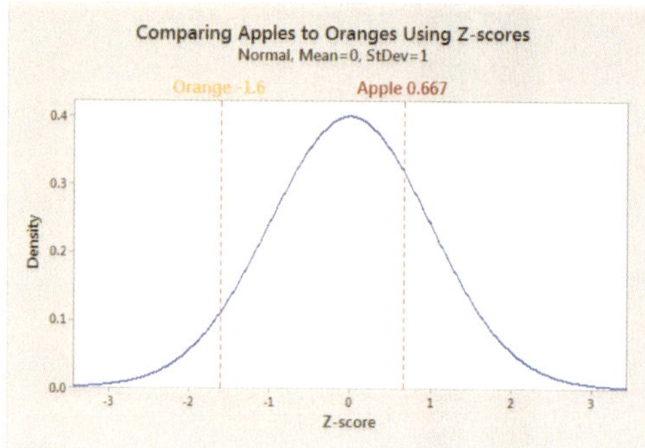

While our apple weighs more than our orange, we are comparing a somewhat heavier than average apple to a downright puny orange.

Finding Areas under the Curve of a Normal Distribution

The normal distribution is a probability distribution. As with any probability distribution, the proportion of the area that falls under the curve between two points on a probability distribution plot indicates the probability that a value will fall within that interval.

However, when you're working with the normal distribution and convert values to standard scores, you can calculate areas by looking up Z-scores in a Standard Normal Distribution Table.

Because there are an infinite number of different normal distributions, publishers can't print a table for each distribution. However, you can transform the values from any normal distribution into Z-scores, and then use a table of standard scores to calculate probabilities.

Using a Table of Z-scores

Let's take the Z-score for our apple (0.667) and use it to determine its weight percentile. A percentile is the proportion of a population that falls below a specific value. Consequently, to determine the percentile, we need to find the area that corresponds to the range of Z-scores that are less than 0.667. In the portion of the table below, the closest Z-score to ours is 0.65, which we'll use.

z	*Height*	*Area*
0.00	39.89	0
0.05	39.84	3.99
0.10	39.69	7.97
0.15	39.45	11.92
0.20	39.10	15.85
0.25	38.67	19.74
0.30	38.14	23.58
0.35	37.52	27.37
0.40	36.83	31.08
0.45	36.05	34.73
0.50	35.21	38.29
0.55	34.29	41.77
0.60	33.32	45.15
0.65	32.30	48.43
0.70	31.23	51.61

The trick with these tables is to use the values in conjunction with the properties of the normal distribution to calculate the probability that you need. The table value indicates that the area of the curve between -0.65 and +0.65 is 48.43%. However, that's not what we want to know. We want the area that is less than a Z-score of 0.65.

We know that the two halves of the normal distribution are mirror images of each other. So, if the area for the interval from -0.65 and +0.65 is 48.43%, then the range from 0 to +0.65 must be half of that: 48.43/2 = 24.215%. Additionally, we know that the area for all scores less than zero is half (50%) of the distribution.

Therefore, the area for all scores up to 0.65 = 50% + 24.215% = 74.215%

Our apple is at approximately the 74th percentile.

Below is a probability distribution plot produced by statistical software that shows the same percentile along with a graphical representation of the corresponding area under the curve. The value is slightly different because we used a Z-score of 0.65 from the table while the software uses the more precise value of 0.667.

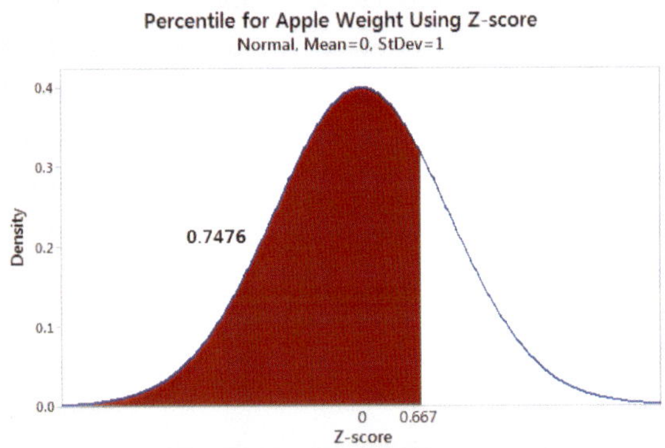

References

- Novak S.Y. (2011) Extreme value methods with applications to finance. London: CRC/ Chapman & Hall/Taylor & Francis. ISBN 9781-43983-5746.
- What-is-probability: stattrek.com, Retrieved 17 July 2018
- Agresti, Alan; Coull, Brent A. (May 1998), "Approximate is better than 'exact' for interval estimation of binomial proportions" (PDF), The American Statistician, 52 (2): 119–126, doi:10.2307/2685469, retrieved 2015-01-05
- What-are-probability-axioms-3126567: thoughtco.com, Retrieved 19 May 2018
- Taboga, Marco. "Lectures on Probability Theory and Mathematical Statistics". statlect.com. Retrieved 18 December 2017.
- Random-variable-and-its-probability-distribution, probability: toppr.com, Retrieved 16 April 2018
- MacKay, David (2003). Information Theory, Inference and Learning Algorithms. Cambridge University Press; First Edition. ISBN 978-0521642989.
- Independent-random-variables, fundamentals-of-probability: statlect.com, Retrieved 30 April 2018
- Kachitvichyanukul, V.; Schmeiser, B. W. (1988). "Binomial random variate generation". Communications of the ACM. 31 (2): 216–222. doi:10.1145/42372.42381.

5

Statistical Data - Types and Collection Methods

The collection of data in statistics occurs through observational and experimental studies or via sampling. There can be many classifications of statistical data. This chapter discusses the varied types of statistical data, such as quantitative and qualitative data, primary and secondary data, etc. as well as the collection methods of survey method, census method, etc.

Statistical Analysis

Statistical data analysis is a procedure of performing various statistical operations. It is a kind of quantitative research, which seeks to quantify the data, and typically, applies some form of statistical analysis. Quantitative data basically involves descriptive data, such as survey data and observational data.

Statistical data analysis generally involves some form of statistical tools, which a layman cannot perform without having any statistical knowledge. There are various software packages to perform statistical data analysis. This software includes Statistical Analysis System (SAS), Statistical Package for the Social Sciences (SPSS), Stat soft, etc.

Data in statistical data analysis consists of variable(s). Sometimes the data is univariate or multivariate. Depending upon the number of variables, the researcher performs different statistical techniques.

If the data in statistical data analysis is multiple in numbers, then several multivariates can be performed. These are factor statistical data analysis, discriminant statistical data analysis, etc. Similarly, if the data is singular in number, then the univariate statistical data analysis is performed. This includes t test for significance, z test, f test, ANOVA one way, etc.

The data in statistical data analysis is basically of 2 types, namely, continuous data and discreet data. The continuous data is the one that cannot be counted. For example, intensity of a light can be measured but cannot be counted. The discreet data is the one that can be counted. For example, the number of bulbs can be counted.

The continuous data in statistical data analysis is distributed under continuous distribution function, which can also be called the probability density function, or simply pdf.

The discreet data in statistical data analysis is distributed under discreet distribution function, which can also be called the probability mass function or simple pmf.

We use the word 'density' in continuous data of statistical data analysis because density cannot be counted, but can be measured. We use the word 'mass' in discreet data of statistical data analysis because mass cannot be counted.

There are various pdf's and pmf's in statistical data analysis. For example, Poisson distribution is the commonly known pmf, and normal distribution is the commonly known pdf.

These distributions in statistical data analysis help us to understand which data falls under which distribution. If the data is about the intensity of a bulb, then the data would be falling in Poisson distribution.

There is a major task in statistical data analysis, which comprises of statistical inference. The statistical inference is mainly comprised of two parts: estimation and tests of hypothesis.

Estimation in statistical data analysis mainly involves parametric data—the data that consists of parameters. On the other hand, tests of hypothesis in statistical data analysis mainly involve non parametric data— the data that consists of no parameters.

Types of Data

Categorical Data

Categorical data represents characteristics. Therefore it can represent things like a person's gender, language etc. Categorical data can also take on numerical values (Example: 1 for female and 0 for male). Note that those numbers don't have mathematical meaning.

Nominal Data

Nominal values represent discrete units and are used to label variables, that have no quantitative value. Just think of them as "labels". Note that nominal data that has no order. Therefore if you would change the order of its values, the meaning would not change. You can see two examples of nominal features below:

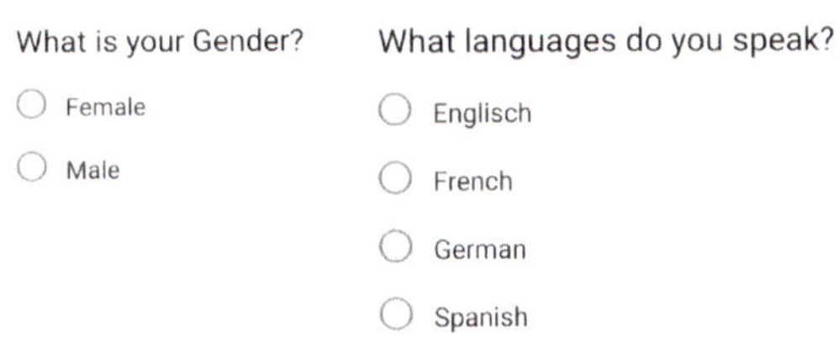

The left feature that describes a persons gender would be called „dichotomous", which is a type of nominal scales that contains only two categories.

Ordinal Data

Ordinal values represent discrete and ordered units. It is therefore nearly the same as nominal data, except that it's ordering matters. You can see an example below:

What Is Your Educational Background?

- 1 - Elementary
- 2 - High School
- 3 - Undegraduate
- 4 - Graduate

Note that the difference between Elementary and High School is different than the difference between High School and College. This is the main limitation of ordinal data, the differences between the values is not really known. Because of that, ordinal scales are usually used to measure non-numeric features like happiness, customer satisfaction and so on.

Numerical Data

1. Discrete Data

We speak of discrete data if its values are distinct and separate. In other words: We speak of discrete data if the data can only take on certain values. This type of data can't be measured but it can be counted. It basically represents information that can be categorized into a classification. An example is the number of heads in 100 coin flips.

You can check by asking the following two questions whether you are dealing with discrete data or not: Can you count it and can it be divided up into smaller and smaller parts?

2. Continuous Data

Continuous Data represents measurements and therefore their values can't be counted but they can be measured. An example would be the height of a person, which you can describe by using intervals on the real number line.

3. Interval Data

Interval values represent ordered units that have the same difference. Therefore we speak of interval data when we have a variable that contains numeric values that are ordered and where we know the exact differences between the values. An example would be a feature that contains temperature of a given place like you can see below:

Temperature?

- - 10
- -5
- 0
- + 5
- + 10
- + 15

The problem with interval values data is that they don't have a „true zero". That means in regards to our example, that there is no such thing as no temperature. With interval data, we can add and subtract, but we cannot multiply, divide or calculate ratios. Because there is no true zero, a lot of descriptive and inferential statistics can't be applied.

Ratio Data

Ratio values are also ordered units that have the same difference. Ratio values are the same as interval values, with the difference that they do have an absolute zero. Good examples are height, weight, length etc.

Statistical Methods

Nominal Data

When you are dealing with nominal data, you collect information through:

Frequencies: The Frequency is the rate at which something occurs over a period of time or within a dataset.

Proportion: You can easily calculate the proportion by dividing the frequency by the total number of events. (e.g how often something happened divided by how often it could happen)

Percentage

Visualization Methods: To visualize nominal data you can use a pie chart or a bar chart.

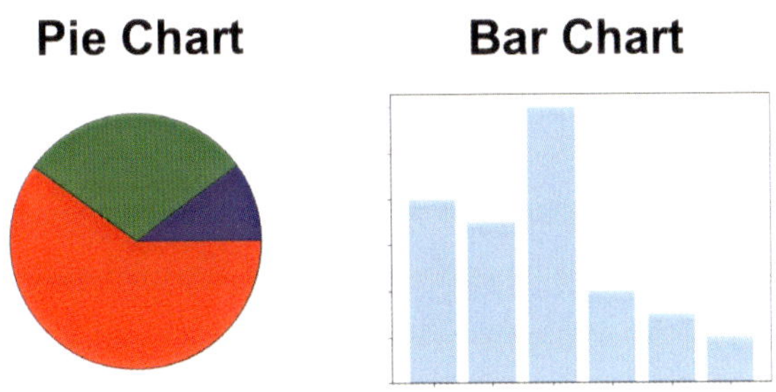

In Data Science, you can use one hot encoding, to transform nominal data into a numeric feature.

Ordinal Data

When you are dealing with ordinal data, you can use the same methods like with nominal data, but you also have access to some additional tools. Therefore you can summarize your ordinal data with frequencies, proportions, percentages. And you can visualize it with pie and bar charts. Additionally, you can use percentiles, median, mode and the interquartile range to summarize your data.

In Data Science, you can use one label encoding, to transform ordinal data into a numeric feature.

Continuous Data

When you are dealing with continuous data, you can use the most methods to describe your data. You can summarize your data using percentiles, median, interquartile range, mean, mode, standard deviation, and range.

Visualization Methods: To visualize continuous data, you can use a histogram or a box-plot. With a histogram, you can check the central tendency, variability, modality, and kurtosis of a distribution. Note that a histogram can't show you if you have any outliers. This is why we also use box-plots.

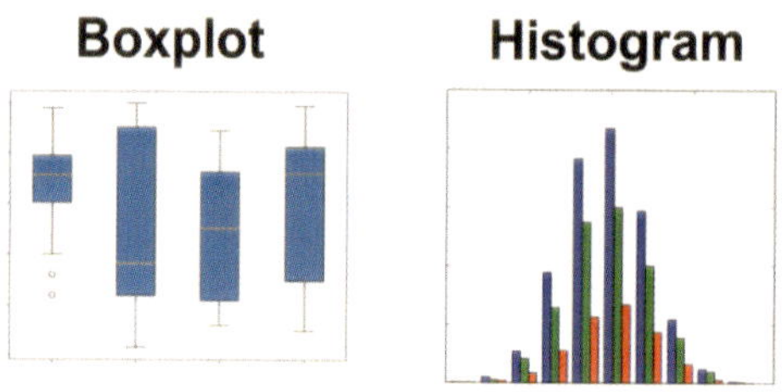

Primary Data

Primary data is data originated for the first time by the researcher through direct efforts and experience, specifically for the purpose of addressing his research problem. Also known as the first hand or raw data. Primary data collection is quite expensive, as the research is conducted by the organisation or agency itself, which requires resources like investment and manpower. The data collection is under direct control and supervision of the investigator.

Sources of Primary Data

The sources of primary data are primary units such as basic experimental units, individuals, households. Following methods are used to collect data from primary units usually and these methods depends on the nature of the primary unit. Published data and the data collected in the past is called secondary data.

- Personal Investigation: The researcher conducts the experiment or survey himself/herself and collected data from it. The collected data is generally accurate and reliable. This method of collecting primary data is feasible only in case of small scale laboratory, field experiments or pilot surveys and is not practicable for large scale experiments and surveys because it take too much time.
- Through Investigators: The trained (experienced) investigators are employed to collect the required data. In case of surveys, they contact the individuals and fill in the questionnaires after asking the required information, where a questionnaire is an inquiry form having a number of questions designed to obtain information from the respondents. This method of collecting data is usually employed by most of the organizations and its gives reasonably accurate information but it is very costly and may be time taking too.

- Through Questionnaire: The required information (data) is obtained by sending a questionnaire (printed or soft form) to the selected individuals (respondents) (by mail) who fill in the questionnaire and return it to the investigator. This method is relatively cheap as compared to "through investigator" method but non-response rate is very high as most of the respondents don't bother to fill in the questionnaire and send it back to investigator.
- Through Local Sources: The local representatives or agents are asked to send requisite information who provide the information based upon their own experience. This method is quick but it gives rough estimates only.
- Through Telephone: The information may be obtained by contacting the individuals on telephone. It's a Quick and provide accurate required information.
- Through Internet: With the introduction of information technology, the people may be contacted through internet and the individuals may be asked to provide the pertinent information. Google survey is widely used as online method for data collection now a day. There are many paid online survey services too.

Some Advantages of using Primary data:

1. The investigator collects data specific to the problem under study.
2. There is no doubt about the quality of the data collected (for the investigator).
3. If required, it may be possible to obtain additional data during the study period.

Some Disadvantages of using Primary data (for reluctant/uninterested investigators):

The investigator has to contend with all the hassles of data collection

- Deciding why, what, how, when to collect
- Getting the data collected (personally or through others)
- Getting funding and dealing with funding agencies
- Ethical considerations (consent, permissions, etc.)

Ensuring the data collected is of a high standard

- All desired data is obtained accurately, and in the format it is required in
- There is no fake/cooked up data
- Unnecessary/useless data has not been included

Cost of obtaining the data is often the major expense in studies

Secondary Data

Secondary data implies second-hand information which is already collected and recorded by any person other than the user for a purpose, not relating to the current research problem.

Sources of Secondary Data

The secondary data may be available from the following sources:

- Government Organizations: Federal and Provincial Bureau of Statistics, Crop Reporting Service-Agriculture Department, Census and Registration Organization etc.
- Semi-Government Organization: Municipal committees, District Councils, Commercial and Financial Institutions like banks etc.
- Teaching and Research Organizations.
- Research Journals and Newspapers.
- Internet

Some Advantages of using Secondary data:

1. The data's already there- no hassles of data collection.
2. It is less expensive.
3. The investigator is not personally responsible for the quality of data.

Some disadvantages of using Secondary data:

1. The investigator cannot decide what is collected (if specific data about something is required, for instance).
2. One can only hope that the data is of good quality.
3. Obtaining additional data (or even clarification) about something is not possible (most often).

Quantitative and Qualitative Data

Qualitative Data

Qualitative data collection is a method in which the characteristics, attributes, properties, qualities, etc. of a phenomenon or thing is described. It is the description of data in a language rather than in numbers. This method does not measure the characteristics but describes them. For example; Favorite color = blue.

It is also sometimes referred to as "categorical data." It does not focus on drawing any inferences. It only deals with data that can be observed like texture, taste, smell, beauty, but is not measured.

Qualitative data, in recent years, has lost reliability to some extent and has come under criticism, but they provide a better description and thus have more validity to them. Research uses a combination of qualitative and quantitative methods because the qualitative data and description backs up the numerical data with the help of better explanations and information.

Quantitative Data

Quantitative data collection is a method in which data that can be numerically counted or expressed is collected. This data is useful for experiments, manipulated analysis, etc. and is represented by histograms, tables, charts, and graphs. It deals with measurements like height, length, volume, area, humidity, temperature, etc.

For example;

Height = 2.8m. Or sometimes they represent the exact number like,

Number of students = 234.

This type of data is associated with some type of scale measurement. The most commonly used scale for this data is a ratio scale. Another general scale measurement is the interval scale.

Quantitative data is criticized for its lack of in-depth description thus it is used by researchers along with qualitative data to back up its reliability with explanations of the qualitative information.

Importance of Quantitative and Qualitative Data

Quantitative and qualitative data provide different outcomes, and are often used together to get a full picture of a population. For example, if data are collected on annual income (quantitative), occupation data (qualitative) could also be gathered to get more detail on the average annual income for each type of occupation.

Quantitative and qualitative data can be gathered from the same data unit depending on whether the variable of interest is numerical or categorical. For example:

Data unit	**Numeric variable**	**= Quantitative data**	**Categorical variable**	**= Qualitative data**
A person	"*How many* children do you have?"	4 children	"In *which country* were your children born?"	Australia
	"*How much* do you earn?"	$60,000 p.a.	"*What* is your occupation?"	Photographer
	"How many hours do you work?"	38 hours per week	"Do you work *full-time or part-time*?"	Full-time
A house	"*How many* square metres is the house?"	200 square metres	"In *which city or town* is the house located?"	Brisbane
A business	"*How many* workers are currently employed?"	264 employees	"*What* is the *industry* of the business?"	Retail
A farm	"*How many* milk cows are located on the farm?	36 cows	"*What* is the main *activity* of the farm?"	Dairy

Uses of Quantitative and Qualitative Data

It is important to identify whether the data are quantitative or qualitative as this affects the statistics that can be produced.

Frequency counts: The number of times an observation occurs (frequency) for a data item (variable) can be shown for both quantitative and qualitative data.

The graphs below arrange the quantitative and qualitative data to show the frequency distribution of the data.

Quantitative Data

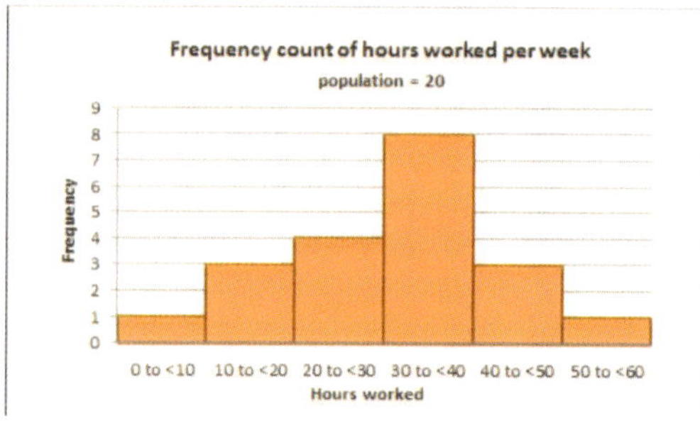

Qualitative Data

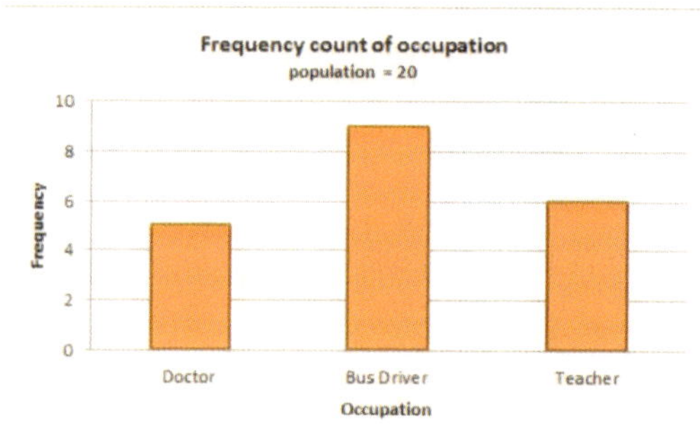

As absolute frequencies can be calculated on quantitative and qualitative data, relative frequencies can also be produced, such as percentages, proportions, rates and ratios. For example, the graphs above show 4 people (20%) worked less than 30 hours per week, and 6 people (30%) are teachers.

Descriptive (Summary) Statistics

Statistics that describe or summarise can be produced for quantitative data and to a lesser extent for qualitative data.

As quantitative data are always numeric they can be ordered, added together, and the frequency of an observation can be counted. Therefore, all descriptive statistics can be calculated using quantitative data.

As qualitative data represent individual (mutually exclusive) categories, the descriptive statistics that can be calculated are limited, as many of these techniques require numeric values which can be logically ordered from lowest to highest and which express a count.

Mode can be calculated, as it it the most frequency observed value. Median, measures of shape, measures of spread such as the range and interquartile range require an ordered data set with a logical low-end value and high-end value. Variance and standard deviation require the mean to be calculated, which is not appropriate for categorical variables as they have no numerical value.

Inferential Statistics

By making inferences about quantitative data from a sample, estimates or projections for the total population can be produced.

Quantitative data can be used to inform broader understandings of a population, or to consider how that population may change or progress into the future.

For example, a simple income projection for an employee in 2015 may be inferred from the rate of change for data collected in 2000, 2005, and 2010.

As shown in the graph below, data collected over time indicates a 5% increase every five years. Therefore, if the rate of increase continues to follow the same pattern, it can be projected that the annual income for that employee in 2015 will be $46,305; which is the 2010 wage of $44,100 increased by an additional 5%.

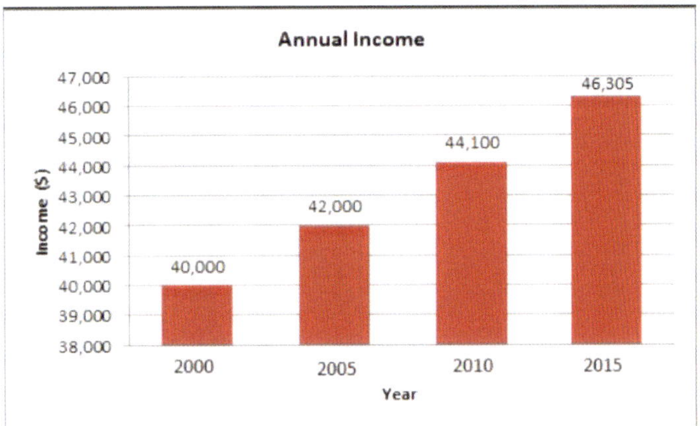

Qualitative data are not compatible with inferential statistics as all techniques are based on numeric values.

Census Method

The Census Method is also called as a Complete Enumeration Survey Method wherein each and every item in the universe is selected for the data collection. The universe might constitute a particular place, a group of people or any specific locality which is the complete set of items and which are of interest in any particular situation.

The census method is most commonly used by the government in connection with the national population, housing census, agriculture census, etc. where the vast knowledge about these fields is required. Whenever the entire population is studied to collect the detailed data about every unit, then the census method is applied.

The census method can be applied in a situation where the separate data for every unit in the population is to be collected, such that the separate actions for each is taken. For example, the preparation of the voter's list for election purposes, income tax assessment, recruitment of personnel, etc. are some of the areas where the census method is adopted. This method can be used where the population is comprised of heterogeneous items, i.e. different characteristics.

Though the census method provides a complete data of the population under study, it is very costly and time-consuming. Often, this method is dropped down because of these constraints and the sampling method, where certain items representative of the larger group, is selected to draw the conclusions.

Advantages and Disadvantages of Census Method

Following are the advantages and disadvantages of census method of data collection

Advantages of Census Method

Intensive Study: Data collection through census method gives opportunity to the investigator to have an intensive study about a problem study about a problem. The investigator gathers a lot of knowledge through this method.

Intensive Study

Higher Degree of Accuracy

Higher Degree of Accuracy In this method there would be higher degree of accuracy in data. No other method is accurate like census method when the universe is small.

Higher Degree of Accuracy

Suitable for Heterogeneous Units

Suitable for Heterogeneous Units: This method is also applicable for units having heterogeneity or difference.

Indispensable in Some Cases

Indispensable in Some Cases: In certain cases this method is very important and suitable to be used for data collection. Without this method the study of a universe remains uncompleted.

Disadvantages of Census Method of Data Collection

Following are the demerits of census method

Inconvenient

Inconvenient: This method is inconvenient because it take much time, money and set of industrial organizations.

Possible in Few Cases

Possible in Few Cases: This method is possible only in few cases of data collection.

Possible in Limited Circumstances

Possible in Limited Circumstances: This method is possible only in limited universe it requires not vast area to study.

Time and Labour Consuming

Time & Labour Consuming: It is much time and labour consuming. It requires a lot of people to involve in data collection.

More Statistical Errors

More Statistical Errors: Data collected through this method would have more statistical errors. Census method of data collection is helpful only in an area where inquiry is limited, more time and finance available and data have high degree of accuracy otherwise not.

Survey Method

A Survey is defined is a research method used for collecting data from a pre-defined group of respondents to gain information and insights on various topics of interest. Surveys have a variety of purposes and can be carried out in many ways depending on the methodology chosen and the objectives to be achieved.

The data is usually obtained through the use of standardized procedures whose purpose is to ensure that each respondent is able to answer the questions at a level playing field to avoid biased opinions that could influence the outcome of the research or study. A survey involves asking people for information through a questionnaire, which can be distributed on paper, although with the arrival of new technologies it is more common to distribute them using digital media such as social networks, email, QR codes or URLs.

Characteristics of a Survey

The need to observe or research facts about a situation leads us to conduct a survey. As we mentioned at the beginning, a survey is a method of gathering information.

Need to Conducting a Survey

First, a sample, also referred to as audience, is needed which should consist of a series of survey respondent's data with required demographic characteristics, who can relevantly answer your survey questions and provide the best insights. Better the quality of your survey sample, better will be your response quality and better your insights.

Surveys come in many different forms and have a variety of purposes, but they have common underlying characteristics. The basic characteristics of a survey are:

Survey Questions: Questionnaire for your Survey

Effective survey questions are the cornerstone for the success of any survey and subsequently, any research study. Whether it an email survey, SMS survey, web intercept survey or a mobile app survey, the single common denominator that determines how effectively you are able to collect accurate and complete survey responses is your survey questions and their types.

Multiple choice questions are the most common type of survey questions, in which, some of the popular question types are: dichotomous question, semantic differential scale question, rank order questions, rating scale questions and open ended questions .

To conduct a survey, it is important to plan the type of survey to ensure you get the optimum number of responses required for your survey. It could be a mix of interviews and survey questions or a questionnaire. Interviews could be telephone interviews, face-to-face interviews, online interviews and questionnaires can be mall surveys or web surveys. The underlying difference between a survey and a questionnaire is that a questionnaire may or may not be delivered in the form of a survey, but a survey always consists of a questionnaire.

Sample and Sample Determination

A sample is a selection of respondents from a population in such a manner that the sample represents the total population as closely as possible. Once you have determined your sample, the total number of individuals in that particular sample is the sample size.

Selecting a sample size depends on the end objective of your research study. It should consist of a series of survey respondents data with required demographic characteristics, who can relevantly answer your survey questions and provide the best insights. If the quality of the sample for survey is good, the response quality and insights are better.

Survey Methods

Survey methodology studies the in-depth sampling of individual units from a population and administering data collection techniques on that sample. It includes instruments or processes that ask different question types to a predefined sample, to conduct data-collection and increase the survey response rate.

The two distinctive member types are in a survey methodology are, professionals in the field that focus on empirical survey errors and others that work to design surveys and reduce them. it is therefore both a scientific field and a profession. The primary tasks of a survey methodologist while administering a survey is to identify and create samples, validate test questions, select the mode to administer questions and validate methods for data collection, statistical analysis and data reporting.

Survey Methods Based on Design

Surveys can be administered by the time they take to complete, the two types are:

- Cross-sectional studies: Cross-sectional study is defined as an observational research type that analyzes data of variables collected at one given point of time across a sample population. population or a pre-defined subset. This study type is also known as cross-sectional analysis, transverse study or prevalence study. The data gathered in a cross-sectional study is from people who are similar in all variables except the one variable which is under study. This variable remains constant throughout the cross-sectional study. This is unlike longitudinal study, where variables in the study can change over the course of research.

- Longitudinal studies: Longitudinal study is an observational study that employs continuous or repeated measures to follow particular individuals over prolonged period of time often years or decades. Longitudinal study collects data that is either qualitative or quantitative in nature. In longitudinal study a survey creator is not interfering with survey respondents. Survey respondents are observed over a period of time ranging from months to even decades to observe any changes in them or their attitude. For example, a researcher wants to find out which disease affects young boys (in the age group of 10-15) then the researcher will observe the individuals over that period to collect meaningful data.

Survey Methods based on Distribution

There are different ways of survey distribution. Some of the most commonly used methods are:

- Email: Sending out an email is the easiest way of conducting a survey. The respondents are targeted and there is higher chance of response due to the the respondents already knowing about your brand. You can use the QuestionPro email management feature to send out and collect survey responses.
- Buy respondents: Buying a sample helps achieve a lot of the response criteria because the people who are being asked to respond have signed up to do so and the qualifying criteria for the research study is met.
- Embed survey in website: Embedding a survey in a website ensures that the number of responses is very high. This can be done while the person enters the website or is exiting it. A non intrusive method of collecting feedback is important to achieve a higher number of responses. The responses received are also honest due to the high brand recall value and the responses are quick to collect and analyze due to them being in a digital format.
- Post to social network: Posting on social networks is another effective way of collecting responses. The survey can be posted as a link and people that follow the brand can take a survey. This method is used when there is no upper cap on the number of survey responses required and is the easiest and fastest way of eliciting responses.
- QR code: QuestionPro QR codes store the URL for the survey. You can print/publish this code in magazines, on signs, business cards, or on just about any object/medium. Users with a camera phone equipped with the correct reader application can scan the image of the QR Code to open the survey in the phone's browser.
- QuestionPro App: The QuestionPro App allows to quickly circulate surveys and the responses can be collected both online and offline.
- API: You can use the API integration of the QuestionPro platform for potential respondents to take your survey.

- SMS: Using SMS surveys are another quick way to collect feedback. This method can be used in the case of quick responses and when the survey is simple, straightforward and not too long. This method is used to increase the open and response rate of collecting feedback.

Surveys can be distributed using one, some or a mix of the above methods depending on the basis of the research objective and the resources being used for any particular survey. Many factors play a part in the mode of distribution of surveys like cost, research study type, flexibility of questions, time to collect responses, statistical analysis to be run on data and willingness of the respondent to take part in the survey.

You can conduct a telephone or email survey and then make a selection of respondents for a face-to-face interview. Survey data are sometimes also obtained through questionnaires filled out by respondents in groups, for example, a school class or a group of shoppers in a shopping center.

You can also classify the surveys by their content, being able to use open or closed questions to know, for example, opinions, attitudes, details of a fact, habits, experiences for a later classification and analysis of the obtained results.

Online Survey

An online survey is set of structured questions that the respondent completes over the internet, generally through filling out a form. An online survey is an easier way to reach out to the respondents as it is less time consuming than the traditional way of gathering information through one to one interaction and also less expensive.

Online surveys can differ in length and format used. The data is collected and stored in a database which is later evaluated by an expert in the field.

As an incentive for respondents to fill out online surveys, businesses offer them with rewards like gift cards, reward points that they can redeem for goods or services later, free airline miles, discount at gas stations etc.

Online surveys with rewards is a win-win situation for both, businesses and respondents. The businesses or organizations get valuable data from a controlled environment, for the market research and for the respondents it's a humble gesture from businesses to participate in the survey and spend their valuable time.

Advantages of Online Survey

Online surveys are one of the less time consuming methods to reach out to your target audiences, but that's not the only advantage of online survey, let us understand what are the other advantages of using online survey:

- Accuracy: In online surveys, usually the margin of error is reduced, as the respondents register their responses by easy selection buttons. Tradition methods require human interference and according to a study human interference increase the margin of error by 10%.

- Easy and quick to analyze: Since all the responses are registered online, it is extremely easy and quick to analyze the data in real time. It is also quick to draw inferences and share the result.
- Ease of participation: In this new age technology oriented universe, most people on this planet have access to the internet. Respondents prefer receiving the survey over the email. Ease of participation greatly increases as the respondents can choose a suitable time and place, according to their convenience to register their responses.
- Great branding exercise: While designing the survey online, organizations or businesses have this opportunity to design their survey to align to their brand. Using logos and similar brand language (color and fonts) in the survey, gives an advantage to the businesses as respondents are able to connect better with the brand.
- Respondents can be honest and flexible at the same time: According to a study researchers have found an increased participation by respondents when deployed with online surveys rather than answering lengthy questions. By designing surveys that ask relevant questions, respondents are honest with their answers and can skip the questions or respondent to a more neutral option in the survey, increasing their flexibility to respond.

Survey Examples

It is important for a researcher to be able to conduct surveys using the right questions and the right medium to administer and track responses. QuestionPro is a platform that helps not only to create but also to deploy different types of survey. We have more than 300+ types of survey template and survey examples including:

- Customer Satisfaction (CSAT) + Net Promoter Score (NPS) Survey: We hear this time and again that customer is king; and it is true. A satisfied customer is a customer that helps your brand and organization grow, through direct means as well as being an advocate for your brand. This survey template talks about the goodwill your brand has created and how referenceable it is.
- Company Communications Evaluation Survey Template: This survey example is usually used to analyze employee perspective about what is important to be communicated regularly, topics that should be included in the newsletter, updates on the bulletin board, the efficiency of an organization's management in communication etc.
- Hardware Product Evaluation Survey Template: Improving hardware product features isn't a very easy proposition due to a lot of elements like raw materials, supply chain and manufacturing lines getting affected by it. Hence, while eliciting feedback for hardware, it is important to be as objective as possible. This survey helps understand exactly what product innovation is required.
- Strategic Planning Survey: Innovation is important to any organization's product or service lines. Hence, implementing customer support and making product or service tweaks when required are extremely important for the sustenance and growth of an organization. This survey template helps organizations chalk out their business strategy.

- Business Demographic Survey: This survey template is designed to carry demographic questions and examples that help gain information on occupation, primary area of business, job function and description, organization's gross income etc.
- Course Evaluation Survey: This survey template helps educational institutions conduct period feedback on their course and if students find it helpful or not, if it's stimulating enough and students see this is as value for money along with accentuated learning.

References

- Statistical-data-analysis: statisticssolutions.com, Retrieved 10 May 2018
- Difference-between-primary-and-secondary-data: keydifferences.com, Retrieved 19 April 2018
- Primary-and-secondary-data-in-statistics: itfeature.com, Retrieved 17 June 2018
- Types-of-data-primary-and-secondary-data: communitymedicine4asses.com, Retrieved 28 May 2018
- What-is-census-method-of-data-collection-advantages-disadvantages, social-research-methodology: studylecturenotes.com, Retrieved 08 July 2018

6

Statistical Hypothesis Testing

A statistical hypothesis is a hypothesis, which can be tested by observing a process modeled using a set of random variables. Statistical hypothesis testing is a form of statistical inference. All the important aspects of statistical hypothesis testing, such as paired difference test, null hypothesis, P-value, etc. have been included in this chapter.

A statistical hypothesis is an assumption about a population parameter. This assumption may or may not be true. Hypothesis testing refers to the formal procedures used by statisticians to accept or reject statistical hypotheses.

Statistical Hypotheses

The best way to determine whether a statistical hypothesis is true would be to examine the entire population. Since that is often impractical, researchers typically examine a random sample from the population. If sample data are not consistent with the statistical hypothesis, the hypothesis is rejected.

There are two types of statistical hypotheses:

- Null hypothesis: The null hypothesis, denoted by H_0, is usually the hypothesis that sample observations result purely from chance.
- Alternative hypothesis: The alternative hypothesis, denoted by H_1 or H_a, is the hypothesis that sample observations are influenced by some non-random cause.

For example, suppose we wanted to determine whether a coin was fair and balanced. A null hypothesis might be that half the flips would result in Heads and half, in Tails. The alternative hypothesis might be that the number of Heads and Tails would be very different. Symbolically, these hypotheses would be expressed as

$$H_o : P = 0.5$$

$$H_a : P \neq 0.5$$

Suppose we flipped the coin 50 times, resulting in 40 Heads and 10 Tails. Given this result, we would be inclined to reject the null hypothesis. We would conclude, based on the evidence, that the coin was probably not fair and balanced.

Hypothesis Tests

Statisticians follow a formal process to determine whether to reject a null hypothesis, based on sample data. This process, called hypothesis testing, consists of four steps.

- State the hypotheses. This involves stating the null and alternative hypotheses. The hypotheses are stated in such a way that they are mutually exclusive. That is, if one is true, the other must be false.
- Formulate an analysis plan. The analysis plan describes how to use sample data to evaluate the null hypothesis. The evaluation often focuses around a single test statistic.
- Analyze sample data. Find the value of the test statistic (mean score, proportion, t statistic, z-score, etc.) described in the analysis plan.
- Interpret results. Apply the decision rule described in the analysis plan. If the value of the test statistic is unlikely, based on the null hypothesis, reject the null hypothesis.

Decision Errors

Two types of errors can result from a hypothesis test.

- Type I error. A Type I error occurs when the researcher rejects a null hypothesis when it is true. The probability of committing a Type I error is called the significance level. This probability is also called alpha, and is often denoted by α.
- Type II error. A Type II error occurs when the researcher fails to reject a null hypothesis that is false. The probability of committing a Type II error is called Beta, and is often denoted by β. The probability of *not* committing a Type II error is called the Power of the test.

Decision Rules

The analysis plan includes decision rules for rejecting the null hypothesis. In practice, statisticians describe these decision rules in two ways - with reference to a P-value or with reference to a region of acceptance.

- P-value. The strength of evidence in support of a null hypothesis is measured by the P-value. Suppose the test statistic is equal to *S*. The P-value is the probability of observing a test statistic as extreme as *S*, assuming the null hypotheis is true. If the P-value is less than the significance level, we reject the null hypothesis.
- Region of acceptance. The region of acceptance is a range of values. If the test statistic falls within the region of acceptance, the null hypothesis is not rejected. The region of acceptance is defined so that the chance of making a Type I error is equal to the significance level.

The set of values outside the region of acceptance is called the region of rejection. If the test statistic falls within the region of rejection, the null hypothesis is rejected. In such cases, we say that the hypothesis has been rejected at the α level of significance.

These approaches are equivalent. Some statistics texts use the P-value approach; others use the region of acceptance approach. On this website, we tend to use the region of acceptance approach.

One-Tailed and Two-Tailed Tests

A test of a statistical hypothesis, where the region of rejection is on only one side of the sampling distribution, is called a one-tailed test. For example, suppose the null hypothesis states that the mean is less than or equal to 10. The alternative hypothesis would be that the mean is greater than 10. The region of rejection would consist of a range of numbers located on the right side of sampling distribution; that is, a set of numbers greater than 10.

A test of a statistical hypothesis, where the region of rejection is on both sides of the sampling distribution, is called a two-tailed test. For example, suppose the null hypothesis states that the mean is equal to 10. The alternative hypothesis would be that the mean is less than 10 or greater than 10. The region of rejection would consist of a range of numbers located on both sides of sampling distribution; that is, the region of rejection would consist partly of numbers that were less than 10 and partly of numbers that were greater than 10.

Paired Difference Test

The test used to compare two sets of measurements to evaluate if their population means differ is a paired difference test. It is a type of location test. It uses additional information about the samples, which are not present in an ordinary unpaired testing situation, in order to increase the statistical power, or to reduce the effects of confounders.

There are specific methods available for carrying out paired difference tests. Such methods are available for normally distributed difference t-test and the paired Z-test, and for the differences that may not be normally distributed, the Wilcoxon signed-rank test.

The paired sample *t*-test, sometimes called the dependent sample *t*-test, is a statistical procedure used to determine whether the mean difference between two sets of observations is zero. In a paired sample *t*-test, each subject or entity is measured twice, resulting in *pairs* of observations. Common applications of the paired sample *t*-test include case-control studies or repeated-measures designs. Suppose you are interested in evaluating the effectiveness of a company training program. One approach you might consider would be to measure the performance of a sample of employees before and after completing the program, and analyze the differences using a paired sample *t*-test.

Hypotheses

Like many statistical procedures, the paired sample *t*-test has two competing hypotheses, the null hypothesis and the alternative hypothesis. The null hypothesis assumes that the true mean difference between the paired samples is zero. Under this model, all observable differences are explained by random variation. Conversely, the alternative hypothesis assumes that the true mean difference between the paired samples is not equal to zero. The alternative hypothesis can take one of several forms depending on the expected outcome. If the direction of the difference does not matter, a two-tailed hypothesis is used. Otherwise, an upper-tailed or lower-tailed hypothesis can be used to increase the power of the test. The null hypothesis remains the same for each type of alternative hypothesis. The paired sample *t*-test hypotheses are formally defined below:

- The null hypothesis (H_0) assumes that the true mean difference (μ_d) is equal to zero.

- The two-tailed alternative hypothesis (H_1) assumes that μ_d is not equal to zero.
- The upper-tailed alternative hypothesis (H_1) assumes that μ_d is greater than zero.
- The lower-tailed alternative hypothesis (H_1) assumes that μ_d is less than zero.

The mathematical representations of the null and alternative hypotheses are defined below:

- $H_0: \mu_d = 0$
- $H_1: \mu_d \neq 0$ (two-tailed)
- $H_1: \mu_d > 0$ (upper-tailed)
- $H_1: \mu_d < 0$ (lower-tailed)

It is important to remember that hypotheses are never about data, they are about the processes which produce the data. In the formulas above, the value of μ_d is unknown. The goal of hypothesis testing is to determine the hypothesis (null or alternative) with which the data are more consistent.

Assumptions

As a parametric procedure (a procedure which estimates unknown parameters), the paired sample *t*-test makes several assumptions. Although *t*-tests are quite robust, it is good practice to evaluate the degree of deviation from these assumptions in order to assess the quality of the results. In a paired sample *t*-test, the observations are defined as the differences between two sets of values, and each assumption refers to these differences, not the original data values. The paired sample *t*-test has four main assumptions:

- The dependent variable must be continuous (interval/ratio).
- The observations are independent of one another.
- The dependent variable should be approximately normally distributed.
- The dependent variable should not contain any outliers.

Level of Measurement

The paired sample *t*-test requires the sample data to be numeric and continuous, as it is based on the normal distribution. Continuous data can take on any value within a range (income, height, weight, etc.). The opposite of continuous data is discrete data, which can only take on a few values (Low, Medium, High, etc.). Occasionally, discrete data can be used to approximate a continuous scale, such as with Likert-type scales.

Independence

Independence of observations is usually not testable, but can be reasonably assumed if the data collection process was random without replacement. In our example, it is reasonable to assume that the participating employees are independent of one another.

Normality

To test the assumption of normality, a variety of methods are available, but the simplest is to inspect the data visually using a tool like a histogram below figure. Real-world data are almost never perfectly normal, so this assumption can be considered reasonably met if the shape looks approximately symmetric and bell-shaped. The data in the example figure below is approximately normally distributed.

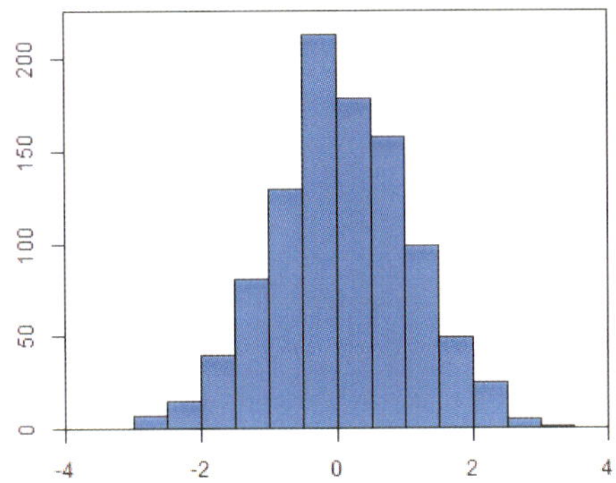

Figure: Histogram of an approximately normally distributed variable.

Outliers

Outliers are rare values that appear far away from the majority of the data. Outliers can bias the results and potentially lead to incorrect conclusions if not handled properly. One method for dealing with outliers is to simply remove them. However, removing data points can introduce other types of bias into the results, and potentially result in losing critical information. If outliers seem to have a lot of influence on the results, a nonparametric test such as the Wilcoxon Signed Rank Test may be appropriate to use instead. Outliers can be identified visually using a boxplot below figure.

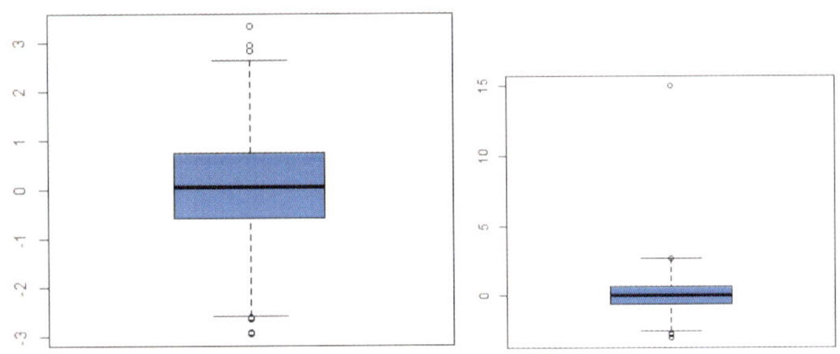

Figure: Boxplots of a variable without outliers (left) and with an outlier (right).

Procedure

The procedure for a paired sample *t*-test can be summed up in four steps. The symbols to be used are defined below:

- D = Differences between two paired samples
- d_i = The ith observation in D
- n = The sample size

- $\bar{d}$ = The sample mean of the differences.
- $\hat{\sigma}$ = The sample standard deviation of the differences.
- T =The critical value of a *t*-distribution with ($n-1$) degrees of freedom.
- t = The *t*-statistic (*t*-test statistic) for a paired sample *t*-test.
- p = The p-value (probability value) for the *t*-statistic.

The four steps are listed below:

- Calculate the sample mean.
  - Undefined control sequence\cfrac.
- Calculate the sample standard deviation.
  - Undefined control sequence\cfrac.
- Calculate the test statistic.
  - Undefined control sequence\cfrac.
- Calculate the probability of observing the test statistic under the null hypothesis. This value is obtained by comparing *t* to a *t*-distribution with (n – 1) degrees of freedom. This can be done by looking up the value in a table, such as those found in many statistical textbooks, or with statistical software for more accurate results.
- $p = 2 \cdot \Pr(T > |t|)$ (two-tailed)
- $p = \Pr(T > t)$ (upper-tailed)
- $p = \Pr(T < t)$ (lower-tailed)

determine whether the results provide sufficient evidence to reject the null hypothesis in favor of the alternative hypothesis.

Interpretation

There are two types of significance to consider when interpreting the results of a paired sample *t*-test, statistical significance and practical significance.

Statistical Significance

Statistical significance is determined by looking at the *p*-value. The *p*-value gives the probability of observing the test results under the null hypothesis. The lower the *p*-value, the lower the probability of obtaining a result like the one that was observed if the null hypothesis was true. Thus, a low *p*-value indicates decreased support for the null hypothesis. However, the possibility that the null hypothesis is true and that we simply obtained a very rare result can never be ruled out completely. The cutoff value for determining statistical significance is ultimately decided on by the researcher, but usually a value of .05 or less is chosen. This corresponds to a 5% (or less) chance of obtaining a result like the one that was observed if the null hypothesis was true.

Practical Significance

Practical significance depends on the subject matter. It is not uncommon, especially with large sample sizes, to observe a result that is statistically significant but not practically significant. In most cases, both types of significance are required in order to draw meaningful conclusions.

Power of the Paired Z-test

Suppose we are using a Z-test to analyze the data, where the variances of the pre-treatment and post-treatment data σ_1^2 and σ_2^2 are known (the situation with a t-test is similar). The unpaired Z-test statistic is

$$\frac{\bar{Y}_2 - \bar{Y}_1}{\sqrt{\sigma_1^2 / n + \sigma_2^2 / n}},$$

The power of the unpaired, one-sided test carried out at level $\alpha = 0.05$ can be calculated as follows:

$$\begin{aligned} P\left(\frac{\bar{Y}_2 - \bar{Y}_1}{\sqrt{\sigma_1^2 / n + \sigma_2^2 / n}} > 1.64\right) &= P\left(\frac{\bar{Y}_2 - \bar{Y}_1}{S} > 1.64\sqrt{\sigma_1^2 / n + \sigma_2^2 / n} / S\right) \\ &= P\left(\frac{\bar{Y}_2 - \bar{Y}_1 - \delta + \delta}{S} > 1.64\sqrt{\sigma_1^2 / n + \sigma_2^2 / n} / S\right) \\ &= P\left(\frac{\bar{Y}_2 - \bar{Y}_1 - \delta}{S} > 1.64\sqrt{\sigma_1^2 / n + \sigma_2^2 / n} / S - \delta / S\right) \\ &= 1 - \Phi(1.64\sqrt{\sigma_1^2 / n + \sigma_2^2 / n} / S - \delta / S), \end{aligned}$$

Where S is the standard deviation of $\bar{D}$, Φ is the standard normal cumulative distribution function, and $\delta = EY_2 - EY_1$ is the true effect of the treatment. The constant 1.64 is the 95th percentile of the standard normal distribution, which defines the rejection region of the test.

By a similar calculation, the power of the paired Z-test is

$$1 - \Phi(1.64 - \delta / S).$$

By comparing the expressions for power of the paired and unpaired tests, one can see that the paired test has more power as long as

$$\sqrt{\sigma_1^2 / n + \sigma_2^2 / n} / S = \sqrt{\frac{\sigma_1^2 + \sigma_2^2}{\sigma_1^2 + \sigma_2^2 - 2\sigma_1\sigma_2\rho}} > 1 \text{ where } \rho := \mathrm{corr}(Y_{i1}, Y_{i2}).$$

This condition is met whenever ρ, the within-pairs correlation, is positive.

A Random Effects Model for Paired Testing

The following statistical model is useful for understanding the paired difference test

$$Y_{ij} = \mu_j + \alpha_i + \epsilon_{ij}$$

where α_i is a random effect that is shared between the two values in the pair, and ε_{ij} is a random noise term that is independent across all data points. The constant values μ_1, μ_2 are the expected values of the two measurements being compared, and our interest is in $\delta = \mu_2 - \mu_1$.

In this model, the α_i capture "stable confounders" that have the same effect on the pre-treatment and post-treatment measurements. When we subtract to form D_i, the α_i cancel out, so do not contribute to the variance. The within-pairs covariance is

$$\text{cov}(Y_{i1}, Y_{i2}) = \text{var}(\alpha_i).$$

This is non-negative, so it leads to better performance for the paired difference test compared to the unpaired test, unless the α_i are constant over i, in which case the paired and unpaired tests are equivalent.

In less mathematical terms, the unpaired test assumes that the data in the two groups being compared are independent. This assumption determines the form for the variance of $\bar{D}$. However when two measurements are made for each subject, it is unlikely that the two measurements are independent. If the two measurements within a subject are positively correlated, the unpaired test overstates the variance of D, making it a conservative test in the sense that its actual type I error probability will be lower than the nominal level, with a corresponding loss of statistical power. In rare cases, the data may be negatively correlated within subjects, in which case the unpaired test becomes anti-conservative. The paired test is generally used when repeated measurements are made on the same subjects, since it has the correct level regardless of the correlation of the measurements within pairs.

Use in Reducing Confounding

Another application of paired difference testing arises when comparing two groups in a set of observational data, with the goal being to isolate the effect of one factor of interest from the effects of other factors that may play a role. For example, suppose teachers adopt one of two different approaches, denoted "A" and "B", to teaching a particular mathematical topic. We may be interested in whether the performances of the students on a standardized mathematics test differ according to the teaching approach. If the teachers are free to adopt approach A or approach B, it is possible that teachers whose students are already performing well in mathematics will preferentially choose method A (or vice versa). In this situation, a simple comparison between the mean performances of students taught with approach A and approach B will likely show a difference, but this difference is partially or entirely due to the pre-existing differences between the two groups of students. In this situation, the baseline abilities of the students serve as a confounding variable, in that they are related to both the outcome (performance on the standardized test), and to the treatment assignment to approach A or approach B.

It is possible to reduce, but not necessarily eliminate, the effects of confounding variables by forming "artificial pairs" and performing a pairwise difference test. These artificial pairs are constructed based on additional variables that are thought to serve as confounders. By pairing students whose values on the confounding variables are similar, a greater fraction of the difference in the value of interest (e.g. the standardized test score in the example discussed above), is due to the factor of interest, and a lesser fraction is due to the confounder. Forming artificial pairs for paired difference testing is an example of a general approach for reducing the effects of confounding when making comparisons using observational data called matching.

As a concrete example, suppose we observe student test scores X under teaching strategies A and B, and each student has either a "high" or "low" level of mathematical knowledge before the two teaching strategies are implemented. However, we do not know which students are in the "high" category and which are in the "low" category. The population mean test scores in the four possible groups are $\begin{array}{ccc} & A & B \\ \text{High} & \mu_{HA} & \mu_{HB} \\ \text{Low} & \mu_{LA} & \mu_{LB} \end{array}$ and the proportions of students in the groups are $\begin{array}{ccc} & A & B \\ \text{High} & p_{HA} & p_{HB} \\ \text{Low} & p_{LA} & p_{LB} \end{array}$ w

where $p_{HA} + p_{HB} + p_{LA} + p_{LB} = 1$.

The "treatment difference" among students in the "high" group is $\mu_{HA} - \mu_{HB}$ and the treatment difference among students in the "low" group is $\mu_{LA} - \mu_{LB}$. In general, it is possible that the two teaching strategies could differ in either direction, or show no difference, and the effects could differ in magnitude or even in sign between the "high" and "low" groups. For example, if strategy B were superior to strategy A for well-prepared students, but strategy A were superior to strategy B for poorly prepared students, the two treatment differences would have opposite signs.

Since we do not know the baseline levels of the students, the expected value of the average test score $\bar{X}_A$ among students in the A group is an average of those in the two baseline levels:

$$E\bar{X}_A = \mu_{HA}\frac{p_{HA}}{p_{HA} + p_{LA}} + \mu_{LA}\frac{p_{LA}}{p_{HA} + p_{LA}},$$

and similarly the average test score $\bar{X}_B$ among students in the B group is

$$E\bar{X}_B = \mu_{HB}\frac{p_{HB}}{p_{HB} + p_{LB}} + \mu_{LB}\frac{p_{LB}}{p_{HB} + p_{LB}}.$$

Thus the expected value of the observed treatment difference $\bar{D} = \bar{X}_A - \bar{X}_B$ is

$$\mu_{HA}\frac{p_{HA}}{p_{HA} + p_{LA}} - \mu_{HB}\frac{p_{HB}}{p_{HB} + p_{LB}} + \mu_{LA}\frac{p_{LA}}{p_{HA} + p_{LA}} - \mu_{LB}\frac{p_{LB}}{p_{HB} + p_{LB}}.$$

A reasonable null hypothesis is that there is no effect of the treatment within either the "high" or "low" student groups, so that $\mu_{HA} = \mu_{HB}$ and $\mu_{LA} = \mu_{LB}$. Under this null hypothesis, the expected value of D will be zero if

$$p_{HA} = (p_{HA} + p_{LA})(p_{HA} + p_{HB})$$

And

$$p_{HB} = (p_{HB} + p_{LB})(p_{HA} + p_{HB}).$$

This condition asserts that the assignment of students to the A and B teaching strategy groups is independent of their mathematical knowledge before the teaching strategies are implemented. If this holds, baseline mathematical knowledge is not a confounder, and conversely, if baseline mathematical knowledge is a confounder, the expected value of $\bar{D}$ will generally different from zero. If the expected value of $\bar{D}$ under the null hypothesis is not equal to zero, then a situation where we reject the null hypothesis could either be due to an actual differential effect between teaching strategies A and B, or it could be due to non-independence in the assignment of students to the A and B groups (even in the complete absence of an effect due to the teaching strategy).

This example illustrates that if we make a direct comparison between two groups when confounders are present, we do not know whether any difference that is observed is due to the grouping itself, or is due to some other factor. If we are able to pair students by an exact or estimated measure of their baseline mathematical ability, then we are only comparing students "within rows" of the table of means given above. Consequently, if the null hypothesis holds, the expected value of $\bar{D}$ will equal zero, and statistical significance levels have their intended interpretation.

Null Hypothesis

A null hypothesis is a precise statement about a population that we try to reject with sample data.

We don't usually believe our null hypothesis (or H_0) to be true. However, we need *some* exact statement as a starting point for statistical significance testing.

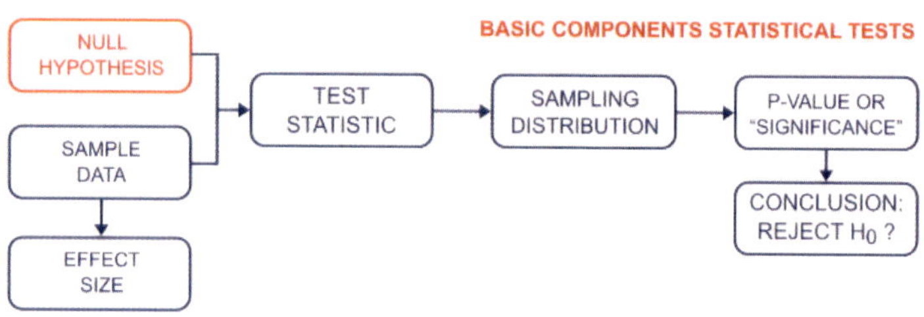

Null Hypothesis Examples

Often -but not always- the null hypothesis states there is no association or difference between variables or subpopulations. Like so, some typical null hypotheses are:

- The correlation between frustration and aggression is zero.
- The average income for men is similar to that for women.
- Nationality is (perfectly) unrelated to music preference.
- The average population income was equal over 2012 through 2016.

"Null" does not Mean "Zero"

A common misunderstanding is that "null" implies "zero". This is often *but not always* the case. For example, a null hypothesis may also state that the correlation between frustration and aggression is 0.5. No zero involved here and-although somewhat unusual- perfectly valid. The "null" in "null hypothesis" derives from "nullify": the null hypothesis is the statement that we're trying to refute, regardless whether it does (not) specify a zero effect.

Null Hypothesis - Limitations

Thus far, we only concluded that the population correlation is probably not zero. That's the only conclusion from our null hypothesis approach and it's not really that interesting. What we really want to know is the population correlation. Our sample correlation of 0.25 seems a reasonable estimate. We call such a single number a point estimate.

Now, a new sample may come up with a different correlation. An interesting question is how much our sample correlations would fluctuate over samples if we'd draw many of them. The figure below shows precisely that, assuming our sample size of N = 100 and our (point) estimate of 0.25 for the population correlation.

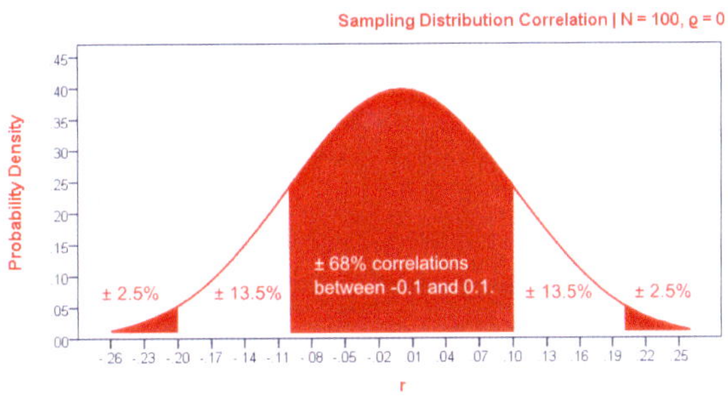

Confidence Intervals

Our sample outcome suggests that some 95% of *many* samples should come up with a correlation between 0.06 and 0.43. This range is known as a confidence interval. Although not precisely correct, it's most easily thought of as the bandwidth that's likely to enclose the population correlation. One thing to note is that the concidence interval is quite wide. It almost contains a zero correlation, exactly the null hypothesis we rejected earlier.

Another thing to note is that our sampling distribution and confidence interval are slightly asymmetrical. They are symmetrical for most other statistics (such as means or beta coefficients) but not correlations.

Terminology

Simple Hypothesis

Any hypothesis which specifies the population distribution completely. For such a hypothesis the sampling distribution of any statistic is a function of the sample size alone.

Composite Hypothesis

Any hypothesis which does *not* specify the population distribution completely. Example: A hypothesis specifying a normal distribution with a specified mean and an unspecified variance.

The simple/composite distinction was made by Neyman and Pearson.

Exact Hypothesis

Any hypothesis that specifies an exact parameter value. Example: $\mu = 100$. Synonym: point hypothesis.

Inexact Hypothesis

Those specifying a parameter range or interval. Examples: $\mu \leq 100$; $95 \leq \mu \leq 105$.

Fisher required an exact null hypothesis for testing.

A one-tailed hypothesis (tested using a one-sided test) is an inexact hypothesis in which the value of a parameter is specified as being either:

- Above or equal to a certain value, or
- Below or equal to a certain value.

A one-tailed hypothesis is said to have directionality.

Fisher's original (lady tasting tea) example was a one-tailed test. The null hypothesis was asymmetric. The probability of guessing all cups correctly was the same as guessing all cups incorrectly, but Fisher noted that only guessing correctly was compatible with the lady's claim.

Goals of Null Hypothesis Tests

There are many types of significance tests for one, two or more samples, for means, variances and proportions, paired or unpaired data, for different distributions, for large and small samples; all have null hypotheses. There are also at least four goals of null hypotheses for significance tests:

- Technical null hypotheses are used to verify statistical assumptions. For example, the residuals between the data and a statistical model cannot be distinguished from random noise. If true, there is no justification for complicating the model.
- Scientific null assumptions are used to directly advance a theory. For example, the angular momentum of the universe is zero. If not true, the theory of the early universe may need revision.
- Null hypotheses of homogeneity are used to verify that multiple experiments are producing consistent results. For example, the effect of a medication on the elderly is consistent with that of the general adult population. If true, this strengthens the general effectiveness conclusion and simplifies recommendations for use.

- Null hypotheses that assert the equality of effect of two or more alternative treatments, for example, a drug and a placebo, are used to reduce scientific claims based on statistical noise. This is the most popular null hypothesis; It is so popular that many statements about significant testing assume such null hypotheses.

Rejection of the null hypothesis is *not necessarily* the real goal of a significance tester. An adequate statistical model may be associated with a failure to reject the null; the model is adjusted until the null is not rejected. The numerous uses of significance testing were well known to Fisher who discussed many in his book written a decade before defining the null hypothesis.

A statistical significance test shares much mathematics with a confidence interval. They are mutually illuminating. A result is often significant when there is confidence in the sign of a relationship (the interval does not include 0). Whenever the sign of a relationship is important, statistical significance is a worthy goal. This also reveals weaknesses of significance testing: A result can be significant without a good estimate of the strength of a relationship; significance can be a modest goal. A weak relationship can also achieve significance with enough data. Reporting both significance and confidence intervals is commonly recommended.

The varied uses of significance tests reduce the number of generalizations that can be made about all applications.

Choice of the Null Hypothesis

The choice of the null hypothesis is associated with sparse and inconsistent advice. Fisher mentioned few constraints on the choice and stated that many null hypotheses should be considered and that many tests are possible for each. The variety of applications and the diversity of goals suggests that the choice can be complicated. In many applications the formulation of the test is traditional. A familiarity with the range of tests available may suggest a particular null hypothesis and test. Formulating the null hypothesis is not automated (though the calculations of significance testing usually are). Sir David Cox has said, "How [the] translation from subject-matter problem to statistical model is done is often the most critical part of an analysis".

A statistical significance test is intended to test a hypothesis. If the hypothesis summarizes a set of data, there is no value in testing the hypothesis on that set of data. Example: If a study of last year's weather reports indicates that rain in a region falls primarily on weekends, it is only valid to test that null hypothesis on weather reports from any *other* year.Testing hypotheses suggested by the data is circular reasoning that proves nothing; It is a special limitation on the choice of the null hypothesis.

A routine procedure is as follows: Start from the scientific hypothesis. Translate this to a statistical alternative hypothesis and proceed: "Because H_a expresses the effect that we wish to find evidence for, we often begin with H_a and then set up H_0 as the statement that the hoped-for effect is not present." This advice is *reversed* for modeling applications where we hope not to find evidence against the null.

A complex case example is as follows: The gold standard in clinical research is the randomized placebo-controlled double-blind clinical trial. But testing a new drug against a (medically ineffective) placebo may be unethical for a serious illness. Testing a new drug against an older medically effective drug raises fundamental philosophical issues regarding the goal of the test and the motivation of the experimenters. The standard "no difference" null hypothesis may reward the pharmaceutical company for gathering inadequate data. "Difference" is a better null hypothesis in this case, but statistical significance is not an adequate criterion for reaching a nuanced conclusion which requires a good numeric estimate of the drug's effectiveness. A "minor" or "simple" proposed change in the null hypothesis ((new vs old) rather than (new vs placebo)) can have a dramatic effect on the utility of a test for complex non-statistical reasons.

Directionality

The choice of null hypothesis (H_0) and consideration of directionality is critical.

Tailedness of the Null-hypothesis Test

Consider the question of whether a tossed coin is fair (i.e. that on average it lands heads up 50% of the time) and an experiment where you toss the coin 5 times. A possible result of the experiment that we consider here is 5 heads. Let outcomes be considered unlikely with respect to an assumed distribution if their probability is lower than a significance threshold of 0.05.

A potential null hypothesis implying a one-tail test is "this coin is not biased toward heads". Beware that, in this context, the word "tail" takes two meanings: either as outcome of a single toss, or as region of extremal values in a probability distribution.

Indeed, with a fair coin the probability of this experiment outcome is $1/2^5=0.031$, which would be even lower if the coin were biased in favour of tails. Therefore, the observations are not likely enough for the null hypothesis to hold, and the test refutes it. Since the coin is ostensibly neither fair nor biased toward tails, the conclusion of the experiment is that the coin is biased towards heads.

Alternatively, a null hypothesis implying a two-tailed test is "this coin is fair". This one null hypothesis could be examined by looking out for either too many tails or too many heads in the experiments. The outcomes that would tend to refuse this null hypothesis are those with a large number of heads or a large number of tails, and our experiment with 5 heads would seem to belong to this class.

However, the probability of 5 tosses of the same kind, irrespective of whether these are head or tails, is twice as much as that of the 5-head occurrence singly considered. Hence, under this two-tailed null hypothesis, the observation receives a probability value of 0.063. Hence again, with the same significance threshold used for the one-tailed test (0.05), the same outcome is not statistically significant. Therefore, the two-tailed null hypothesis will be preserved in this case, not supporting the conclusion reached with the single-tailed null hypothesis, that the coin is biased towards heads.

This example illustrates that the conclusion reached from a statistical test may depend on the precise formulation of the null and alternative hypotheses.

Fisher said, "the null hypothesis must be exact, that is free of vagueness and ambiguity, because it must supply the basis of the 'problem of distribution,' of which the test of significance is the solution", implying a more restrictive domain for H_0. According to this view, the null hypothesis must be numerically exact—it must state that a particular quantity or difference is equal to a particular number. In classical science, it is most typically the statement that there is no effect of a particular treatment; in observations, it is typically that there is no difference between the value of a particular measured variable and that of a prediction.

Most statisticians believe that it is valid to state direction as a part of null hypothesis, or as part of a null hypothesis/alternative hypothesis pair. However, the results are not a full description of all the results of an experiment, merely a single result tailored to one particular purpose. For example, consider an H_0 that claims the population mean for a new treatment is an improvement on a well-established treatment with population mean = 10 (known from long experience), with the one-tailed alternative being that the new treatment's mean > 10. If the sample evidence obtained through x-bar equals −200 and the corresponding t-test statistic equals −50, the conclusion from the test would be that there is no evidence that the new treatment is better than the existing one: it would not report that it is markedly worse, but that is not what this particular test is looking for. To overcome any possible ambiguity in reporting the result of the test of a null hypothesis, it is best to indicate whether the test was two-sided and, if one-sided, to include the direction of the effect being tested.

The statistical theory required to deal with the simple cases of directionality dealt with here, and more complicated ones, makes use of the concept of an unbiased test.

The directionality of hypotheses is not always obvious. The explicit null hypothesis of Fisher's Lady tasting tea example was that the Lady had no such ability, which led to a symmetric probability distribution. The one-tailed nature of the test resulted from the one-tailed alternate hypothesis (a term not used by Fisher). The null hypothesis became implicitly one-tailed. The logical negation of the Lady's one-tailed claim was also one-tailed. (Claim: Ability > 0; Stated null: Ability = 0; Implicit null: Ability ≤ 0).

Pure arguments over the use of one-tailed tests are complicated by the variety of tests. Some tests (for instance the χ^2 goodness of fit test) are inherently one-tailed. Some probability distributions are asymmetric. The traditional tests of 3 or more groups are two-tailed.

Advice concerning the use of one-tailed hypotheses has been inconsistent and accepted practice varies among fields. The greatest objection to one-tailed hypotheses is their potential subjectivity. A non-significant result can sometimes be converted to a significant result by the use of a one-tailed hypothesis (as the fair coin test, at the whim of the analyst). The flip side of the argument: One-sided tests are less likely to ignore a real effect. One-tailed tests can suppress the publication of data that differs in sign from predictions. Objectivity was a goal of the developers of statistical tests.

It is a common practice to use a one-tailed hypotheses by default. However, "If you do not have a specific direction firmly in mind in advance, use a two-sided alternative. Moreover, some users of statistics argue that we should *always* work with the two-sided alternative."

One alternative to this advice is to use three-outcome tests. It eliminates the issues surrounding directionality of hypotheses by testing twice, once in each direction and combining the results to produce three possible outcomes. Variations on this approach have a history, being suggested perhaps 10 times since 1950.

Disagreements over one-tailed tests flow from the philosophy of science. While Fisher was willing to ignore the unlikely case of the Lady guessing all cups of tea incorrectly (which may have been appropriate for the circumstances), medicine believes that a proposed treatment that kills patients is significant in every sense and should be reported and perhaps explained. Poor statistical reporting practices have contributed to disagreements over one-tailed tests. Statistical significance resulting from two-tailed tests is insensitive to the sign of the relationship; Reporting significance alone is inadequate. "The treatment has an effect" is the uninformative result of a two-tailed test. "The treatment has a beneficial effect" is the more informative result of a one-tailed test. "The treatment has an effect, reducing the average length of hospitalization by 1.5 days" is the most informative report, combining a two-tailed significance test result with a numeric estimate of the relationship between treatment and effect. Explicitly reporting a numeric result eliminates a philosophical advantage of a one-tailed test. An underlying issue is the appropriate form of an experimental science without numeric predictive theories: A model of numeric results is more informative than a model of effect signs (positive, negative or unknown) which is more informative than a model of simple significance (non-zero or unknown); in the absence of numeric theory signs may suffice.

P- value

The P-value provides a measure of this distance. The P-value (in this situation) is the probability to the right of our test statistic calculated using the null distribution. The further out the test statistic is in the tail, the smaller the P-value, and the stronger the evidence against the null hypothesis in favor of the alternative.

The P-value can be interpreted in terms of a hypothetical repetition of the study. Suppose the null hypothesis is true and a new dataset is obtained independently of the first dataset but using the same sampling procedure. If the new dataset is used to calculate a new value of the test statistic (same formula but new data), what is the probability that the new value will be further out in the tail (assuming a one-tailed test) than the original value? This probability is the P-value.

The P-value is often incorrectly interpreted as the probability that the null hypothesis is true. Try not to make this mistake. In a frequentist interpretation of probability, there is nothing random about whether the hypothesis is true, the randomness is in the process generating the data. One can interpret "the probability that the null hypothesis is true" using subjective probability, a measure of one's belief that the null hypothesis is true. One can then calculate this subjective probability by specifying a prior probability (subjective belief before looking at the data) that the null hypothesis is true, and then use the data and the model to update one's subjective probability. This is called the Bayesian approach because Bayes' Theorem is used to update subjective probabilities to reflect new information.

When reporting a P-value to persons unfamiliar with statistics, it is often necessary to use descriptive language to indicate the strength of the evidence. I tend to use the following sort of language. Obviously the cut-offs are somewhat arbitrary and another person might use different language.

$P > 0.10$	No evidence against the null hypothesis. The data appear to be consistent with the null hypothesis.
$0.05 < P < 0.10$	Weak evidence against the null hypothesis in favor of the alternative.
$0.01 < P < 0.05$	Moderate evidence against the null hypothesis in favor of the alternative.
$0.001 < P < 0.01$	Strong evidence against the null hypothesis in favor of the alternative.
$P < 0.001$	Very strong evidence against the null hypothesis in favor of the alternative.

In using this kind of language, one should keep in mind the difference between statistical significance and practical significance. In a large study one may obtain a small P-value even though the magnitude of the effect being tested is too small to be of importance It is a good idea to support a P-value with a confidence interval for the parameter being tested.

A P-value can also be reported more formally in terms of a fixed level α test. Here α is a number selected independently of the data, usually 0.05 or 0.01, more rarely 0.10. We reject the null hypothesis at level α if the P-value is smaller than α, otherwise we fail to reject the null hypothesis at level α. There is essentially no difference between a P-value of 0.051 and 0.049. In some situations it may be necessary to proceed with some course of action based on our belief in whether the null or alternative hypothesis is true. More often, it seems better to report the P-value as a measure of evidence.

A fixed level α test can be calculated without first calculating a P-value. This is done by comparing the test statistic with a critical value of the null distribution corresponding to the level α. This is usually the easiest approach when doing hand calculations and using statistical tables, which provide percentiles for a relatively small set of probabilities. Most statistical software produces P-values which can be compared directly with α. There is no need to repeat the calculation by hand.

Suppose we are comparing a new medical treatment with a standard treatment, the control. The null hypothesis is that of no treatment effect (no difference between treatment and control). The alternative hypothesis is that the treatment effect (mean difference of treatment minus control using some outcome variable) is positive. We want to have good chance of reporting a small P-value assuming the alternative hypothesis is true and the magnitude of the effect is large enough to be of practical importance. The power of a level α test is defined to be the probability that the null hypothesis will be rejected at level α (i.e., the P-value will be less than α) assuming the alternative hypothesis is true. The power generally depends on the variability of the data (lower variance, higher power), the sample size (higher n, higher power), and the magnitude of the effect (larger effect, higher power).

Assessing Normality using the Ryan-Joiner Test.

Null hypothesis: the data $\{x_1, ..., x_n\}$ are a random sample of size n from a normal distribution.

Alternative hypothesis: the data are a random sample from some other distribution.

Test statistic: r = the correlation between the data and the normal scores.

The normal scores are defined by the following graph.

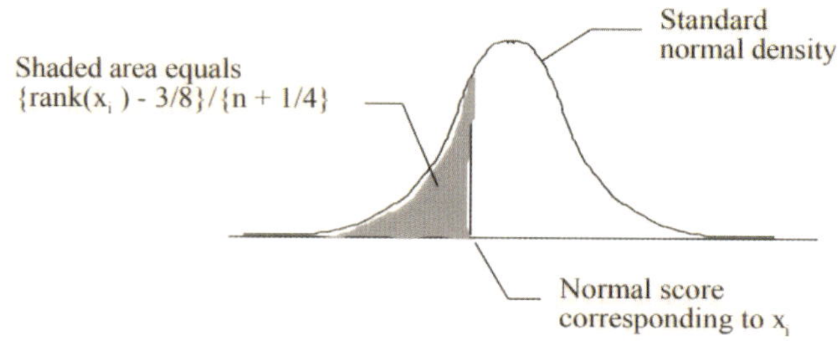

Rationale: If the data are a sample from a normal distribution then the normal probability plot (plot of normal scores against the data) will be close to a straight line, and the correlation r will be close to 1. If the data are sampled from a non-normal distribution then the plot may show a marked deviation from a straight line, resulting in a smaller correlation r. Smaller values of r are therefore regarded as stronger evidence against the null hypothesis.

Null distribution of r: I do not know whether this distribution has a name. We might call it the Ryan-Joiner distribution, corresponding to the name of the test. The density will be skewed to the left, with most of the probability close to 1, as in the picture below.

P-value: The probability to the left of the observed correlation r calculated using the null distribution; i.e., the area under the density to the left of r. You do not need to know how to calculate this. Minitab does the calculation for you.

Interpretation: If you want to use simple descriptive language, you can use the table above. The strength of evidence is described directly in terms of the P-value.

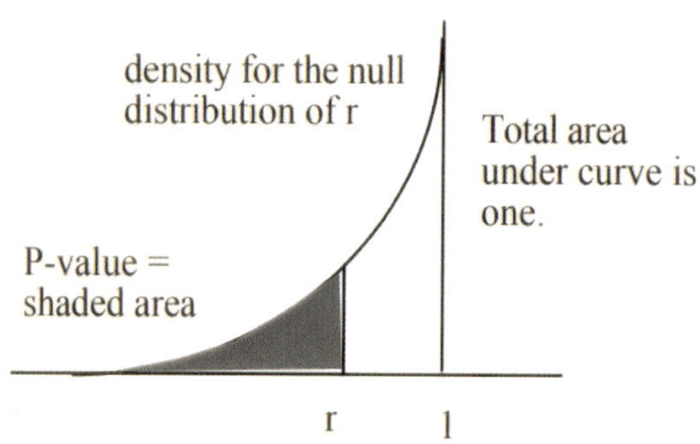

Examples

Here a few simple examples follow, each illustrating a potential pitfall.

One Roll of a Pair of Dice

Suppose a researcher rolls a pair of dice once and assumes a null hypothesis that the dice are fair, not loaded or weighted toward any specific number/roll/result; uniform. The test statistic is "the sum of the rolled numbers" and is one-tailed. The researcher rolls the dice and observes that both dice show 6, yielding a test statistic of 12. The *p*-value of this outcome is 1/36 (because under the assumption of the null hypothesis, the test statistic is uniformly distributed) or about 0.028 (the highest test statistic out of 6×6 = 36 possible outcomes). If the researcher assumed a significance level of 0.05, this result would be deemed significant and the hypothesis that the dice are fair would be rejected.

In this case, a single roll provides a very weak basis (that is, insufficient data) to draw a meaningful conclusion about the dice. This illustrates the danger with blindly applying *p*-value without considering the experiment design.

Five Heads in a Row

Suppose a researcher flips a coin five times in a row and assumes a null hypothesis that the coin is fair. The test statistic of "total number of heads" can be one-tailed or two-tailed: a one-tailed test corresponds to seeing if the coin is biased towards heads, but a two-tailed test corresponds to seeing if the coin is biased either way. The researcher flips the coin five times and observes heads each time (HHHHH), yielding a test statistic of 5. In a one-tailed test, this is the upper extreme of all possible outcomes, and yields a *p*-value of $(1/2)^5 = 1/32 \approx 0.03$. If the researcher assumed a significance level of 0.05, this result would be deemed significant and the hypothesis that the coin is fair would be rejected. In a two-tailed test, a test statistic of zero heads (TTTTT) is just as extreme and thus the data of HHHHH would yield a *p*-value of $2\times(1/2)^5 = 1/16 \approx 0.06$, which is not significant at the 0.05 level.

This demonstrates that specifying a direction (on a symmetric test statistic) halves the *p*-value (increases the significance) and can mean the difference between data being considered significant or not.

Sample Size Dependence

Suppose a researcher flips a coin some arbitrary number of times (n) and assumes a null hypothesis that the coin is fair. The test statistic is the total number of heads and is a two-tailed test. Suppose the researcher observes heads for each flip, yielding a test statistic of n and a *p*-value of $2/2^n$. If the coin was flipped only 5 times, the *p*-value would be $2/32 = 0.0625$, which is not significant at the 0.05 level. But if the coin was flipped 10 times, the *p*-value would be $2/1024 \approx 0.002$, which is significant at the 0.05 level.

In both cases the data suggest that the null hypothesis is false (that is, the coin is not fair somehow), but changing the sample size changes the *p*-value. In the first case, the sample size is not large enough to allow the null hypothesis to be rejected at the 0.05 level (in fact, the *p*-value can never be below 0.05 for the coin example).

This demonstrates that in interpreting *p*-values, one must also know the sample size, which complicates the analysis.

Alternating Coin Flips

Suppose a researcher flips a coin ten times and assumes a null hypothesis that the coin is fair. The test statistic is the total number of heads and is two-tailed. Suppose the researcher observes alternating heads and tails with every flip (HTHTHTHTHT). This yields a test statistic of 5 and a *p*-value of 1 (completely unexceptional), as that is the expected number of heads.

Suppose instead that the test statistic for this experiment was the "number of alternations" (that is, the number of times when H followed T or T followed H), which is one-tailed. That would yield a test statistic of 9, which is extreme and has a *p*-value of $2/2^9 = 1/256 \approx 0.0039$. That would be considered extremely significant, well beyond the 0.05 level. These data indicate that, in terms of one test statistic, the data set is extremely unlikely to have occurred by chance, but it does not suggest that the coin is biased towards heads or tails.

By the first test statistic, the data yield a high *p*-value, suggesting that the number of heads observed is not unlikely. By the second test statistic, the data yield a low *p*-value, suggesting that the pattern of flips observed is very, very unlikely. There is no "alternative hypothesis" (so only rejection of the null hypothesis is possible) and such data could have many causes. The data may instead be forged, or the coin may be flipped by a magician who intentionally alternated outcomes.

This example demonstrates that the *p*-value depends completely on the test statistic used and illustrates that *p*-values can only help researchers to reject a null hypothesis, not consider other hypotheses.

Coin Flipping

As an example of a statistical test, an experiment is performed to determine whether a coin flip is fair (equal chance of landing heads or tails) or unfairly biased (one outcome being more likely than the other).

Suppose that the experimental results show the coin turning up heads 14 times out of 20 total flips. The null hypothesis is that the coin is fair, and the test statistic is the number of heads. If a right-tailed test is considered, the *p*-value of this result is the chance of a fair coin landing on heads *at least* 14 times out of 20 flips. That probability can be computed from binomial coefficients as

$$\begin{aligned}&\text{Prob}(14 \text{ heads}) + \text{Prob}(15 \text{ heads}) + \cdots + \text{Prob}(20 \text{ heads})\\ &= \frac{1}{2^{20}}\left[\binom{20}{14} + \binom{20}{15} + \cdots + \binom{20}{20}\right] = \frac{60{,}460}{1{,}048{,}576} \approx 0.058\end{aligned}$$

This probability is the *p*-value, considering only extreme results that favor heads. This is called a one-tailed test. However, the deviation can be in either direction, favoring either heads or tails. The two-tailed *p*-value, which considers deviations favoring either heads or tails, may instead be calculated. As the binomial distribution is symmetrical for a fair coin, the two-sided *p*-value is simply twice the above calculated single-sided *p*-value: the two-sided *p*-value is 0.115.

In the above example:

- Null hypothesis (H_0): The coin is fair, with Prob(heads) = 0.5.
- Test statistic: Number of heads.
- Level of significance: 0.05.
- Observation O: 14 heads out of 20 flips.

- Two-tailed p-value of observation O given H_0 = 2*min(Prob(no. of heads ≥ 14 heads), Prob(no. of heads ≤ 14 heads))= 2*min(0.058, 0.978) = 2*0.058 = 0.115.

The Prob(no. of heads ≤ 14 heads) = 1 - Prob(no. of heads ≥ 14 heads) + Prob(no. of head = 14) = 1 - 0.058 + 0.036 = 0.978; however, symmetry of the binomial distribution makes it an unnecessary computation to find the smaller of the two probabilities. Here, the calculated p-value exceeds 0.05, so the observation is consistent with the null hypothesis, as it falls within the range of what would happen 95% of the time were the coin in fact fair. Hence, the null hypothesis at the 5% level is not rejected. Although the coin did not fall evenly, the deviation from the expected outcome is small enough to be consistent with chance.

However, had one more head been obtained, the resulting p-value (two-tailed) would have been 0.0414 (4.14%). The null hypothesis is rejected when a 5% cut-off is used.

Type I and Type II Errors

Type I error, also known as a "false positive": the error of rejecting a null hypothesis when it is actually true. In other words, this is the error of accepting an alternative hypothesis (the real hypothesis of interest) when the results can be attributed to chance. Plainly speaking, it occurs when we are observing a difference when in truth there is none (or more specifically - no statistically significant difference). So the probability of making a type I error in a test with rejection region R is $P(R/H_0 \text{ is true})$.

Type II error, also known as a "false negative": the error of not rejecting a null hypothesis when the alternative hypothesis is the true state of nature. In other words, this is the error of failing to accept an alternative hypothesis when you don't have adequate power. Plainly speaking, it occurs when we are failing to observe a difference when in truth there is one. So the probability of making a type II error in a test with rejection region R is $1 - P(R/H_a \text{ is true})$. The power of the test can be $P(R/H_a \text{ is true})$.

Hypothesis testing is the art of testing if variation between two sample distributions can just be explained through random chance or not. If we have to conclude that two distributions vary in a meaningful way, we must take enough precaution to see that the differences are not just through random chance. At the heart of Type I error is that we don't want to make an unwarranted hypothesis so we exercise a lot of care by minimizing the chance of its occurrence. Traditionally we try to set Type I error as .05 or .01 - as in there is only a 5 or 1 in 100 chance that the variation that we are seeing is due to chance. This is called the 'level of significance'. Again, there is no guarantee that 5 in 100 is rare enough so significance levels need to be chosen carefully. For example, a factory where a six sigma quality control system has been implemented requires that errors never add up to more than the probability of being six standard deviations away from the mean (an incredibly rare event). Type I error is generally reported as the p-value.

Multiple Hypothesis Testing

In Statistics, multiple testing refers to the potential increase in Type I error that occurs when statistical tests are used repeatedly, for example while doing multiple comparisons to test null hypotheses stating that the averages of several disjoint populations are equal to each other (homogeneous).

Intuitively, even if a particular outcome of an experiment is very unlikely to happen, the fact that the experiment is repeated multiple times will increase the probability that the outcome appears at least once. As an example, if a coin is tossed 10 times and lands 10 times on tail, it will usually be considered evidence that the coin is biased, because the probability of observing such a series is very low for a fair coin ($2^{-10} \approx 10^{-3}$). However, if the same series of ten tails in a row appears as part of 10,000 tosses with the same coin, it is more likely to be seen as a random fluctuation in the long series of tosses.

If the significance level for a given experiment is α, the experimentwise significance level will increase exponentially (significance decreases) as the number of tests increases. More precisely, assuming all tests are independent, if n tests are performed, the experimentwise significance level will be given by $1-(1 - \alpha)^n \approx n\alpha$ when α is small Thus, in order to retain the same overall rate of false positives in a series of multiple tests, the standards for each test must be more stringent. Intuitively, reducing the size of the allowable error (alpha) for each comparison by the number of comparisons will result in an overall alpha which does not exceed the desired limit, and this can be mathematically proved true. For instance, to obtain the usual alpha of 0.05 with ten tests, requiring an alpha of .005=0.05/10 / = α n for each test can be demonstrated to result in an overall alpha which does not exceed 0.05. This technique is known as Bonferroni correction.

However, it can also be demonstrated that this technique may be conservative (depending on the correlation structure among tests), i.e. will in actuality result in a true alpha of significantly less than 0.05; therefore raising the rate of false negatives, failing to identify an unnecessarily high percentage of actual significant differences in the data. For this reason, there has been a great deal of attention paid to developing better techniques for multiple testing, such that the overall rate of false positives can be maintained without inflating the rate of false negatives unnecessarily.

False Discovery Rate

For large-scale multiple testing (for example, as is very common in genomics when using technologies such as DNA microarrays) one can instead control the false discovery rate (FDR), defined to be the expected proportion of false positives among all significant tests.

False discovery rate (FDR) controls the expected proportion of incorrectly rejected null hypotheses (type I errors) in a list of rejected hypotheses. It is a less conservative comparison procedure with greater power than family wise error rate (FWER) control, at a cost of increasing the likelihood of obtaining type I errors. (Bonferroni correction controls FWER; FWER = P(the number of type I errors ≥ 1)).

The q-value is defined to be the FDR analogue of the p-value. The q-value of an individual hypothesis test is the minimum FDR at which the test may be called significant.

To estimate the q-value and FDR, we need following notations:

- m is the number of tests
- m_0 is the number of true null hypotheses
- m - m_0 is the number of false null hypotheses
- U is the number of true negatives (the tests not rejected when the null is true)
- V is the number of false positives (the tests rejected when the null is true)
- T is the number of false negatives (the tests not rejected when the null is not true)
- S is the number of true positives (the tests rejected when the null is not true)
- R = V + S is the total number of tests rejected
- $H_1 \ldots H_m$ the null hypotheses being tested (so there are m tests)
- In m hypothesis tests of which m_0 are true null hypotheses, R is an observable random variable, and S, T, U, and V are all unobservable random variables.

The false discovery rate (FDR) is given by $E\left(\frac{V}{V+S}\right) = E\left(\frac{V}{R}\right)$ and one wants to keep this value below a threshold α:

The Simes procedure ensures that its expected value $E\left(\frac{V}{R}\right)$ is less than a given α. This procedure is only valid when the m tests are independent. Let $H_1 \ldots H_m$ be the null hypotheses and $P_1 \ldots P_m$ their corresponding p values. Order these values in increasing order and denote them by $P_{(1)} \ldots P_{(m)}$. For a given α, find the largest k such that $P_{(k)} \le \frac{k}{m} a$. Then reject all $H_{(i)}$ for i = 1,...,k. The q-value of $H_{(k)}$ can be estimated by $mP_{(k)} / k$. It is also the FDR if we reject all the null hypotheses with p-values $\le P_k$.

The positive discover rate (pFDR) is given by

$$E\left(\frac{V}{R} \middle| R > 0\right) = P(\text{null is true} \mid \text{observed statistic in the rejection region})$$

$$= \frac{P(\text{observed statistic in the rejection region} \mid \text{null is true}) \mathrm{P}(\text{null is true})}{P(\text{observed statistic in the rejection region})}$$

If we reject the null when the p-value $\le P(k)$,

- P(observed statistic in the rejection region| $H_{(k)}$ is true) = $P_{(k)}$,
- P(null is true) $= \frac{W(\lambda)}{(1-\lambda)m}$, where $W(\lambda)$ = the number of hypotheses with pvalues ≥ λ.

  This equation is based on the consideration that the large p-values are most likely to come from the null, uniformly distributed p-values. So usually a relatively large λ will be used.
- P(observed statistic in the rejection region) = $1 - W\left(P_{(k)}\right) / m$

The q-value of $H_{(k)}$ controlling the pFDR then can be estimated by

$$\frac{P_{(k)}W(\lambda)}{(1-\lambda)\left(m-W\left(P_{(k)}\right)\right)}.$$

It is also the estimated pFDR if we reject all the null hypotheses with p-values $\leq P_{(k)}$,

Application Domains

Statistical tests always involve a trade-off between:

1. The acceptable level of false positives (in which a non-match is declared to be a match) and
2. The acceptable level of false negatives (in which an actual match is not detected).

A threshold value can be varied to make the test more restrictive or more sensitive, with the more restrictive tests increasing the risk of rejecting true positives, and the more sensitive tests increasing the risk of accepting false positives.

Inventory Control

An automated inventory control system that rejects high-quality goods of a consignment commits a type I error, while a system that accepts low-quality goods commits a type II error.

Computers

The notions of false positives and false negatives have a wide currency in the realm of computers and computer applications, as follows.

Computer Security

Security vulnerabilities are an important consideration in the task of keeping computer data safe, while maintaining access to that data for appropriate users. Moulton, stresses the importance of:

- Avoiding the type I errors (or false positives) that classify *authorized users* as *imposters.*
- Avoiding the type II errors (or false negatives) that classify *imposters* as *authorized users.*

Spam Filtering

A false positive occurs when spam filtering or spam blocking techniques wrongly classify a legitimate email message as spam and, as a result, interferes with its delivery. While most anti-spam tactics can block or filter a high percentage of unwanted emails, doing so without creating significant false-positive results is a much more demanding task.

A false negative occurs when a spam email is not detected as spam, but is classified as *non-spam.* A low number of false negatives is an indicator of the efficiency of spam filtering.

Malware

The term "false positive" is also used when antivirus software wrongly classifies a harmless file as a virus. The incorrect detection may be due to heuristics or to an incorrect virus signature in a database. Similar problems can occur with antitrojan or antispyware software.

Optical Character Recognition

Detection algorithms of all kinds often create false positives. Optical character recognition (OCR) software may detect an "a" where there are only some dots that *appear* to be an "a" to the algorithm being used.

Security Screening

False positives are routinely found every day in airport security screening, which are ultimately visual inspection systems. The installed security alarms are intended to prevent weapons being brought onto aircraft; yet they are often set to such high sensitivity that they alarm many times a day for minor items, such as keys, belt buckles, loose change, mobile phones, and tacks in shoes.

The ratio of false positives (identifying an innocent traveller as a terrorist) to true positives (detecting a would-be terrorist) is, therefore, very high; and because almost every alarm is a false positive, the positive predictive value of these screening tests is very low.

The relative cost of false results determines the likelihood that test creators allow these events to occur. As the cost of a false negative in this scenario is extremely high (not detecting a bomb being brought onto a plane could result in hundreds of deaths) whilst the cost of a false positive is relatively low (a reasonably simple further inspection) the most appropriate test is one with a low statistical specificity but high statistical sensitivity (one that allows a high rate of false positives in return for minimal false negatives).

Biometrics

Biometric matching, such as for fingerprint recognition, facial recognition or iris recognition, is susceptible to type I and type II errors. The null hypothesis is that the input *does* identify someone in the searched list of people, so:

- The probability of type I errors is called the "false reject rate" (FRR) or false non-match rate (FNMR),
- While the probability of type II errors is called the "false accept rate" (FAR) or false match rate (FMR).

If the system is designed to rarely match suspects then the probability of type II errors can be called the "false alarm rate". On the other hand, if the system is used for validation (and acceptance is the norm) then the FAR is a measure of system security, while the FRR measures user inconvenience level.

Medical Screening

In the practice of medicine, there is a significant difference between the applications of *screening* and *testing*.

- *Screening* involves relatively *cheap* tests that are given to large populations, none of whom manifest any clinical indication of disease (e.g., Pap smears).
- *Testing* involves far more *expensive*, often invasive, procedures that are given only to those who manifest some clinical indication of disease, and are most often applied to confirm a suspected diagnosis.

For example, most states in the USA require newborns to be screened for phenylketonuria and hypothyroidism, among other congenital disorders. Although they display a high rate of false positives, the screening tests are considered valuable because they greatly increase the likelihood of detecting these disorders at a far earlier stage.

The simple blood tests used to *screen* possible blood donors for HIV and hepatitis have a significant rate of false positives; however, physicians use much more expensive and far more precise *tests* to determine whether a person is actually infected with either of these viruses.

Perhaps the most widely discussed false positives in medical screening come from the breast cancer screening procedure mammography. The US rate of false positive mammograms is up to 15%, the highest in world. One consequence of the high false positive rate in the US is that, in any 10-year period, half of the American women screened receive a false positive mammogram. False positive mammograms are costly, with over $100 million spent annually in the U.S. on follow-up testing and treatment. They also cause women unneeded anxiety. As a result of the high false positive rate in the US, as many as 90–95% of women who get a positive mammogram do *not* have the condition. The lowest rate in the world is in the Netherlands, 1%. The lowest rates are generally in Northern Europe where mammography films are read twice and a high threshold for additional testing is set (the high threshold decreases the power of the test).

The ideal population screening test would be cheap, easy to administer, and produce *zero* false-negatives, if possible. Such tests usually produce more false-positives, which can subsequently be sorted out by more sophisticated (and expensive) testing.

Medical Testing

False negatives and false positives are significant issues in medical testing. False negatives may provide a falsely reassuring message to patients and physicians that disease is absent, when it is actually present. This sometimes leads to inappropriate or inadequate treatment of both the patient and their disease. A common example is relying on cardiac stress tests to detect coronary atherosclerosis, even though cardiac stress tests are known to only detect limitations of coronary artery blood flow due to advanced stenosis.

False negatives produce serious and counter-intuitive problems, especially when the condition being searched for is common. If a test with a *false negative rate* of only 10%, is used to test a population with a *true occurrence rate* of 70%, many of the negatives detected by the test will be false.

False positives can also produce serious and counter-intuitive problems when the condition being searched for is rare, as in screening. If a test has a *false positive rate* of one in ten thousand, but only one in a million samples (or people) is a *true positive*, most of the positives detected by that test will be false. The probability that an observed positive result is a false positive may be calculated using Bayes' theorem.

Type III and Type IV Error

A type III error is where you correctly reject the null hypothesis, but it's rejected for the wrong reason. This compares to a Type I error (incorrectly rejecting the null hypothesis) and a Type II error (not rejecting the null when you should). Type III errors are not considered serious, as they do mean you arrive at the correct decision. They usually happen because of random chance and are a rare occurrence.

You can also think of a Type III error as giving the right answer (i.e. correctly rejecting the null) to the wrong question. Either way, you're still arriving at the correct conclusion for the wrong reason. When we say the "wrong question", that normally means you've formulated your hypotheses incorrectly. In other words, both your null and alternate hypotheses may be poorly worded or completely incorrect.

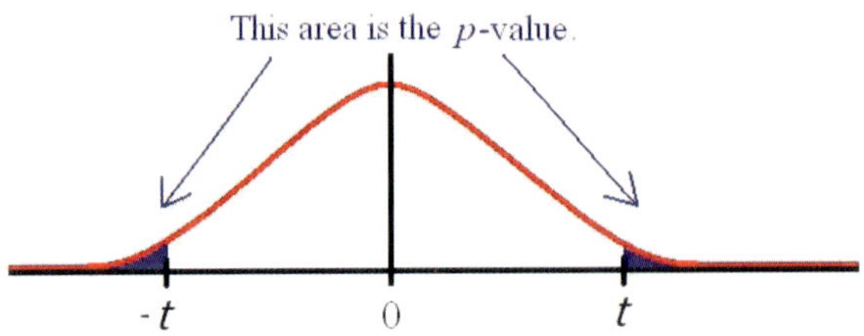

A two-tailed t-distribution

Type III errors can generally be avoided by running a two-tailed test instead of a one-tailed test. A one-tailed test has a higher power if your hypothesized direction is correct. However, if your direction is wrong, the one-tailed test will return the probability of a Type III error (only you won't realize this!). For example, let's say you hypothesize that there is a difference between the means of two samples, and that the mean difference is lower. You test this theory by running a left-tailed test. The test returns a small p-value and you (correctly) reject the null hypothesis that the means are the same. However, unknown to you, the means *are* different: it's just that one set is *higher* (i.e. you should have run a right-tailed test), not lower.

Type III errors aren't limited to differences between means. The can happen in every type of statistical test (e.g., correlations, proportions, variances etc).

Type IV Error

A Type IV error is directly related to a Type IV error; it's actually a specific type of Type III error. When you correctly reject the null hypothesis, but make a mistake interpreting the results, you have committed a Type IV error. Some common reasons that Type IV errors happen include:

- Aggregation bias (the wrong assumption that "what is true for the group is true for the individual").
- Running the wrong test for your data.
- Collinearity among predictors.

References

- Conover, W.J. (1999), "Chapter 3.4: The Sign Test", Practical Nonparametric Statistics(Third ed.), Wiley, pp. 157–176, ISBN 0-471-16068-7
- Hypothesis-testing: stattrek.com, Retrieved 10 May 2018
- Wasserstein, Ronald L.; Lazar, Nicole A. (7 March 2016). "The ASA's Statement on p-Values: Context, Process, and Purpose". The American Statistician. 70 (2): 129–133. doi:10.1080/00031305.2016.1154108. Retrieved 30 October 2016.
- Manova-analysis-paired-sample-t-test: statisticssolutions.com, Retrieved 19 July 2018
- Ranstam, J. (August 2012). "Why the P-value culture is bad and confidence intervals a better alternative". Osteoarthritis and Cartilage. 20 (8): 805–808. doi:10.1016/j.joca.2012.04.001. Retrieved 7 March 2017.
- Null-hypothesis: spss-tutorials.com, Retrieved 26 May 2018
- Marden, John I. (December 2000). "Hypothesis Testing: From p Values to Bayes Factors". Journal of the American Statistical Association. 95 (452): 1316. doi:10.2307/2669779.
- Sprent, P. (1989), Applied Nonparametric Statistical Methods (Second ed.), Chapman & Hall, ISBN 0-412-44980-3
- Type-iii-error-in-statistical-tests: statisticshowto.datasciencecentral.com, Retrieved 16 March 2018
- Bhattacharya, Bhaskar; Habtzghi, DeSale (2002). "Median of the p value under the alternative hypothesis". The American Statistician. American Statistical Association. 56(3): 202–6. doi:10.1198/000313002146. Retrieved 19 February 2016.

7

Regression Analysis

Regression analysis is a framework of statistical processes for the estimation of the relationships between variables in statistical modeling. This is an important chapter, which will analyze the principles and types of regression analysis, such as linear regression, nonlinear regression, nonparametric regression, semiparametric regression analysis, etc.

The Regression Analysis is a statistical tool used to determine the probable change in one variable for the given amount of change in another. This means, the value of the unknown variable can be estimated from the known value of another variable.

The degree to which the variables are correlated to each other depends on the Regression Line. The regression line is a single line that best fits the data, i.e. all the points plotted are connected via a line in the manner that the distance from the line to the points is the smallest.

The regression also tells about the relationship between the two or more variables, then what is the difference between regression and correlation? Well, there are two important points of differences between Correlation and Regression. These are:

- The Correlation Coefficient measures the "degree of relationship "between variables, say X and Y whereas the Regression Analysis studies the "nature of relationship" between the variables.
- Correlation coefficient does not clearly indicate the cause-and-effect relationship between the variables, i.e. it cannot be said with certainty that one variable is the cause, and the other is the effect. Whereas, the Regression Analysis clearly indicates the cause-and-effect relationship between the variables.

The regression analysis is widely used in all the scientific disciplines. In economics, it plays a significant role in measuring or estimating the relationship among the economic variables. For example, the two variables – price (X) and demand (Y) are closely related to each other, so we can find out the probable value of X from the given value of Y and similarly the probable value of Y can be found out from the given value of X.

Types of Regression Analysis

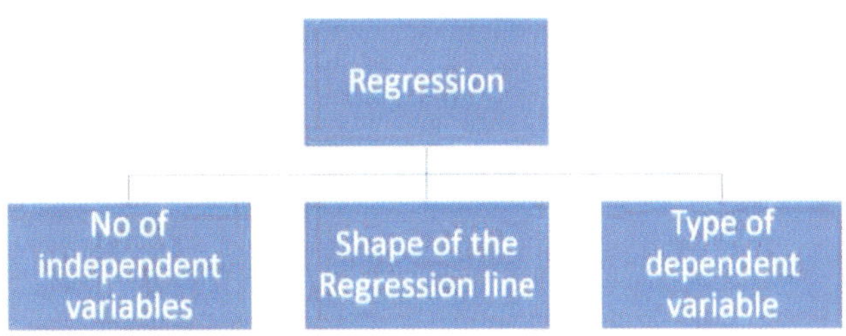

There are various kinds of regression techniques available to make predictions. These techniques are mostly driven by three metrics (number of independent variables, type of dependent variables and shape of regression line).

Linear Regression Analysis

Linear regression is a basic and commonly used type of predictive analysis. The overall idea of regression is to examine two things:

(1) Does a set of predictor variables do a good job in predicting an outcome (dependent) variable?

(2) Which variables in particular are significant predictors of the outcome variable, and in what way do they–indicated by the magnitude and sign of the beta estimates–impact the outcome variable? These regression estimates are used to explain the relationship between one dependent variable and one or more independent variables. The simplest form of the regression equation with one dependent and one independent variable is defined by the formula y = c + b*x, where y = estimated dependent variable score, c = constant, b = regression coefficient, and x = score on the independent variable.

Naming the Variables:

There are many names for a regression's dependent variable. It may be called an outcome variable, criterion variable, endogenous variable, or regressand. The independent variables can be called exogenous variables, predictor variables, or regressors.

Three major uses for regression analysis are (1) determining the strength of predictors, (2) forecasting an effect, and (3) trend forecasting.

First, the regression might be used to identify the strength of the effect that the independent variable(s) have on a dependent variable. Typical questions are what is the strength of relationship between dose and effect, sales and marketing spending, or age and income.

Second, it can be used to forecast effects or impact of changes. That is, the regression analysis helps us to understand how much the dependent variable changes with a change in one or more independent variables.

Third, regression analysis predicts trends and future values. The regression analysis can be used to get point estimates. A typical question is, "what will the price of gold be in 6 months?"

Types of Linear Regression

Simple Linear Regression

Simple linear regression is a statistical method that allows us to summarize and study relationships between two continuous (quantitative) variables:

- One variable, denoted x, is regarded as the predictor, explanatory, or independent variable.
- The other variable, denoted y, is regarded as the response, outcome, or dependent variable.

Because the other terms are used less frequently today, we'll use the "predictor" and "response" terms to refer to the variables encountered in this course. The other terms are mentioned only to make you aware of them should you encounter them in other arenas. Simple linear regression gets its adjective "simple," because it concerns the study of only one predictor variable.

Types of Relationships

Before proceeding, we must clarify what types of relationships we won't study in this course, namely, deterministic (or functional) relationships. Here is an example of a deterministic relationship.

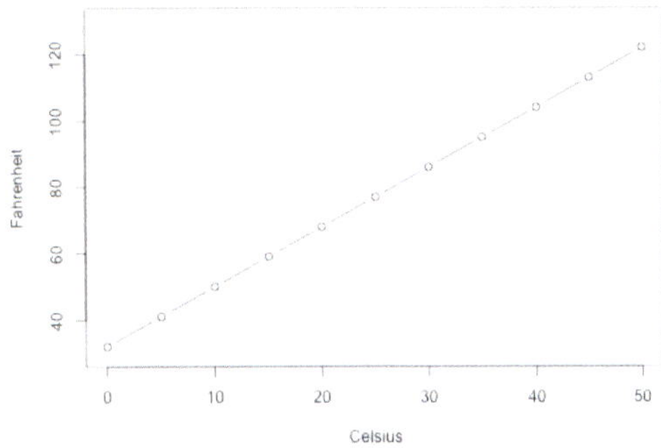

Note that the observed (x, y) data points fall directly on a line. As you may remember, the relationship between degrees Fahrenheit and degrees Celsius is known to be:

$$\text{Fahr} = \frac{9}{5}\text{Cels} + 32$$

That is, if you know the temperature in degrees Celsius, you can use this equation to determine the temperature in degrees Fahrenheit *exactly*.

Here are some examples of other deterministic relationships that students from previous semesters have shared:

- Circumference = $\pi \times$ diameter.
- Hooke's Law: $Y = \alpha + \beta X$, where Y = amount of stretch in a spring, and X = applied weight.
- Ohm's Law: $I = V/r$, where V = voltage applied, r = resistance, and I = current.
- Boyle's Law: For a constant temperature, $P = \alpha/V$, where P = pressure, α = constant for each gas, and V = volume of gas.

For each of these deterministic relationships, the equation *exactly* describes the relationship between the two variables. This course does not examine deterministic relationships. Instead, we are interested in statistical relationships, in which the relationship between the variables is not perfect.

Multiple Linear Regression Model

A linear regression model that contains more than one predictor variable is called a *multiple linear regression model*. The following model is a multiple linear regression model with two predictor variables, x_1 and x_2.

$$Y = \beta_0 + \beta_1 x_1 + \beta_2 x_2 + \in$$

The model is linear because it is linear in the parameters β_0, β_1 and β_2. The model describes a plane in the three-dimensional space of Y, x_1 and x_2. The parameter β_0 is the intercept of this plane. Parameters β_1 and β_2 are referred to as *partial regression coefficients*. Parameter β_1 represents the change in the mean response corresponding to a unit change in x_1 when x_2 is held constant. Parameter β_2 represents the change in the mean response corresponding to a unit change in x_2 when x_1 is held constant. Consider the following example of a multiple linear regression model with two predictor variables, x_1 and x_2:

$$Y = 30 + 5x_1 + 7x_2 + \in$$

This regression model is a first order multiple linear regression model. This is because the maximum power of the variables in the model is 1. (The regression plane corresponding to this model is shown in the figure below.) Also shown is an observed data point and the corresponding random error, $\in$. The true regression model is usually never known (and therefore the values of the random error terms corresponding to observed data points remain unknown). However, the regression model can be estimated by calculating the parameters of the model for an observed data set.

One of the following figures shows the contour plot for the regression model the above equation. The contour plot shows lines of constant mean response values as a function of x_1 and x_2. The contour lines for the given regression model are straight lines as seen on the plot. Straight contour lines result for first order regression models with no interaction terms.

A linear regression model may also take the following form:

$$Y = \beta_0 + \beta_1 x_1 + \beta_2 x_2 + \beta_{12} x_1 x_2 + \in$$

A cross-product term, $x_1 x_2$, is included in the model. This term represents an interaction effect between the two variables x_1 and x_2. Interaction means that the effect produced by a change in the predictor variable on the response depends on the level of the other predictor variable(s). As an example of a linear regression model with interaction, consider the model given by the equation $Y = 30 + 5x_1 + 7x_2 + 3x_1 x_2 + \in$. The regression plane and contour plot for this model are shown in the following two figures, respectively.

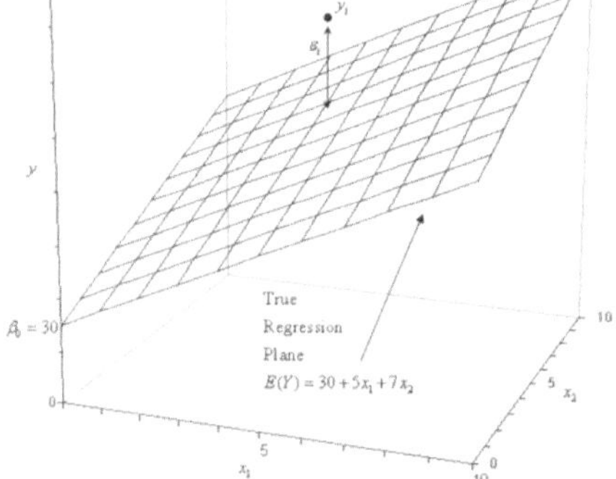

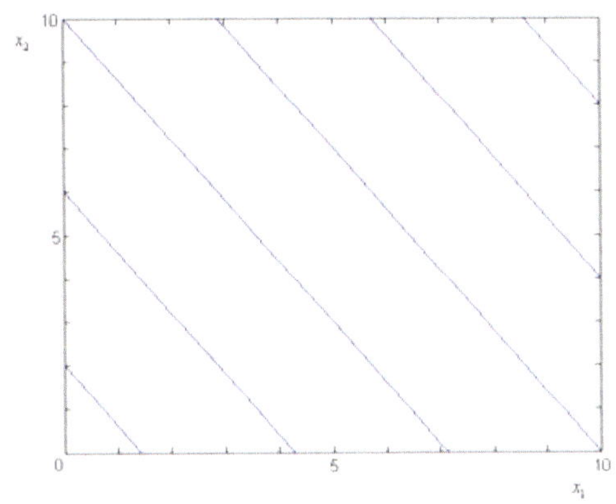

Now consider the regression model shown next:

$$Y = \beta_0 + \beta_1 x_1 + \beta_2 x_1^2 + \beta_3 x_1^3 + \in$$

This model is also a linear regression model and is referred to as a *polynomial regression model*. Polynomial regression models contain squared and higher order terms of the predictor variables making the response surface curvilinear. As an example of a polynomial regression model with an interaction term consider the following equation:

$$Y = 500 + 5x_1 + 7x_2 - 3x_1^2 - 5x_2^2 + 3x_1 x_2 + \in$$

This model is a *second order* model because the maximum power of the terms in the model is two. The regression surface for this model is shown in the following figure. Such regression models are used in RSM to find the optimum value of the response Y. Notice that, although the shape of the regression surface is curvilinear, the regression model is still linear because the model is linear in the parameters. The contour plot for this model is shown in the second of the following two figures:

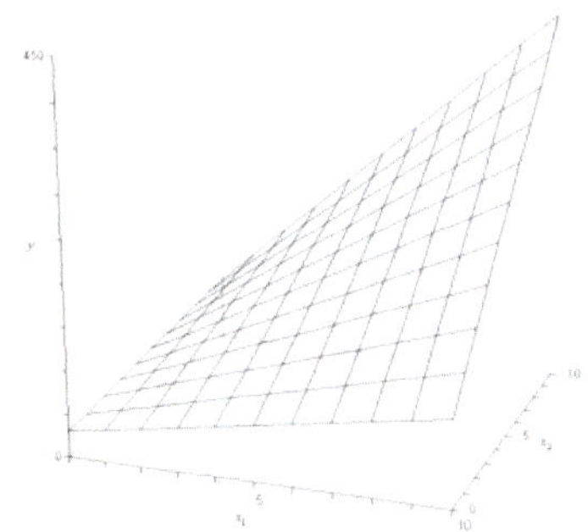

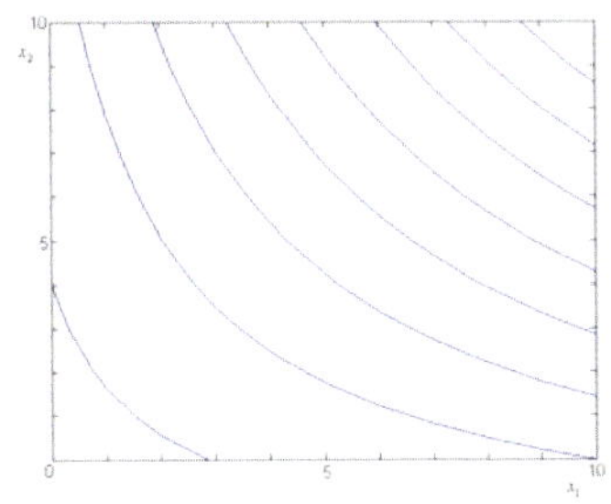

All multiple linear regression models can be expressed in the following general form:

$$Y = \beta_0 + \beta_1 x_1 + \beta_2 x_2 + \ldots + \beta_k x_k + \in$$

where k denotes the number of terms in the model. For example, the model can be written in the general form using $x_3 = x_1^2$, $x_4 = x_2^3$ and $x_5 = x_1 x_2$ as follows:

$$Y = 500 + 5x_1 + 7x_2 - 3x_3 - 5x_4 + 3x_5 + \in$$

Logistic Regression

In statistics, the logistic model (or logit model) is a widely used statistical model that, in its basic form, uses a logistic function to model a binary dependent variable; many more complex extensions exist. In regression analysis, logistic regression (or logit regression) is estimating the parameters of a logistic model; it is a form of binomial regression. Mathematically, a binary logistic model has a dependent variable with two possible values, such as pass/fail, win/lose, alive/dead or healthy/sick; these are represented by an indicator variable, where the two values are labeled "0" and "1". In the logistic model, the log-odds (the logarithm of the odds) for the value labeled "1" is a linear combination of one or more independent variables ("predictors"); the independent variables can each be a binary variable (two classes, coded by an indicator variable) or a continuous variable (any real value). The corresponding probability of the value labeled "1" can vary between 0 (certainly the value "0") and 1 (certainly the value "1"), hence the labeling; the function that converts log-odds to probability is the logistic function, hence the name. The unit of measurement for the log-odds scale is called a *logit*, from *logistic unit*, hence the alternative names. Analogous models with a different sigmoid function instead of the logistic function can also be used, such as the probit model; the defining characteristic of the logistic model is that increasing one of the independent variables multiplicatively scales the odds of the given outcome at a *constant* rate, with each dependent variable having its own parameter; for a binary independent variable this generalizes the odds ratio.

Logistic regression was developed by statistician David Cox in 1958. The binary logistic regression model has extensions to more than two levels of the dependent variable: categorical outputs with more than two values are modelled by multinomial logistic regression, and if the multiple categories are ordered, by ordinal logistic regression, for example the proportional odds ordinal logistic model. The model itself simply models probability of output in terms of input, and does not perform statistical classification (it is not a classifier), though it can be used to make a classifier, for instance by choosing a cutoff value and classifying inputs with probability greater than the cutoff as one class, below the cutoff as the other; this is a common way to make a binary classifier. The coefficients are generally not computed by a closed-form expression, unlike linear least squares.

Applications

Logistic regression is used in various fields, including machine learning, most medical fields, and social sciences. For example, the Trauma and Injury Severity Score (TRISS), which is widely used to predict mortality in injured patients, was originally developed by Boyd et al. using logistic regression. Many other medical scales used to assess severity of a patient have been developed using logistic regression. Logistic regression may be used to predict the risk of developing a given disease (e.g. diabetes; coronary heart disease), based on observed characteristics of the patient (age, sex, body mass index, results of various blood tests, etc.). Another example might be to predict whether an Indian voter will vote BJP or Trinamool Congress or Left Front or Congress, based on age, income, sex, race, state of residence, votes in previous elections, etc. The technique can also be used in engineering, especially for predicting the probability of failure of a given process, system or product. It is also used in marketing applications such as prediction of a customer's propensity to purchase a product or halt a subscription, etc. In economics it can be used to predict the likelihood of a person's choosing to be in the labor force, and a business application would be to predict the likelihood of a homeowner defaulting on a mortgage. Conditional random fields, an extension of logistic regression to sequential data, are used in natural language processing.

Examples

Logistic Model

One may begin to understand logistic regression by first considering a logistic model with given parameters, then seeing how the coefficients can be estimated ("regressed") from data. Consider a model with two predictors, x_1 and x_2; these may be continuous variables (taking a real number as value), or indicator functions for binary variables (taking value 0 or 1). Then the general form of the log-odds (here denoted by *l*) is:

$$l = \beta_0 + \beta_1 x_1 + \beta_2 x_2,$$

where the coefficients β_i are the parameters of the model. Note that this is a linear model: the log-odds *l* are a linear combination of the predictors x_1 and x_2 , including a constant term β_0. The corresponding odds are the exponent:

$$o = b^{\beta_0 + \beta_1 x_1 + \beta_2 x_2}$$

where *b* is the base of the logarithm and exponent. This is now a non-linear model, since the odds are not a linear combination of the predictors.

Odds of $o = o:1$ for an event are the same as probability of $o/(o+1)$, since happening *o* of the time and not happening 1 time corresponds to happening *o* times out of a total of *o* + 1 events, so the corresponding probability is:

$$p = \frac{b^{\beta_0 + \beta_1 x_1 + \beta_2 x_2}}{b^{\beta_0 + \beta_1 x_1 + \beta_2 x_2} + 1} = \frac{1}{1 + b^{-(\beta_0 + \beta_1 x_1 + \beta_2 x_2)}}$$

The base *b* is usually taken to be *e*, but for exposition here use 10, so the odds are more understandable. Suppose the coefficients are $-3, 1, 2$, so the model is:

$$l = -3 + 1 \cdot x_1 + 2 \cdot x_2,$$

This can be interpreted as follows:

- $\beta_0 = -3$ is the *y*-intercept, and can thus be interpreted as the log-odds when the predictors are all zero; here the log-odds are -3, so the odds for are $10^{-3} = 10^{-3}:1 = 1:1000$, and the probability is $1/(1000+1) = 1/1001 \approx 0.001 = 0.1\%$.

- Alternatively, grouping the weighted predictors as a single contribution to the total logit, $l' = 1 \cdot x_1 + 2 \cdot x_2$, so $l = l' - 3$, the quantity $-\beta_0$ is the *x*-intercept, and can be interpreted as the weighted predictors corresponding to even odds: if $l' = 1 \cdot x_1 + 2 \cdot x_2 = 3$, then the log-odds are 0, the odds for are $10^0 = 10^0:1 = 1:1$, and the probability is $1/(1+1) = 1/2 = 50\%$.
- $\beta_1 = 1$ means that increasing x_1 by 1 increases the log-odds by $1 \cdot 1$, so it *multiplies* the odds by 10 10; this is sometimes referred to as the "effect" of the predictor x_1.
- $\beta_2 = 2$ means that increasing x_2 by 1 increases the log-odds by $2 \cdot 1$, so it *multiplies* the odds by $10^2 = 100$ Thus the effect of x_2 on the log-odds is twice as great as the effect of x_1.

Such a model can be used for various purposes. For example, given an individual datum with values of predictors x_1, x_2, one can estimate the log-odds (hence odds and probability) of the outcome by putting in the values in the formula. Alternatively, consider comparing two medical treatments, one of which decreases x_1 by 3 and the other that decreases x_2 by 2. Assuming this model is valid, the first treatment reduces the log-odds of the outcome by $1 \cdot 3 = 3$, so it reduces the odds by a factor of $10^3 = 1000$ but the other treatment reduces the log-odds by $2 \cdot 2 = 4$, so it reduces the odds by a factor of $10^4 = 10000$, and thus is more effective, all else equal.

In order to estimate the parameters of such a logistic model and compute how well it fits the data, one must do a logistic regression.

Probability of Passing an Exam Versus Hours of Study

Suppose we wish to answer the following question:

A group of 20 students spend between 0 and 6 hours studying for an exam. How does the number of hours spent studying affect the probability that the student will pass the exam?

The reason for using logistic regression for this problem is that the values of the dependent variable, pass and fail, while represented by "1" and "0", are not cardinal numbers. If the problem was changed so that pass/fail was replaced with the grade 0–100 (cardinal numbers), then simple regression analysis could be used.

The table shows the number of hours each student spent studying, and whether they passed (1) or failed (0).

Hours	0.50	0.75	1.00	1.25	1.50	1.75	1.75	2.00	2.25	2.50	2.75	3.00	3.25	3.50	4.00	4.25	4.50	4.75	5.00	5.50
Pass	0	0	0	0	0	0	1	0	1	0	1	0	1	0	1	1	1	1	1	1

The graph shows the probability of passing the exam versus the number of hours studying, with the logistic regression curve fitted to the data.

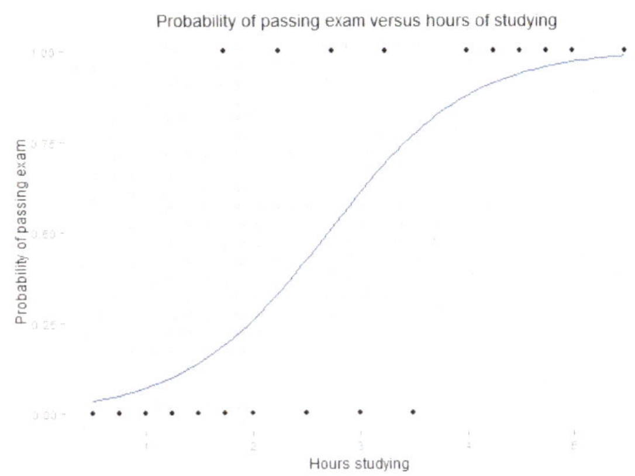

Graph of a logistic regression curve showing probability of passing an exam versus hours studying

The logistic regression analysis gives the following output.

	Coefficient	Std.Error	z-value	P-value (Wald)
Intercept	−4.0777	1.7610	−2.316	0.0206
Hours	1.5046	0.6287	2.393	0.0167

The output indicates that hours studying is significantly associated with the probability of passing the exam ($p = 0.0167$, Wald test). The output also provides the coefficients for $\text{Intercept} = -4.0777$ and $\text{Hours} = 1.5046$. These coefficients are entered in the logistic regression equation to estimate the odds (equivalently, probability) of passing the exam:

$$\text{Log-odds of passing exam} = 1.5046 \cdot \text{Hours} - 4.0777 = 1.5046 \cdot (\text{Hours} - 2.71)$$

$$\text{Odds of passing exam} = \exp\left(1.5046 \cdot \text{Hours} - 4.0777\right) = \exp\left(1.5046 \cdot (\text{Hours} - 2.71)\right)$$

$$\text{Probability of passing exam} = \frac{1}{1 + \exp\left(-\left(1.5046 \cdot \text{Hours} - 4.0777\right)\right)}$$

One additional hour of study is estimated to increase log-odds of passing by 1.5046, so multiplying odds of passing by $\exp(1.5046) \approx 4.5$. The form with the x-intercept (2.71) shows that this estimates even odds (log-odds 0, odds 1, probability 1/2) for a student who studies 2.71 hours.

For example, for a student who studies 2 hours, entering the value $\text{Hours} = 2$ in the equation gives the estimated probability of passing the exam of 0.26:

$$\text{Probability of passing exam} = \frac{1}{1 + \exp\left(-\left(1.5046 \cdot 2 - 4.0777\right)\right)} = 0.26$$

Similarly, for a student who studies 4 hours, the estimated probability of passing the exam is 0.87:

$$\text{Probability of passing exam} = \frac{1}{1 + \exp\left(-\left(1.5046 \cdot 4 - 4.0777\right)\right)} = 0.87$$

This table shows the probability of passing the exam for several values of hours studying.

Hours of study	Passing exam		
	Log-odds	Odds	Probability
1	−2.57	0.076 ≈ 1:13.1	0.07
2	−1.07	0.34 ≈ 1:2.91	0.26
3	0.44	1.55	0.61
4	1.94	6.96	0.87
5	3.45	31.4	0.97

The output from the logistic regression analysis gives a p-value of $p = 0.0167$, which is based on the Wald z-score. Rather than the Wald method, the recommended method to calculate the p-value for logistic regression is the likelihood-ratio test (LRT), which for this data gives $p = 0.0006$.

Ordinal regression is a statistical technique that is used to predict behavior of ordinal level dependent variables with a set of independent variables. The dependent variable is the order response category variable and the independent variable may be categorical or continuous.

Ordered Regression

Assumptions:

- One dependent variable, we cannot use multiple dependent variables.
- Parallel lines assumption: There is one regression equation for each category except the last category. The last category probability can be predicted as 1-second last category probability.
- Adequate cell count: As per the rule of thumb, 80% of cells must have more than 5 counts. No cell should have Zero count. The greater the cell with less count, the less reliable the chi-square test will be.

Multinomial Logistic Regression

Multinomial logistic regression is used when you have a categorical dependent variable with two or more unordered levels (i.e. two or more discrete outcomes). It is practically identical to logistic regression, except that you have multiple possible outcomes instead of just one.

For example, children's food choices are influenced by their parents' choices and the children's pastimes (e.g. sports enthusiast vs. gamer). You could study the relationship between a child's food choices with their parents' choices and children's pastimes. The dependent variable levels would be the different food choices (fast food, healthy choices, protein packed, vegan etc.). Or you might study how workers' education levels and time on the job affect promotions. The independent variables would be education levels and time on the job, and the levels of the dependent variable might be promotion to team-leader roles, sales positions, or management positions.

One level of the dependent variable is chosen as the *reference category*. This is typically the most common or the most frequent category. In the first example above, this might be "fast food". The probability of being in any of the other categories is compared to the probability of being in the reference category. These relative probabilities are the predicted log odds.

Running Multinomial Logistic Regression

This type of regression is usually performed with software. Essentially, the software will run a series of individual binomial logistic regressions for M – 1 categories (one calculation for each category, minus the reference category). When M = 2, multinomial logistic regression, ordered logistic regression, and logistic regression are equal.

Before the advent of computer software, you would have run these individual regressions and then compared the results. The software takes away that chore, and estimates parameters simultaneously, resulting in more efficiency.

Assumptions include:

- The model is specified correctly with no extraneous variables.
- Cases are independent.
- There is no multicollinearity between the independent variables.

Interpreting Results

Multinomial logistic regression works the same way as other types of regression: you're looking for a relationship between the independent and dependent variables. The output will give you sets of coefficients for each variable. The output for each software package will vary.

Nonlinear Regression

Nonlinear regression uses nonlinear regression equations, which take the form:

$$Y = f(X,\beta) + \varepsilon$$

Where:

- X = a vector of p predictors,
- β = a vector of k parameters,
- f(-) = a known regression function,
- ε = an error term.

The formal definition is that if your regression equation looks like the one above, it's nonlinear regression. However, this is actually a lot more difficult than it sounds. Take the following nonlinear regression equations:

- The Michaelis -Menten model: $f(x,\beta)=(\beta_1 x)/(\beta_2+x)$.
- $Y = \beta_0 + (0.4 - \beta_0)e^{-\beta_1(x_i-5)} + \varepsilon_i$.

These both meet the requirement of fitting the form $Y = f(X,\beta) + \varepsilon$, but that isn't immediately obvious without some in-depth knowledge of algebra and regression analysis.

The good news is there is a much simpler, more intuitive definition of nonlinear regression:

If your model uses an equation in the form $Y = a_0 + b_1X_1$, it's a linear regression model. If not, it's nonlinear.

It's *much* easier to spot a linear regression equation, as it's *always* going to take the form $Y = a_0 + b_1X_1$ *.

Nonparametric Regression

Nonparametric regression, like linear regression, estimates mean outcomes for a given set of covariates. Unlike linear regression, nonparametric regression is agnostic about the functional form between the outcome and the covariates and is therefore not subject to misspecification error.

In nonparametric regression, you do not specify the functional form. You specify the dependent variable—the outcome—and the covariates. You specify y, x_1, x_2, and x_3 to fit.

$$y = g(x_1, x_2, x_3) + \epsilon$$

The method does not assume that g()is linear; it could just as well be

$$y = \beta_1 x_1 + \beta_2 x_2^2 + \beta_3 x_1^3 x_2 + \beta_4 x_3 + \epsilon$$

The method does not even assume the function is linear in the parameters. It could just as well be

$$y = \beta_1 x_1^{\beta 2} + \cos(x_2 x_3) + \epsilon$$

Kernel Regression

Kernel regression is a modeling tool which belongs to the family of smoothing methods. Unlike linear regression which is both used to explain phenomena and for prediction (understanding a phenomenon to be able to predict it afterwards), Kernel regression is mostly used for prediction. The structure of the model is variable and complex, the latter working like a filter or black box. There are many variations of Kernel regression in existence.

As with any modeling method, a learning sample of size nlearn is used to estimate the parameters of the model. A sample of size nvalid can then be used to evaluate the quality of the model. Lastly, the model can be applied to a prediction sample of size npred, for which the values of the dependent variable Y are unknown.

The characteristics of Kernel Regression are:

- The use of a kernel function, to weigh the observations of the learning sample, depending on their "distance" from the predicted observation.

  The kernel functions available in XLSTAT are:

  - Uniform
  - Triangle
  - Epanechnikov
  - Quartic
  - Triweight
  - Tricube
  - Gaussian
  - Cosine
- The bandwidth associated to each variable. It is involved in calculating the kernel and the weights of the observations, and differentiates or rescales the relative weights of the variables while at the same time reducing or augmenting the impact of observations of the learning sample, depending on how far they are from the observation to predict.
- The polynomial degree used when fitting the model to the observations of the learning sample. Two strategies are suggested in order to restrict the size of the learning sample taken into account for the estimation of the parameters of the polynomial: Moving window and k nearest neighbors.

Lowess Regression

Locally weighted regression and smoothing scatter plots or LOWESS regression was introduced to create smooth curves through scattergrams.

LOWESS regression is very similar to Kernel regression as it is also based on polynomial regression and requires a kernel function to weight the observations.

Gaussian Process Regression or Kriging

In Gaussian process regression, also known as Kriging, a Gaussian prior is assumed for the regression curve. The errors are assumed to have a multivariate normal distribution and the regression curve is estimated by its posterior mode. The Gaussian prior may depend on unknown hyperparameters, which are usually estimated via empirical Bayes.

Smoothing splines have an interpretation as the posterior mode of a Gaussian process regression.

Nonparametric Multiplicative Regression

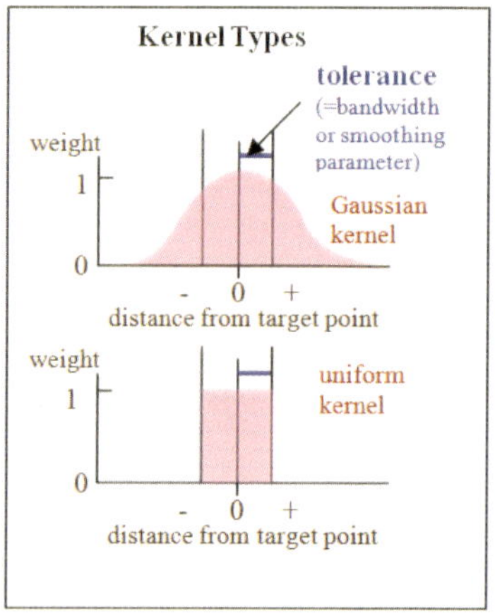

Two kinds of kernels used with kernel smoothers for nonparametric regression.

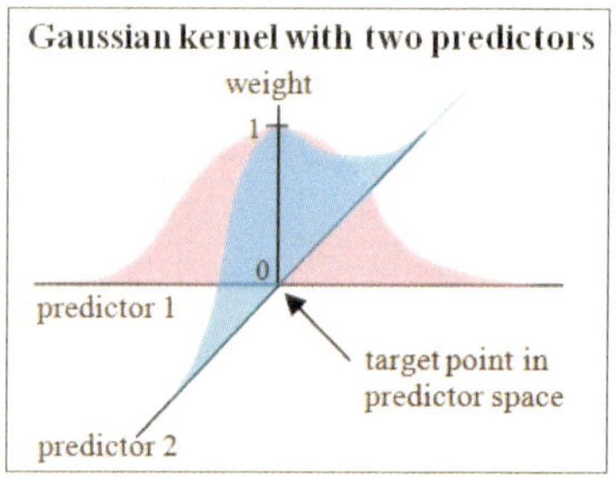

Use of Gaussian kernels for nonparametric multiplicative regression with two predictors. The weights from the kernel function for each predictor are multiplied to obtain a weight for a given data point in estimating a response variable (dependent variable) at a target point in the predictor space.

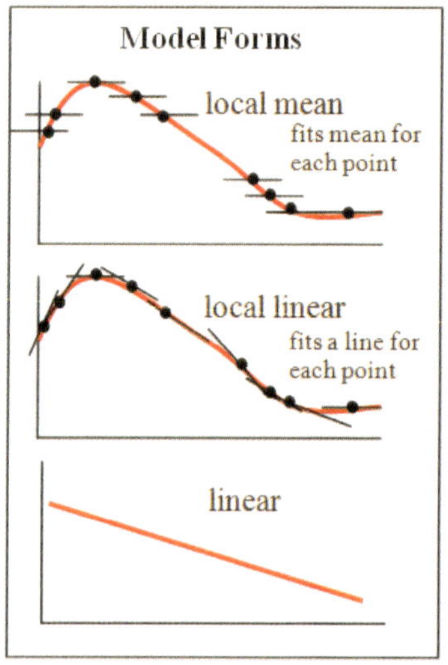

Two commonly used forms of a local model used in nonparametric regression, contrasted with a simple linear model.

Nonparametric multiplicative regression (NPMR) is a form of nonparametric regression based on multiplicative kernel estimation. Like other regression methods, the goal is to estimate a response (dependent variable) based on one or more predictors (independent variables). NPMR can be a good choice for a regression method if the following are true:

- The shape of the response surface is unknown.
- The predictors are likely to interact in producing the response; in other words, the shape of the response to one predictor is likely to depend on other predictors.
- The response is either a quantitative or binary (0/1) variable.

This is a smoothing technique that can be cross-validated and applied in a predictive way.

NPMR Behaves Like an Organism

NPMR has been useful for modeling the response of an organism to its environment. Organismal response to environment tends to be nonlinear and have complex interactions among predictors. NPMR allows you to model automatically the complex interactions among predictors in much the same way that organisms integrate the numerous factors affecting their performance.

A key biological feature of an NPMR model is that failure of an organism to tolerate any single dimension of the predictor space results in overall failure of the organism. For example, assume that a plant needs a certain range of moisture in a particular temperature range. If either temperature or moisture fall outside the tolerance of the organism, then the organism dies. If it is too hot, then no amount of moisture can compensate to result in survival of the plant. Mathematically this works with NPMR because the product of the weights for the target point is zero or near zero if any of the weights for individual predictors (moisture or temperature) are zero or near zero. Note further that in this simple example, the second condition listed above is probably true: the response of the plant to moisture probably depends on temperature and vice versa.

Optimizing the selection of predictors and their smoothing parameters in a multiplicative model is computationally intensive. With a large pool of predictors, the computer must search through a huge number of potential models in search for the best model. The best model has the best fit, subject to overfitting constraints or penalties.

Local Model

NPMR can be applied with several different kinds of local models. By "local model" we mean the way that data points near a target point in the predictor space are combined to produce an estimate for the target point. The most common choices for the local models are the local mean estimator, a local linear estimator, or a local logistic estimator. In each case the weights can be extended multiplicatively to multiple dimensions.

In words, the estimate of the response is a local estimate (for example a local mean) of the observed values, each value weighted by its proximity to the target point in the predictor space, the weights being the product of weights for individual predictors. The model allows interactions, because weights for individual predictors are combined by multiplication rather than addition.

Overfitting Controls

Understanding and using these controls on overfitting is essential to effective modeling with nonparametric regression. Nonparametric regression models can become overfit either by including too many predictors or by using small smoothing parameters (also known as bandwidth or tolerance). This can make a big difference with special problems, such as small data sets or clumped distributions along predictor variables.

The methods for controlling overfitting differ between NPMR and the generalized linear modeling (GLMs). The most popular overfitting controls for GLMs are the Akaike information criterion (AIC) and the Bayesian information criterion (BIC) for model selection. The AIC and BIC depend on the number of parameters in a model. Because NPMR models do not have explicit parameters as such, these are not directly applicable to NPMR models. Instead, one can control overfitting by setting a minimum average neighborhood size, minimum data:predictor ratio, and a minimum improvement required to add a predictor to a model.

Nonparametric regression models sometimes use an AIC based on the "effective number of parameters". This penalizes a measure of fit by the trace of the smoothing matrix—essentially how much each data point contributes to estimating itself, summed across all data points. If, however, you use leave-one-out cross validation in the model fitting phase, the trace of the smoothing matrix is always zero, corresponding to zero parameters for the AIC. Thus, NPMR with cross-validation in the model fitting phase already penalizes the measure of fit, such that the error rate of the training data set is expected to approximate the error rate in a validation data set. In other words, the training error rate approximates the prediction (extra-sample) error rate.

Related Techniques

NPMR is essentially a smoothing technique that can be cross-validated and applied in a predictive way. Many other smoothing techniques are well known, for example smoothing splines and wavelets. The optimal choice of a smoothing method depends on the specific application. Nonparametric regression models always fits for larger data.

Regression Trees

Decision tree learning algorithms can be applied to learn to predict a dependent variable from data. Although the original Classification And Regression Tree (CART) formulation applied only to predicting univariate data, the framework can be used to predict multivariate data including time series.

Semiparametric Regression

A semiparametric model is a regression model with both a finite- and an infinite-dimensional component.

A *finite-dimensional component* is spanned by some list of vectors (a vector is an object that has both magnitude and direction). The two-dimensional and three-dimensional spaces we deal with in everyday geometry are examples of finite-dimensional spaces, but so is a hypothetical 4568-dimensional space. *Infinite dimensional spaces* are spaces that have an infinite, and possibly ill-defined, number of dimensions and possibilities. They aren't spanned by any finite list of vectors.

In contrast to parametric models, which are well-defined in the finite-dimensional space, and non-parametric models, where the parameters can all span an infinite space, a semiparametric model has a component that is finite-dimensional (i.e. it's easy to research and understand), and another that is infinite-dimensional (i.e. beyond the range of ordinary statistical methods).

In research and statistics projects involving semi-parametric models the emphasis is almost always on the parametric component of the model. That's because this is the part which lends itself well to research.

Use of Semiparametric Models

Too often parametric models, while being easy to understand and easy to work with, fail to give a fair representation of what is happening in the real world. Non-parametric models may be better representations but do not lend themselves well to analysis. A semiparametric model allows you to have the best of both worlds: a model that is understandable and can be manipulated while still offering a fair representation of the messiness that is involved in real life.

Examples of Semiparametric Models

One example of a semi-parametric model is the Cox Proportional Hazards Model. This model is very useful in studies of the time remaining before an end or failure; it's used when doing research on the time remaining before a patient dies, or before a light bulb burns out. It's defined as:

$$F(t) = 1 - \exp(-\int_0^t \lambda_0(u) e^{\beta' x} du)$$

Here x is what we call the covariate vector, and our unknown parameters are B and $\lambda_0(u)$. B is finite-dimensional and is often the subject of research; $\lambda_0(u)$. on the other hand, is a unknown, non-negative function of time, and the space of $\lambda_0(u)$. possibilities is infinite. We call it a 'nuisance parameter'; a parameter that can't be entirely omitted in a rigorous treatment of a problem but is not of immediate interest– typically because it is difficult or impossible to study.

Methods

Many different semiparametric regression methods have been proposed and developed. The most popular methods are the partially linear, index and varying coefficient models.

Partially Linear Models

A partially linear model is given by

$$Y_i = X_i'\beta + g(Z_i) + u_i, \quad i = 1, \ldots, n,$$

where Y_i is the dependent variable, X_i is a $p \times 1$ vector of explanatory variables, β is a $p \times 1$ vector of unknown parameters and $Z_i \in \mathrm{R}^q$. The parametric part of the partially linear model is given by the parameter vector β while the nonparametric part is the unknown function $g(Z_i)$. The data is assumed to be i.i.d. with $E(u_i \mid X_i, Z_i) = 0$ and the model allows for a conditionally heteroskedastic error process $E(u_i^2 \mid x,z) = \sigma^2(x,z)$ of unknown form. This type of model was proposed by Robinson (1988) and extended to handle categorical covariates by Racine and Li.

This method is implemented by obtaining a $\sqrt{n}$ consistent estimator of β and then deriving an estimator of $g(Z_i)$ from the nonparametric regression of $Y_i - X_i'\hat{\beta}$ on z using an appropriate nonparametric regression method.

Index Models

A single index model takes the form

$$Y = g(X'\beta_o) + u,$$

where Y, X and β_o are defined as earlier and the error term u satisfies $E(u \mid X) = 0$. The single index model takes its name from the parametric part of the model $x'\beta$ which is a *scalar* single index. The nonparametric part is the unknown function $g(\cdot)$.

Ichimura's Method

The single index model method developed by Ichimura is as follows. Consider the situation in which y is continuous. Given a known form for the function $g(\cdot)$, β_o could be estimated using the nonlinear least squares method to minimize the function

$$\sum_{i=1} \left(Y_i - g(X_i'\beta)\right)^2.$$

Since the functional form of $g(\cdot)$ is not known, we need to estimate it. For a given value for β an estimate of the function

$$G(X_i'\beta) = E(Y_i \mid X_i'\beta) = E\left[g(X_i'\beta_o) \mid X_i'\beta\right]$$

using kernel method. Ichimura proposes estimating $g(X_i'\beta)$ with

$$\hat{G}_{-i}(X_i'\beta),$$

the leave-one-out nonparametric kernel estimator of $G(X_i'\beta)$.

Klein and Spady's Estimator

If the dependent variable y is binary and X_i and u_i are assumed to be independent, Klein and Spady propose a technique for estimating β using maximum likelihood methods. The log-likelihood function is given by

$$L(\beta) = \sum_i (1 - Y_i)\ln\left(1 - \hat{g}_{-i}(X_i'\beta)\right) + \sum_i Y_i \ln\left(\hat{g}_{-i}(X_i'\beta)\right),$$

where $\hat{g}_{-i}(X_i'\beta)$ is the leave-one-out estimator.

Smooth Coefficient/Varying Coefficient Models

Hastie and Tibshirani propose a smooth coefficient model given by

$$Y_i = \alpha(Z_i) + X_i'\beta(Z_i) + u_i = (1 + X_i')\begin{pmatrix}\alpha(Z_i)\\ \beta(Z_i)\end{pmatrix} + u_i = W_i'\gamma(Z_i) + u_i,$$

Where X_i is a $k \times 1$ vector and $\beta(z)$ is a vector of unspecified smooth functions of z.

$\gamma(\cdot)$ may be expressed as

$$\gamma(Z_i) = \left(E\left[W_i W_i' \mid Z_i\right]\right)^{-1} E\left[W_i Y_i \mid Z_i\right].$$

Regression Models

Poisson Regression

To define the basic version of a Poisson regression model, suppose that we have observations $y_1,\ldots,y_n$ for the response variable $y_1,\ldots,y_n$ assumed to be independently distributed Poisson variates with means $\mu_1,\ldots,\mu_n$, i.e.

$$f(y_i|\mu_i) = \frac{\mu_i^{y_i}}{y_i!} = \exp(-\mu_i)$$

The systematic component of the model is specified by some regression function η, depending on regression parameters $\beta_1,\ldots,\beta_k$, with each component relating values $x_{i1},\ldots,x_{ik}$ of explanatory variables to respective means, i.e.

$$\mu_i = \eta_i(\beta) = \eta i(x_i 1,\ldots,x_{ik};\ \beta_1,\ldots,\beta_k).$$

Below Table Distribution by months prior to interview of stressful events reported from subjects: 147 subjects reporting exactly one stressful event in the period from 1 to 18 months prior to interview.

Months 1 2 3 4 5 6 7 8 9 10 11 12 13 14 15 16 17 18
Number 15 11 14 17 5 11 10 4 8 10 7 9 11 3 6 1 1 4

Often, this relationship is such that some monotone transformation g of the means is connected to a linear predictor of explanatory variables,

$$g(\mu_i) = \sum_{j=1}^{k} x_{ij}\beta_j.$$

In this situation g is called the link function and the model defined in this manner is an instance of a generalized linear model. For $\eta_i(\beta) = \left(\sum_{j=1}^{k} xi_j\beta_j\right)$ we have thc familiar log linear model,

$$\log \mu_i = \sum_{j=1}^{k} x_{ij}\beta_j .$$

For the model specified by the stochastic component $f(y_i|\mu_i) = \frac{\mu_i^{y_i}}{y_i!} = \exp(-\mu_i)$ and regression function $\mu_i = \eta_i(\beta) = \eta i(x_i 1,...,x_{ik};\ \beta_1,...,\beta_k).$, the log likelihood function is written as

$$\ell_y(\beta) = \sum_{i=1}^{n} \{ y_i \log[\eta_i(\beta)] - \eta_i(\beta) - log(y_i!) \}.$$

It may be worthwhile noting that this reduces to a k-parameter exponential family log likelihood,

$$\ell_y(\beta) = \sum_{j=1}^{k}\left(\sum_{i=1}^{n} x_{ij} y_i\right)\beta_i - \sum_{i=1}^{n} \exp\left(\sum_{j=1}^{k} x_{ij}\beta_j\right)$$
$$- \sum_{i=1}^{n} \log(y_i!),$$

with jointly sufficient statistics $\sum_{i=1}^{n} x_{ij} y_i, j = 1,.....,k,$ if the model is log linear.

Some Special Cases

Loglinear Models for Contingency Tables Suppose, in obvious notation, y_{ij} with indices i = 1,...,I and j = 1,...,J form a two-dimensional contingency table, according to some classifying factors A and B having I and J categories, respectively. A common method for analyzing data of this kind is to assume that cell frequencies y_{ij} are independently distributed as Poisson and to use loglinear models, where in an analysis-of-variance-like fashion logarithms of expected cell frequencies μ_{ij} are assumed to be sums of several effects, e.g. for the multiplicative model,

$$\log(\mu_{ij}) = \beta_o + \beta_i^A + \beta_j^B,$$

subject to some constraints on the β_s. Sums of independent Poisson variates are again distributed as Poisson with means equal to the sum of respective means. Row totals Y_{i+}, column totals Y_{+j}, and grand total Y_{++} are, therefore, Poisson variates with means $\mu_{i+} = \mu_{i1} + \cdots + \mu_{iJ}$, $\mu_{+j} = \mu_{1j} + \cdots + \mu_{Ij}$, and $\mu_{++} = \sum_{i,j} \mu_{ij}$, respectively. Under the assumption of the multiplicative model these quantities are related by

$$\mu_{ij} = \frac{\mu_i + \mu + j}{\mu + +}$$

showing that the joint distribution of the contingency table is, in a multiplicative manner, completely determined by the marginal distributions.

The Poisson model assumption implies that marginals are random. If, instead, the total is fixed by the sampling design, it may be more appropriate to assume a multinomial distribution for the table. Formally, the multinomial model can be inferred from the Poisson model by conditioning on the total y_{++}. For the probability πij of an observation falling into row i and column j, we then have $\pi_{ij} = \mu_{ij}/\mu_{++}$, and from assuming the multiplicative model $\log(\mu_{ij}) = \beta_o + \beta_i^A + \beta_j^B$, it follows that

$$\pi_{ij} = \pi_i + \pi_{+j} ,$$

Where π_{i+} and π_{+j} are the marginal probabilities of an observation falling into row i and column j, respectively. Hence, row and column variables A and B are independently distributed.

Likewise, if row totals are fixed, then each row may be assumed to be multinomially distributed. Again, this can be inferred from the Poisson model by conditioning on the row totals, and the multiplicative model $\log(\mu_{ij}) = \beta_o + \beta_i^A + \beta_j^B$, implies identical distributions for the rows – a condition usually called homogeneity. It may be worthwhile noting that maximum-likelihood estimates for the parameters in the Poisson models are identical to those obtained for some other sampling designs, such as the multinomial designs just mentioned, making this class of model particularly interesting and useful.

Loglinear models for two- and higher-dimensional contingency tables, used to describe the association and interaction structure connecting the variables, are

Table: Number of recurrences of superficial bladder cancer for 31 male patients with grade 2, stage T1, solitary primary tumors and respective times under observation (in months) by size of primary tumor. Subset of data analyzed in

Size	Recurrences	Time under observation
≤3cm	1	2, 3, 6, 8, 9, 10, 11, 13, 14, 16, 21, 22, 24, 26, 27
	2	7, 13, 15, 18, 23
	3	20
	4	24
>3cm	1	1, 5, 17, 18, 25
	2	18, 25
	3	4
	4	19

Usually, the goal is to find a parsimonious model that fits the data well and allows meaningful substantive interpretation. Most commonly, this search is restricted to hierarchical models.

Multiplicative Rate Models

If occurrences of some kind of event are counted over time, then often interest lies in the rate at which events occur. The rate describes the instantaneous risk for an event to happen at a given point in time. To be more specific, the probability of observing exactly one event in the interval ranging from t to t + h, divided by its length h, is assumed to tend to some value λ(t), as h tends to 0. λ(t), as a function of time t, is called the rate or intensity function.

An important special case, termed the Poisson process, assumes that waiting times between successive events are independent and exponentially distributed with common mean 1/λ. Here, the rate function is constant over time, λ(t) ≡ λ. Furthermore, the number Y (t) of events that occur up to time t is distributed as Poisson with mean $\mu = \lambda t$. Note that the mean of Y (t)/t equals the rate λ. This suggests a Poisson regression approach

$$\log \lambda = \log\left(\frac{\mu}{t}\right) = \alpha + \beta x$$

for modeling the dependence of the rate function on an explanatory variable X. This can be rewritten as

$$\log\mu = \alpha + \beta x + \log t,$$

with log(t) as an offset, i.e. a variable in the linear predictor, the corresponding regression parameter of which is set equal to 1. Observe that this defines a multiplicative model for the rate function,

$$\lambda = e^{\alpha}\left(e^{\beta}\right)^{x},$$

with a baseline rate $\lambda_0 = \exp(\alpha)$ and proportionality factor $\exp(\beta x)$.

For illustrative purposes a subset of the data analyzed in is reprinted in above Table. For 31 male patients, who have been treated for superficial bladder cancer, the number of recurrent tumors has been recorded for some time after removal of the primary tumor. Defining X to be 1 for larger primary tumors (>3 cm) and 0 otherwise, and assuming a Poisson process with rate, yields parameter estimates $\hat{\alpha} = -1.95$ and $\hat{\beta} = 0.385$. The (baseline) rate for smaller tumors is (estimated as) 0.142, the rate for larger tumors being 1.47 times larger. In terms of waiting times between recurrences, means are estimated as 7.06 and 4.80 months, respectively.

Now suppose that we have recorded, for n individuals, time under observation, t_i, and the number y_i of events occurred. Observation times are assumed to be nonrandom and counts to be mutually independent. We also have a set of explanatory variables $x_{i1},\ldots,x_{ik}$ available for each subject. Under the assumption of proportional rates, $\lambda_i = \lambda_0 \exp\sum_{j=1}^{k} x_{ij}\beta_j$ we have

$$\mu_i = \lambda_i \times t_i = \lambda_0 \exp\left[\log(t_i) + \sum_{j=1}^{k} x_{ij}\beta_j\right]$$
$$= \lambda_0 t_i \prod_{j=1}^{k} \exp\left(x_{ij}\beta_j\right),$$

i.e. a loglinear model for the mean of the Poisson process, involving the logarithm of observation times as an "explanatory" variable, with the associated regression parameter fixed at a value of 1.

If the process is such that it can be characterized by a time-varying rate function $\lambda(t)$, it is called a nonhomogeneous Poisson process. Writing

$$\Lambda(t) = \int_0^t \lambda(u)\,du$$

for the integrated rate or intensity function, the number of occurrences of the event in period until time point t is again distributed as Poisson, but with mean equal to $\Lambda(t)$. Note that events in nonoverlapping time intervals are independent, but waiting times between successive events are, contrary to the homogeneous process with constant rate, neither identically distributed nor independent. In this situation model

$$\mu_i = \lambda_i \times t_i = \lambda_0 \exp\left[\log(t_i) + \sum_{j=1}^{k} x_{ij}\beta_j\right] = \lambda_0 t_i \prod_{j=1}^{k} \exp(x_{ij}\beta_j),$$

can be modified, using a baseline rate function λ_0 $(t|\alpha)$, possibly depending on some additional parameter α, to give

$$\mu_i = \exp\left\{ log\left[\Lambda_0(t_i \mid \alpha)\right] + \sum_{j=1}^{k} x_{ij}\beta_j \right\}.$$

Choosing $\Lambda_0(t|\alpha)$ to be t,tα, or exp(αt) corresponds to an exponential, a Weibull, or an extreme value intensity function, respectively, and results in a loglinear model for the Y_{is}. Disregarding constant terms, the likelihood function for this model is

$$L(\alpha,\beta) = \prod_{i=1}^{n}\left[\Lambda_0(t_i \mid \alpha)\exp\left(\sum_{j=1}^{k} x_{ij}\beta_j\right)\right]^{y_i} \times \exp\left[-\Lambda_0(t_i \mid \alpha)\exp\left(\sum_{j=1}^{k} x_{ij}\beta_j\right)\right].$$

If times for occurrences of each event were known, a multiplicative term, depending on the parameter α, would be added to above euation.

There is a close connection to relative risk models, which are very frequently used in epidemiology. This class of models assumes that risk factors interact in a multiplicative way.

Maximum Likelihood-based Parameter Estimation

Given a set of parameters θ and an input vector x, the mean of the predicted Poisson distribution, as stated above, is given by

$$\lambda = \mathrm{E}(Y \mid x) = e^{\theta' x}$$

and thus, the Poisson distribution's probability mass function is given by

$$p(y \mid x;\theta) = \frac{\lambda^y}{y!} e^{-\lambda} = \frac{e^{y\theta' x} e^{-e^{\theta' x}}}{y!}.$$

Now suppose we are given a data set consisting of m vectors $x_i \in \mathbb{R}^{n+1}, i = 1,\ldots,m$, along with a set of m values $y_1,\ldots,y_m \in \mathbb{N}$. Then, for a given set of parameters θ, the probability of attaining this particular set of data is given by

$$p(y_1,\ldots,y_m \mid x_1,\ldots,x_m;\theta) = \prod_{i=1}^{m} \frac{e^{y_i\theta' x_i} e^{-e^{\theta' x_i}}}{y_i!}.$$

By the method of maximum likelihood, we wish to find the set of parameters θ that makes this probability as large as possible. To do this, the equation is first rewritten as a likelihood function in terms of θ:

$$L(\theta \mid X,Y) = \prod_{i=1}^{m} \frac{e^{y_i \theta' x_i} e^{-e^{\theta' x_i}}}{y_i!}.$$

Note that the expression on the right hand side has not actually changed. A formula in this form is typically difficult to work with; instead, one uses the *log-likelihood*:

$$\ell(\theta \mid X,Y) = \log L(\theta \mid X,Y) = \sum_{i=1}^{m} \left(y_i \theta' x_i - e^{\theta' x_i} - \log(y_i!) \right).$$

Notice that the parameters θ only appear in the first two terms of each term in the summation. Therefore, given that we are only interested in finding the best value for θ we may drop the $y_i!$ and simply write

$$\ell(\theta \mid X,Y) = \sum_{i=1}^{m} \left(y_i \theta' x_i - e^{\theta' x_i} \right).$$

To find a maximum, we need to solve an equation $\frac{\partial \ell(\theta \mid X,Y)}{\partial \theta} = 0$ which has no closed-form solution. However, the negative log-likelihood, $-\ell(\theta \mid X,Y)$, is a convex function, and so standard convex optimization techniques such as gradient descent can be applied to find the optimal value of θ.

Poisson Regression in Practice

Poisson regression may be appropriate when the dependent variable is a count, for instance of events such as the arrival of a telephone call at a call centre. The events must be independent in the sense that the arrival of one call will not make another more or less likely, but the probability per unit time of events is understood to be related to covariates such as time of day.

"Exposure" and Offset

Poisson regression may also be appropriate for rate data, where the rate is a count of events divided by some measure of that unit's *exposure* (a particular unit of observation). For example, biologists may count the number of tree species in a forest: events would be tree observations, exposure would be unit area, and rate would be the number of species per unit area. Demographers may model death rates in geographic areas as the count of deaths divided by person−years. More generally, event rates can be calculated as events per unit time, which allows the observation window to vary for each unit. In these examples, exposure is respectively unit area, person−years and unit time. In Poisson regression this is handled as an offset, where the exposure variable enters on the right-hand side of the equation, but with a parameter estimate (for log(exposure)) constrained to 1.

$$\log(\mathrm{E}(Y \mid x)) = \log(\text{exposure}) + \theta' x$$

which implies

$$\log(\mathrm{E}(Y \mid x)) - \log(\text{exposure}) = \log\left(\frac{\mathrm{E}(Y \mid x)}{\text{exposure}}\right) = \theta' x$$

Offset in the case of a GLM in R can be achieved using the offset() function:

```
glm(y ~ offset(log(exposure)) + x, family=poisson(link=log) )
```

Overdispersion and Zero Inflation

A characteristic of the Poisson distribution is that its mean is equal to its variance. In certain circumstances, it will be found that the observed variance is greater than the mean; this is known as overdispersion and indicates that the model is not appropriate. A common reason is the omission of relevant explanatory variables, or dependent observations. Under some circumstances, the problem of overdispersion can be solved by using quasi-likelihood estimation or a negative binomial distribution instead.

Ver Hoef and Boveng described the difference between quasi-Poisson (also called overdispersion with quasi-likelihood) and negative binomial (equivalent to gamma-Poisson) as follows: If $E(Y) = \mu$, the quasi-Poisson model assumes $\text{var}(Y) = \theta\mu$ while the gamma-Poisson assumes $\text{var}(Y) = \mu(1 + \kappa\mu)$, where θ is the quasi-Poisson overdispersion parameter, and κ is the shape parameter of the negative binomial distribution. For both models, parameters are estimated using Iteratively reweighted least squares. For quasi-Poisson, the weights are μ/θ. For negative binomial, the weights are $\mu/(1 + \kappa\mu)$. With large μ and substantial extra-Poisson variation, the negative binomial weights are capped at $1/\kappa$. Ver Hoef and Boveng discussed an example where they selected between the two by plotting mean squared residuals vs. the mean.

Another common problem with Poisson regression is excess zeros: if there are two processes at work, one determining whether there are zero events or any events, and a Poisson process determining how many events there are, there will be more zeros than a Poisson regression would predict. An example would be the distribution of cigarettes smoked in an hour by members of a group where some individuals are non-smokers.

Other generalized linear models such as the negative binomial model or zero-inflated model may function better in these cases.

Use in Survival Analysis

Poisson regression creates proportional hazards models, one class of survival analysis.

Ordinal Regression

In statistics, ordinal regression (also called "ordinal classification") is a type of regression analysis used for predicting an ordinal variable, i.e. a variable whose value exists on an arbitrary scale where only the relative ordering between different values is significant. It can be considered an intermediate problem between regression and classification. Examples of ordinal regression are ordered logit and ordered probit. Ordinal regression turns up often in the social sciences, for example in the modeling of human levels of preference (on a scale from, say, 1–5 for "very poor" through "excellent"), as well as in information retrieval. In machine learning, ordinal regression may also be called ranking learning.

Linear Models for Ordinal Regression

Ordinal regression can be performed using a generalized linear model (GLM) that fits both a coefficient vector and a set of *thresholds* to a dataset. Suppose one has a set of observations, represented by length-p vectors x_1 through x_n, with associated responses y_1 through y_n, where each y_i is an ordinal variable on a scale 1, ..., K. To this data, one fits a length-p coefficient vector w and a set of thresholds $\theta_1, ..., \theta_{K-1}$ with the property that $\theta_1 < \theta_2 < ... < \theta_{K-1}$. This set of thresholds divides the real number line into K disjoint segments, corresponding to the K response levels.

The model can now be formulated as

$$\Pr(y \le i \mid \mathbf{x}) = \sigma(\theta_i - \mathbf{w} \cdot \mathbf{x})$$

or, the cumulative probability of the response y being at most i is given by a function σ (the inverse link function) applied to a linear function of x. Several choices exist for σ; the logistic function

$$\sigma(\theta_i - \mathbf{w} \cdot \mathbf{x}) = \frac{1}{1 + e^{-(\theta_i - \mathbf{w} \cdot \mathbf{x})}}$$

gives the ordered logit model, while using the probit function gives the ordered probit model. A third option is to use an exponential function

$$\sigma(\theta_i - \mathbf{w} \cdot \mathbf{x}) = \exp(-\exp(\theta_i - \mathbf{w} \cdot \mathbf{x}))$$

which gives the proportional hazards model.

Latent Variable Model

The probit version of the above model can be justified by assuming the existence of a real-valued latent variable (unobserved quantity) y^*, determined by

$$y^* = \mathbf{w} \cdot \mathbf{x} + \varepsilon$$

where ε is normally distributed with zero mean and unit variance, conditioned on x. The response variable y results from an "incomplete measurement" of y^*, where one only determines the interval into which y^* falls:

$$y = \begin{cases} 1 \text{ if } y^* \le \theta_1, \\ 2 \text{ if } \theta_1 < y^* \le \theta_2, \\ 3 \text{ if } \theta_2 < y^* \le \theta_3 \\ \vdots \\ K \text{ if } \theta_{K-1} < y^*. \end{cases}$$

Defining $\theta_0 = -\infty$ and $\theta_K = \infty$, the above can be summarized as $y = k$ if and only if $\theta_{k-1} < y^* \leq \theta_k$.

From these assumptions, one can derive the conditional distribution of y as

$$\begin{aligned} P(y=k \mid \mathbf{x}) &= P(\theta_{k-1} < y^* \leq \theta_k \mid \mathbf{x}) \\ &= P(\theta_{k-1} < \mathbf{w}\cdot\mathbf{x} + \varepsilon \leq \theta_k) \\ &= \Phi(\theta_k - \mathbf{w}\cdot\mathbf{x}) - \Phi(\theta_{k-1} - \mathbf{w}\cdot\mathbf{x}) \end{aligned}$$

where Φ is the cumulative distribution function of the standard normal distribution, and takes on the role of the inverse link function σ. The log-likelihood of the model for a single training example x_i, y_i can now be stated as

$$\log L\left(\mathrm{w},\theta \middle| x_i, y_i\right) = \sum_{k=1}^{k} [y_i = k] \log\left[\Phi\left(\theta_k - \mathrm{w}.\mathrm{x}_i\right) - \Phi\left(\theta_{k-1} - \mathrm{w}.\mathrm{x}_i\right)\right]$$

(using the Iverson bracket $[y_i = k]$.) The log-likelihood of the ordered logit model is analogous, using the logistic function instead of Φ.

Alternative Models

In machine learning, alternatives to the latent-variable models of ordinal regression have been proposed. An early result was PRank, a variant of the perceptron algorithm that found multiple parallel hyperplanes separating the various ranks; its output is a weight vector w and a sorted vector of $K-1$ thresholds θ, as in the ordered logit/probit models. The prediction rule for this model is to output the smallest rank k such that wx $< \theta_k$.

Other methods rely on the principle of large-margin learning that also underlies support vector machines.

Another approach is given by Rennie and Srebro, who, realizing that "even just evaluating the likelihood of a predictor is not straight-forward" in the ordered logit and ordered probit models, propose fitting ordinal regression models by adapting common loss functions from classification (such as the hinge loss and log loss) to the ordinal case.

Software

ORCA (Ordinal Regression and Classification Algorithms) is a Octave/MATLAB framework including a wide set of ordinal regression methods.

Ordinary Least Squares (OLS) Regression Model

In statistics, ordinary least squares (OLS) is a type of linear least squares method for estimating the unknown parameters in a linear regression model. OLS chooses the parameters of a linear function of a set of explanatory variables by the principle of least squares: minimizing the sum of the squares of the differences between the observed dependent variable (values of the variable being predicted) in the given dataset and those predicted by the linear function.

Geometrically, this is seen as the sum of the squared distances, parallel to the axis of the dependent variable, between each data point in the set and the corresponding point on the regression surface – the smaller the differences, the better the model fits the data. The resulting estimator can be expressed by a simple formula, especially in the case of a simple linear regression, in which there is a single regressor on the right side of the regression equation.

The OLS estimator is consistent when the regressors are exogenous, and optimal in the class of linear unbiased estimators when the errors are homoscedastic and serially uncorrelated. Under these conditions, the method of OLS provides minimum-variance mean-unbiased estimation when the errors have finite variances. Under the additional assumption that the errors are normally distributed, OLS is the maximum likelihood estimator.

OLS is used in fields as diverse as economics (econometrics), political science, psychology and engineering (control theory and signal processing).

Linear Model

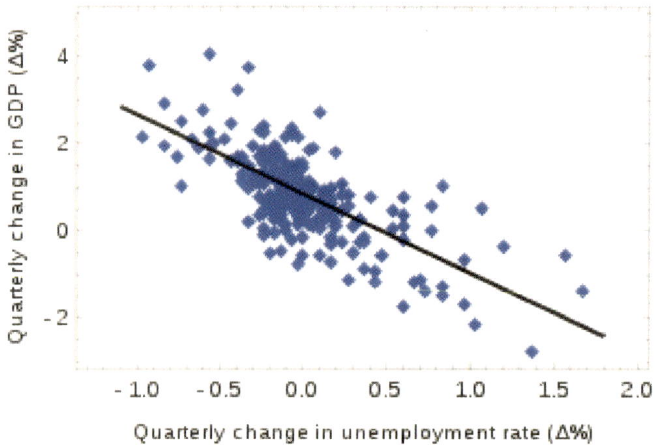

Okun's law in macroeconomics states that in an economy the GDP growth should depend linearly on the changes in the unemployment rate. Here the ordinary least squares method is used to construct the regression line describing this law.

Suppose the data consist of n observations $\{y_i, x_i\}_{i=1}^{n}$. Each observation i includes a scalar response y_i and a column vector x_i of values of p predictors (regressors) x_{ij} for $j = 1, ..., p$. In a linear regression model, the response variable, y_i, is a linear function of the regressors:

$$y_i = \beta_1 x_{i1} + \beta_2 x_{i2} + \cdots + \beta_p x_{ip} + \varepsilon_i,$$

or in vector form,

$$y_i = x_i^T \beta + \varepsilon_i,$$

where β is a $p\times 1$ vector of unknown parameters; the ε_i's are unobserved scalar random variables (errors) which account for influences upon the responses y_i from sources other than the explanators x_i; and x_i is a column vector of the ith observations of all the explanatory variables. This model can also be written in matrix notation as

$$y = X\beta + \varepsilon,$$

where y and ε are $n\times 1$ vectors of the values of the response variable and the errors for the various observations, and X is an $n\times p$ matrix of regressors, also sometimes called the design matrix, whose row i is x_i^T and contains the ith observations on all the explanatory variables.

As a rule, the constant term is always included in the set of regressors X, say, by taking $x_{i1} = 1$ for all $i = 1, \ldots, n$. The coefficient β_1 corresponding to this regressor is called the *intercept*.

There may be some relationship between the regressors. For instance, the third regressor may be the square of the second regressor. In this case (assuming that the first regressor is constant) we have a quadratic model in the second regressor. But this is still considered a linear model because it is linear in the βs.

Matrix/vector Formulation

Consider an overdetermined system of m linear equations in n unknown coefficients, $\beta_1, \beta_2, \ldots, \beta_n$, with $m > n$.

$$\sum_{j=1}^{n} X_{ij}\beta_j = y_i, (i = 1, 2, \ldots, m),$$

For a linear model as above, not all of X contains information on the data points. The first column is populated with ones, $X_{i1} = 1$, only the other columns contain actual data, and n = number of regressors + 1. This can be written in matrix form as

$$\mathbf{X}\boldsymbol{\beta} = \mathbf{y},$$

where

$$\mathbf{X} = \begin{bmatrix} X_{11} & X_{12} & \cdots & X_{1n} \\ X_{21} & X_{22} & \cdots & X_{2n} \\ \vdots & \vdots & \ddots & \vdots \\ X_{m1} & X_{m2} & \cdots & X_{mn} \end{bmatrix}, \quad \boldsymbol{\beta} = \begin{bmatrix} \beta_1 \\ \beta_2 \\ \vdots \\ \beta_n \end{bmatrix}, \quad \mathbf{y} = \begin{bmatrix} y_1 \\ y_2 \\ \vdots \\ y_m \end{bmatrix}.$$

Such a system usually has no solution, so the goal is instead to find the coefficients $\hat{\mathbf{a}}$ which fit the equations "best", in the sense of solving the quadratic minimization problem

$$\hat{\boldsymbol{\beta}} = \arg\min_{\boldsymbol{\beta}} S(\boldsymbol{\beta}),$$

where the objective function S is given by

$$S(\boldsymbol{\beta}) = \sum_{i=1}^{m} \left| y_i - \sum_{j=1}^{n} X_{ij}\beta_j \right|^2 = \left\| \mathbf{y} - \mathbf{X}\boldsymbol{\beta} \right\|^2.$$

A justification for choosing this criterion is given in properties below. This minimization problem has a unique solution, provided that the n columns of the matrix $\mathbf{X}$ are linearly independent, given by solving the normal equations

$$(X^T X)\hat{\beta} = X^T y.$$

The matrix $\mathbf{X}^T\mathbf{X}$ is known as the Gramian matrix of $\mathbf{X}$, which possesses several nice properties such as being a positive semi-definite matrix, and the matrix $\mathbf{X}^T\mathbf{y}$ is known as the moment matrix of regressand by regressors. Finally, $\hat{\beta}$ is the coefficient vector of the least-squares hyperplane, expressed as

$$\hat{\beta} = (X^T X)^{-1} X^T y.$$

Estimation

Suppose b is a "candidate" value for the parameter vector β. The quantity $y_i - x_i^T b$, called the residual for the i-th observation, measures the vertical distance between the data point $(x_i\ y_i)$ and the hyperplane $y = x^T b$, and thus assesses the degree of fit between the actual data and the model. The sum of squared residuals (SSR) (also called the error sum of squares (ESS) or residual sum of squares (RSS)) is a measure of the overall model fit:

$$S(b) = \sum_{i=1}^{n} (y_i - x_i^T b)^2 = (y - Xb)^T (y - Xb),$$

where T denotes the matrix transpose, and the rows of X, denoting the values of all the independent variables associated with a particular value of the dependent variable, are $X_i = x_i^T$. The value of b which minimizes this sum is called the OLS estimator for $\boldsymbol{\beta}$. The function $S(b)$ is quadratic in b with positive-definite Hessian, and therefore this function possesses a unique global minimum at $b = \hat{\beta}$, , which can be given by the explicit formula:

$$\hat{\beta} = \operatorname{argmin}_{b \in \mathbb{R}^p} S(b) = (X^T X)^{-1} X^T y \,.$$

The product $N=X^T X$ is a normal matrix and its inverse, $Q=N^{-1}$, is the *cofactor matrix* of β, closely related to its covariance matrix, C_β. The matrix $(X^T X)^{-1} X^T = Q X^T$ is called the Moore–Penrose pseudoinverse matrix of X. This formulation highlights the point that estimation can be carried out if, and only if, there is no perfect multicollinearity between the explanatory variables (which would cause the normal matrix to have no inverse).

After we have estimated β, the fitted values (or predicted values) from the regression will be

$$\hat{y} = X\hat{\beta} = Py,$$

where $P = X(X^T X)^{-1} X^T$ is the projection matrix onto the space V spanned by the columns of X. This matrix P is also sometimes called the hat matrix because it "puts a hat" onto the variable y. Another matrix, closely related to P is the *annihilator* matrix $M = I_n - P$; this is a projection matrix onto the space orthogonal to V. Both matrices P and M are symmetric and idempotent (meaning that $P^2 = P$), and relate to the data matrix X via identities $PX = X$ and $MX = 0$. Matrix M creates the residuals from the regression:

$$\hat{\varepsilon} = y - \hat{y} = y - X\hat{\beta} = My = M(X\beta + \varepsilon) = (MX)\beta + M\varepsilon = M\varepsilon.$$

Using these residuals we can estimate the value of σ^2, called the reduced chi-squared:

$$s^2 = \frac{\hat{\varepsilon}^{\mathrm{T}}\hat{\varepsilon}}{n-p} = \frac{(My)^{\mathrm{T}}My}{n-p} = \frac{y^{\mathrm{T}}M^{\mathrm{T}}My}{n-p} = \frac{y^{\mathrm{T}}My}{n-p} = \frac{S(\hat{\beta})}{n-p}, \qquad \hat{\sigma}^2 = \frac{n-p}{n}s^2$$

The numerator, $n-p$, is the statistical degrees of freedom. The first quantity, s^2, is the OLS estimate for σ^2, whereas the second, $\hat{\sigma}^2$, is the MLE estimate for σ^2. The two estimators are quite similar in large samples; the first one is always unbiased, while the second is biased but minimizes the mean squared error of the estimator. In practice s^2 is used more often, since it is more convenient for the hypothesis testing. The square root of s^2 is called the regression standard error, standard error of the regression, or standard error of the equation.

It is common to assess the goodness-of-fit of the OLS regression by comparing how much the initial variation in the sample can be reduced by regressing onto X. The coefficient of determination R^2 is defined as a ratio of "explained" variance to the "total" variance of the dependent variable y:

$$R^2 = \frac{\sum(\hat{y}_i - \bar{y})^2}{\sum(y_i - \bar{y})^2} = \frac{y^{\mathrm{T}}P^{\mathrm{T}}LPy}{y^{\mathrm{T}}Ly} = 1 - \frac{y^{\mathrm{T}}My}{y^{\mathrm{T}}Ly} = 1 - \frac{\mathrm{RSS}}{\mathrm{TSS}}$$

where TSS is the total sum of squares for the dependent variable, $L = I_n - \mathbf{1}\mathbf{1}^T/n$, and $\mathbf{1}$ is an $n\times 1$ vector of ones. (L is a "centering matrix" which is equivalent to regression on a constant; it simply subtracts the mean from a variable.) In order for R^2 to be meaningful, the matrix X of data on regressors must contain a column vector of ones to represent the constant whose coefficient is the regression intercept. In that case, R^2 will always be a number between 0 and 1, with values close to 1 indicating a good degree of fit.

The variance in the prediction of the independent variable as a function of the dependent variable is given in the Polynomial least squares.

Simple Linear Regression Model

If the data matrix X contains only two variables, a constant and a scalar regressor x_i, then this is called the "simple regression model". This case is often considered in the beginner statistics classes, as it provides much simpler formulas even suitable for manual calculation. The parameters are commonly denoted as (α, β):

$$y_i = \alpha + \beta x_i + \varepsilon_i.$$

The least squares estimates in this case are given by simple formulas

$$\hat{\beta} = \frac{\sum x_i y_i - \frac{1}{n}\sum x_i \sum y_i}{\sum x_i^2 - \frac{1}{n}(\sum x_i)^2} = \frac{\mathrm{Cov}[x,y]}{\mathrm{Var}[x]}$$

$$\hat{\alpha} = \bar{y} - \hat{\beta}\bar{x},$$

where Var(.) and Cov(.) are sample parameters.

Alternative Derivations

The least squares estimator $\hat{\beta}$ was obtained as a value that minimizes the sum of squared residuals of the model. However it is also possible to derive the same estimator from other approaches. In all cases the formula for OLS estimator remains the same: $\hat{\beta} = (X^T X)^{-1} X^T y$; the only difference is in how we interpret this result.

Projection

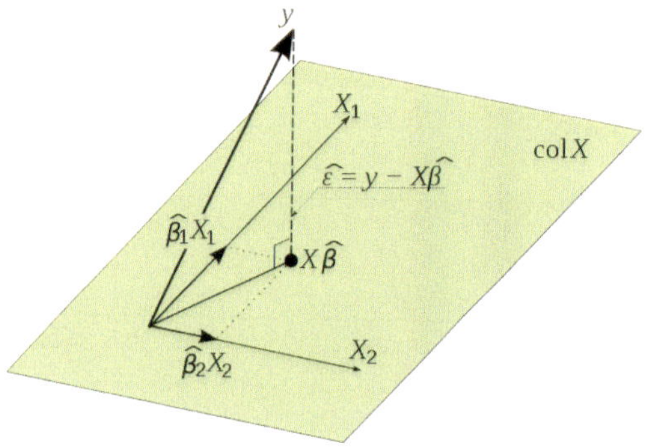

OLS estimation can be viewed as a projection onto the linear space spanned by the regressors. (Here each of X_1 and X_2 refers to a column of the data matrix.)

For mathematicians, OLS is an approximate solution to an over determined system of linear equations $X\beta \approx y$, where β is the unknown. Assuming the system cannot be solved exactly (the number of equations n is much larger than the number of unknowns p), we are looking for a solution that could provide the smallest discrepancy between the right- and left- hand sides. In other words, we are looking for the solution that satisfies

$$\hat{\beta} = \arg\min_{\beta} \| y - X\beta \|,$$

where $\| \; \|$ is the standard L^2 norm in the n-dimensional Euclidean space R^n. The predicted quantity $X\beta$ is just a certain linear combination of the vectors of regressors. Thus, the residual vector $y - X\beta$ will have the smallest length when y is projected orthogonally onto the linear subspace spanned by the columns of X. The OLS estimator $\hat{\beta}$ in this case can be interpreted as the coefficients of vector decomposition of $\hat{y} = Py$ along the basis of X.

In other words, the gradient equations at the minimum can be written as:

$$(\mathbf{y} - X\hat{\boldsymbol{\beta}})^{\mathrm{T}} X = 0.$$

A geometrical interpretation of these equations is that the vector of residuals, $\mathbf{y} - X\hat{\boldsymbol{\beta}}$ is orthogonal to the column space of X, since the dot product $(\mathbf{y} - X\hat{\boldsymbol{\beta}}) \cdot X\mathbf{v}$ is equal to zero for *any* conformal vector, v. This means that $\mathbf{y} - X\hat{\boldsymbol{\beta}}$ is the shortest of all possible vectors $\mathbf{y} - X\boldsymbol{\beta}$, that is, the variance of the residuals is the minimum possible. This is illustrated at the right.

Introducing $\hat{\boldsymbol{\gamma}}$ and a matrix K with the assumption that a matrix $[X \; K]$ is non-singular and $K^{\mathrm{T}} X = 0$ (cf. Orthogonal projections), the residual vector should satisfy the following equation:

$$\hat{\mathbf{r}} \triangleq \mathbf{y} - X\hat{\boldsymbol{\beta}} = K\hat{\boldsymbol{\gamma}}.$$

The equation and solution of linear least squares are thus described as follows:

$$\mathbf{y} = \begin{bmatrix} X & K \end{bmatrix} \begin{pmatrix} \hat{\beta} \\ \hat{\gamma} \end{pmatrix},$$

$$\begin{pmatrix} \hat{\beta} \\ \hat{\gamma} \end{pmatrix} = \begin{bmatrix} X & K \end{bmatrix}^{-1} \mathbf{y} = \begin{bmatrix} (X^{\mathrm{T}}X)^{-1}X^{\mathrm{T}} \\ (K^{\mathrm{T}}K)^{-1}K^{\mathrm{T}} \end{bmatrix} \mathbf{y}.$$

Another way of looking at it is to consider the regression line to be a weighted average of the lines passing through the combination of any two points in the dataset. Although this way of calculation is more computationally expensive, it provides a better intuition on OLS.

Maximum Likelihood

The OLS estimator is identical to the maximum likelihood estimator (MLE) under the normality assumption for the error terms. This normality assumption has historical importance, as it provided the basis for the early work in linear regression analysis by Yule and Pearson. From the properties of MLE, we can infer that the OLS estimator is asymptotically efficient (in the sense of attaining the Cramér–Rao bound for variance) if the normality assumption is satisfied.

Generalized Method of Moments

In iid case the OLS estimator can also be viewed as a GMM estimator arising from the moment conditions

$$\mathrm{E}\big[x_i(y_i - x_i^T\beta)\big] = 0.$$

These moment conditions state that the regressors should be uncorrelated with the errors. Since x_i is a p-vector, the number of moment conditions is equal to the dimension of the parameter vector β, and thus the system is exactly identified. This is the so-called classical GMM case, when the estimator does not depend on the choice of the weighting matrix.

Note that the original strict exogeneity assumption $E[\varepsilon_i | x_i] = 0$ implies a far richer set of moment conditions than stated above. In particular, this assumption implies that for any vector-function f, the moment condition $\mathrm{E}[f(x_i)\cdot\varepsilon_i] = 0$ will hold. However it can be shown using the Gauss–Markov theorem that the optimal choice of function f is to take $f(x) = x$, which results in the moment equation posted above.

Properties

Assumptions

There are several different frameworks in which the linear regression model can be cast in order to make the OLS technique applicable. Each of these settings produces the same formulas and same results. The only difference is the interpretation and the assumptions which have to be imposed in order for the method to give meaningful results. The choice of the applicable framework depends mostly on the nature of data in hand, and on the inference task which has to be performed.

One of the lines of difference in interpretation is whether to treat the regressors as random variables, or as predefined constants. In the first case (random design) the regressors x_i are random and sampled together with the y_i's from some population, as in an observational study. This approach allows for more natural study of the asymptotic properties of the estimators. In the other interpretation (fixed design), the regressors X are treated as known constants set by a design, and y is sampled conditionally on the values of X as in an experiment. For practical purposes, this distinction is often unimportant, since estimation and inference is carried out while conditioning on X.

Classical Linear Regression Model

The classical model focuses on the "finite sample" estimation and inference, meaning that the number of observations n is fixed. This contrasts with the other approaches, which study the asymptotic behavior of OLS, and in which the number of observations is allowed to grow to infinity.

- Correct specification. The linear functional form must coincide with the form of the actual data-generating process.
- Strict exogeneity. The errors in the regression should have conditional mean zero:

  $$\mathrm{E}[\varepsilon \mid X] = 0.$$

  The immediate consequence of the exogeneity assumption is that the errors have mean zero: $\mathrm{E}[\varepsilon] = 0$, and that the regressors are uncorrelated with the errors: $E\left[X^T\varepsilon\right] = 0$.

  The exogeneity assumption is critical for the OLS theory. If it holds then the regressor variables are called *exogenous*. If it doesn't, then those regressors that are correlated with the error term are called *endogenous*, and then the OLS estimates become invalid. In such case the method of instrumental variables may be used to carry out inference.

- No linear dependence. The regressors in X must all be linearly independent. Mathematically, this means that the matrix X must have full column rank almost surely:

  $$\Pr\left[\operatorname{rank}(X) = p\right] = 1.$$

  Usually, it is also assumed that the regressors have finite moments up to at least the second moment. Then the matrix $Q_{xx} = E\left[X^T X / n\right]$ is finite and positive semi-definite.

  When this assumption is violated the regressors are called linearly dependent or perfectly multicollinear. In such case the value of the regression coefficient β cannot be learned, although prediction of y values is still possible for new values of the regressors that lie in the same linearly dependent subspace.

- Spherical errors:

  $$\operatorname{Var}[\varepsilon \mid X] = \sigma^2 I_n,$$

where I_n is the identity matrix in dimension n, and σ^2 is a parameter which determines the variance of each observation. This σ^2 is considered a nuisance parameter in the model, although usually it is also estimated. If this assumption is violated then the OLS estimates are still valid, but no longer efficient.

It is customary to split this assumption into two parts:

- Homoscedasticity: $E\left[\varepsilon_i^2 | X\right] = \sigma^2$, which means that the error term has the same variance σ^2 in each observation. When this requirement is violated this is called heteroscedasticity, in such case a more efficient estimator would be weighted least squares. If the errors have infinite variance then the OLS estimates will also have infinite variance (although by the law of large numbers they will nonetheless tend toward the true values so long as the errors have zero mean). In this case, robust estimation techniques are recommended.
- No autocorrelation: the errors are uncorrelated between observations: $E\left[\varepsilon_i\varepsilon_j | X\right] = 0 \, fori \neq j$. This assumption may be violated in the context of time series data, panel data, cluster samples, hierarchical data, repeated measures data, longitudinal data, and other data with dependencies. In such cases generalized least squares provides a better alternative than the OLS. Another expression for autocorrelation is *serial correlation*.
- Normality. It is sometimes additionally assumed that the errors have normal distribution conditional on the regressors:

  $$\varepsilon | X \sim \mathcal{N}(0, \sigma^2 I_n).$$

  This assumption is not needed for the validity of the OLS method, although certain additional finite-sample properties can be established in case when it does (especially in the area of hypotheses testing). Also when the errors are normal, the OLS estimator is equivalent to the maximum likelihood estimator (MLE), and therefore it is asymptotically efficient in the class of all regular estimators. Importantly, the normality assumption applies only to the error terms; contrary to a popular misconception, the response (dependent) variable is not required to be normally distributed.

Independent and Identically Distributed (IID)

In some applications, especially with cross-sectional data, an additional assumption is imposed — that all observations are independent and identically distributed. This means that all observations are taken from a random sample which makes all the assumptions listed earlier simpler and easier to interpret. Also this framework allows one to state asymptotic results (as the sample size $n\rightarrow\infty$), which are understood as a theoretical possibility of fetching new independent observations from the data generating process. The list of assumptions in this case is:

- Id observations: (x_i, y_i) is independent from, and has the same distribution as, (x_j, y_j) for all $i \neq j$;
- No perfect multicollinearity: $Q_{xx} = \mathrm{E}[x_i x_i^{\mathrm{T}}]$ is a positive-definite matrix;
- Exogeneity: $\mathrm{E}[\varepsilon_i | x_i] = 0$;
- Homoscedasticity: $\mathrm{Var}[\varepsilon_i | x_i] = \sigma^2$.

Time Series Model

- The stochastic process $\{x_i, y_i\}$ is stationary and ergodic; if $\{x_i, y_i\}$ is nonstantionary, OLS results are often spurious unless $\{x_i, y_i\}$ is co-integrating.
- The regressors are *predetermined*: $E[x_i\varepsilon_i] = 0$ for all $i = 1, \ldots, n$;
- The $p\times p$ matrix $Q_{xx} = E[x_i x_i^T]$ is of full rank, and hence positive-definite;
- $\{x_i\varepsilon_i\}$ is a martingale difference sequence, with a finite matrix of second moments $Q_{xx\varepsilon^2} = E[\varepsilon_i^2 x_i x_i^T]$.

Finite Sample Properties

First of all, under the *strict exogeneity* assumption the OLS estimators $\hat{\beta}$ and s^2 are unbiased, meaning that their expected values coincide with the true values of the parameters:

$$E[\hat{\beta} \mid X] = \beta, \quad E[s^2 \mid X] = \sigma^2.$$

If the strict exogeneity does not hold (as is the case with many time series models, where exogeneity is assumed only with respect to the past shocks but not the future ones), then these estimators will be biased in finite samples.

The *variance-covariance matrix* (or simply *covariance matrix*) of $\hat{\beta}$ is equal to

$$\operatorname{Var}[\hat{\beta} \mid X] = \sigma^2 (X^T X)^{-1} = \sigma^2 Q.$$

In particular, the standard error of each coefficient $\hat{\beta}_j$ is equal to square root of the j-th diagonal element of this matrix. The estimate of this standard error is obtained by replacing the unknown quantity σ^2 with its estimate s^2. Thus,

$$\widehat{\text{s.e.}}(\hat{\beta}_j) = \sqrt{s^2 (X^T X)^{-1}_{jj}}$$

It can also be easily shown that the estimator $\hat{\beta}$ is uncorrelated with the residuals from the model:

$$\operatorname{Cov}[\hat{\beta}, \hat{\varepsilon} \mid X] = 0.$$

The Gauss–Markov theorem states that under the *spherical errors* assumption (that is, the errors should be uncorrelated and homoscedastic) the estimator $\hat{\beta}$ is efficient in the class of linear unbiased estimators. This is called the best linear unbiased estimator (BLUE). Efficiency should be understood as if we were to find some other estimator $\hat{\beta}$ which would be linear in y and unbiased, then

$$\operatorname{Var}[\tilde{\beta} \mid X] - \operatorname{Var}[\hat{\beta} \mid X] \geq 0$$

In the sense that this is a nonnegative-definite matrix. This theorem establishes optimality only in the class of linear unbiased estimators, which is quite restrictive. Depending on the distribution of the error terms ε, other, non-linear estimators may provide better results than OLS.

Assuming Normality

The properties listed so far are all valid regardless of the underlying distribution of the error terms. However, if you are willing to assume that the *normality assumption* holds (that is, that $\varepsilon \sim N(0, \sigma^2 I_n)$), then additional properties of the OLS estimators can be stated.

The estimator $\hat{\beta}$ is normally distributed, with mean and variance as given before:

$$\hat{\beta} \sim \mathcal{N}\left(\beta, \sigma^2 (X^{\mathrm{T}} X)^{-1}\right)$$

where Q is the cofactor matrix. This estimator reaches the Cramér–Rao bound for the model, and thus is optimal in the class of all unbiased estimators. Note that unlike the Gauss–Markov theorem, this result establishes optimality among both linear and non-linear estimators, but only in the case of normally distributed error terms.

The estimator s^2 will be proportional to the chi-squared distribution:

$$s^2 \sim \frac{\sigma^2}{n-p} \cdot \chi^2_{n-p}$$

The variance of this estimator is equal to $2\sigma^4 / (n-p)$, which does not attain the Cramér–Rao bound of $2\sigma^4/n$. However it was shown that there are no unbiased estimators of σ^2 with variance smaller than that of the estimator s^2. If we are willing to allow biased estimators, and consider the class of estimators that are proportional to the sum of squared residuals (SSR) of the model, then the best (in the sense of the mean squared error) estimator in this class will be $\tilde{\sigma}^2 = SSR / (n-p+2)$, which even beats the Cramér–Rao bound in case when there is only one regressor ($p = 1$).

Moreover, the estimators $\hat{\beta}$ and s^2 are independent, the fact which comes in useful when constructing the t- and F-tests for the regression.

Influential Observations

As was mentioned before, the estimator $\hat{\beta}$ is linear in y, meaning that it represents a linear combination of the dependent variables y_i. The weights in this linear combination are functions of the regressors X, and generally are unequal. The observations with high weights are called influential because they have a more pronounced effect on the value of the estimator.

To analyze which observations are influential we remove a specific j-th observation and consider how much the estimated quantities are going to change (similarly to the jackknife method). It can be shown that the change in the OLS estimator for β will be equal to

$$\hat{\beta}^{(j)} - \hat{\beta} = -\frac{1}{1-h_j}(X^{\mathrm{T}} X)^{-1} x_j^{\mathrm{T}} \hat{\varepsilon}_j,$$

where $h_j = x_j^T \left(X^T X\right)^{-1} x_j$ is the j-th diagonal element of the hat matrix P, and x_j is the vector of regressors corresponding to the j-th observation. Similarly, the change in the predicted value for j-th observation resulting from omitting that observation from the dataset will be equal to

$$\hat{y}_j^{(j)} - \hat{y}_j = x_j^{\mathrm{T}} \hat{\beta}^{(j)} - x_j^T \hat{\beta} = -\frac{h_j}{1-h_j} \hat{\varepsilon}_j$$

From the properties of the hat matrix, $0 \le h_j \le 1$, and they sum up to p, so that on average $h_j \approx p/n$. These quantities h_j are called the leverages, and observations with high h_j are called leverage points. Usually the observations with high leverage ought to be scrutinized more carefully, in case they are erroneous, or outliers, or in some other way atypical of the rest of the dataset.

Partitioned Regression

Sometimes the variables and corresponding parameters in the regression can be logically split into two groups, so that the regression takes form

$$y = X_1\beta_1 + X_2\beta_2 + \varepsilon,$$

where X_1 and X_2 have dimensions $n\times p_1$, $n\times p_2$, and β_1, β_2 are $p_1\times 1$ and $p_2\times 1$ vectors, with $p_1 + p_2 = p$.

The Frisch–Waugh–Lovell theorem states that in this regression the residuals $\hat{\varepsilon}$ and the OLS estimate $\hat{\beta}_2$ will be numerically identical to the residuals and the OLS estimate for β_2 in the following regression:

$$M_1 y = M_1 X_2 \beta_2 + \eta,$$

where M_1 is the annihilator matrix for regressors X_1.

The theorem can be used to establish a number of theoretical results. For example, having a regression with a constant and another regressor is equivalent to subtracting the means from the dependent variable and the regressor and then running the regression for the de-meaned variables but without the constant term.

Constrained Estimation

Suppose it is known that the coefficients in the regression satisfy a system of linear equations

$$A: \quad Q^T\beta = c,$$

where Q is a $p\times q$ matrix of full rank, and c is a $q\times 1$ vector of known constants, where $q < p$. In this case least squares estimation is equivalent to minimizing the sum of squared residuals of the model subject to the constraint A. The constrained least squares (CLS) estimator can be given by an explicit formula:

$$\hat{\beta}^c = \hat{\beta} - (X^TX)^{-1}Q\left(Q^T(X^TX)^{-1}Q\right)^{-1}(Q^T\hat{\beta} - c).$$

This expression for the constrained estimator is valid as long as the matrix X^TX is invertible. It was assumed from the beginning of this topic that this matrix is of full rank, and it was noted that when the rank condition fails, β will not be identifiable. However it may happen that adding the restriction A makes β identifiable, in which case one would like to find the formula for the estimator. The estimator is equal to

$$\hat{\beta}^c = R(R^TX^TXR)^{-1}R^TX^Ty + \left(I_p - R(R^TX^TXR)^{-1}R^TX^TX\right)Q(Q^TQ)^{-1}c,$$

where R is a $p\times(p-q)$ matrix such that the matrix $[Q\ R]$ is non-singular, and $R^T Q = 0$. Such a matrix can always be found, although generally it is not unique. The second formula coincides with the first in case when $X^T X$ is invertible.

Large Sample Properties

The least squares estimators are point estimates of the linear regression model parameters β. However, generally we also want to know how close those estimates might be to the true values of parameters. In other words, we want to construct the interval estimates.

Since we haven't made any assumption about the distribution of error term ε_i, it is impossible to infer the distribution of the estimators $\hat{\beta}$ and $\hat{\sigma}^2$. Nevertheless, we can apply the central limit theorem to derive their *asymptotic* properties as sample size n goes to infinity. While the sample size is necessarily finite, it is customary to assume that n is "large enough" so that the true distribution of the OLS estimator is close to its asymptotic limit.

We can show that under the model assumptions, the least squares estimator for β is consistent (that is $\hat{\beta}$ converges in probability to β) and asymptotically normal:[proof]

$$(\hat{\beta}-\beta)\xrightarrow{d}\mathcal{N}\left(0,\sigma^2 Q_{xx}^{-1}\right),$$

where $Q_{xx}=X^T X$

Intervals

Using this asymptotic distribution, approximate two-sided confidence intervals for the j-th component of the vector $\hat{\beta}$ can be constructed as

$$\beta_j \in \left[\hat{\beta}_j \pm q_{1-\frac{\alpha}{2}}^{\mathcal{N}(0,1)}\sqrt{\hat{\sigma}^2\left[Q_{xx}^{-1}\right]_{jj}}\,\right] \text{ at the } 1-\alpha \text{ confidence level,}$$

where q denotes the quantile function of standard normal distribution, and $[\cdot]_{jj}$ is the j-th diagonal element of a matrix.

Similarly, the least squares estimator for σ^2 is also consistent and asymptotically normal (provided that the fourth moment of ε_i exists) with limiting distribution

$$(\hat{\sigma}^2-\sigma^2)\xrightarrow{d}\mathcal{N}\left(0,\mathrm{E}\left[\varepsilon_i^4\right]-\sigma^4\right).$$

These asymptotic distributions can be used for prediction, testing hypotheses, constructing other estimators, etc. As an example consider the problem of prediction. Suppose x_0 is some point within the domain of distribution of the regressors, and one wants to know what the response variable would have been at that point. The mean response is the quantity $y_0 = x_0^{\mathrm{T}}\beta$, whereas the predicted response is $\hat{y}_0 = x_0^{\mathrm{T}}\hat{\beta}$. Clearly the predicted response is a random variable, its distribution can be derived from that of $\hat{\beta}$:

$$\left(\hat{y}_0-y_0\right)\xrightarrow{d}\mathcal{N}\left(0,\sigma^2 x_0^{\mathrm{T}}Q_{xx}^{-1}x_0\right),$$

which allows construct confidence intervals for mean response y_0 to be constructed:

$$y_0 \in \left[x_0^T \hat{\beta} \pm q_{1-\frac{\alpha}{2}}^{\mathcal{N}(0,1)} \sqrt{\hat{\sigma}^2 x_0^T Q_{xx}^{-1} x_0} \right] \text{ at the } 1-\alpha \text{ confidence level.}$$

Hypothesis Testing

Two hypothesis tests are particularly widely used. First, one wants to know if the estimated regression equation is any better than simply predicting that all values of the response variable equal its sample mean (if not, it is said to have no explanatory power). The null hypothesis of no explanatory value of the estimated regression is tested using an F-test. If the calculated F-value is found to be large enough to exceed its critical value for the pre-chosen level of significance, the null hypothesis is rejected and the alternative hypothesis, that the regression has explanatory power, is accepted. Otherwise, the null hypothesis of no explanatory power is accepted.

Second, for each explanatory variable of interest, one wants to know whether its estimated coefficient differs significantly from zero—that is, whether this particular explanatory variable in fact has explanatory power in predicting the response variable. Here the null hypothesis is that the true coefficient is zero. This hypothesis is tested by computing the coefficient's t-statistic, as the ratio of the coefficient estimate to its standard error. If the t-statistic is larger than a predetermined value, the null hypothesis is rejected and the variable is found to have explanatory power, with its coefficient significantly different from zero. Otherwise, the null hypothesis of a zero value of the true coefficient is accepted.

In addition, the Chow test is used to test whether two subsamples both have the same underlying true coefficient values. The sum of squared residuals of regressions on each of the subsets and on the combined data set are compared by computing an F-statistic; if this exceeds a critical value, the null hypothesis of no difference between the two subsets is rejected; otherwise, it is accepted.

Example with Real Data

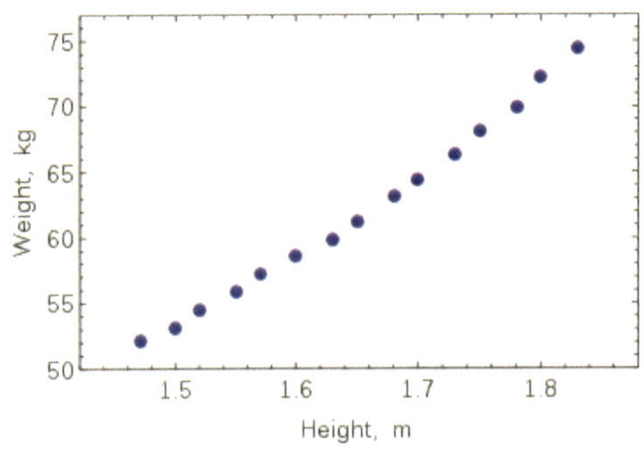

Scatterplot of the data, the relationship is slightly curved but close to linear

The following data set gives average heights and weights for American women aged 30–39.

Height (m)	1.47	1.50	1.52	1.55	1.57	1.60	1.63	1.65	1.68	1.70	1.73	1.75	1.78	1.80	1.83
Weight (kg)	52.21	53.12	54.48	55.84	57.20	58.57	59.93	61.29	63.11	64.47	66.28	68.10	69.92	72.19	74.46

When only one dependent variable is being modeled, a scatterplot will suggest the form and strength of the relationship between the dependent variable and regressors. It might also reveal outliers, heteroscedasticity, and other aspects of the data that may complicate the interpretation of a fitted regression model. The scatterplot suggests that the relationship is strong and can be approximated as a quadratic function. OLS can handle non-linear relationships by introducing the regressor height. The regression model then becomes a multiple linear model:

$$w_i = \beta_1 + \beta_2 h_i + \beta_3 h_i^2 + \varepsilon_i.$$

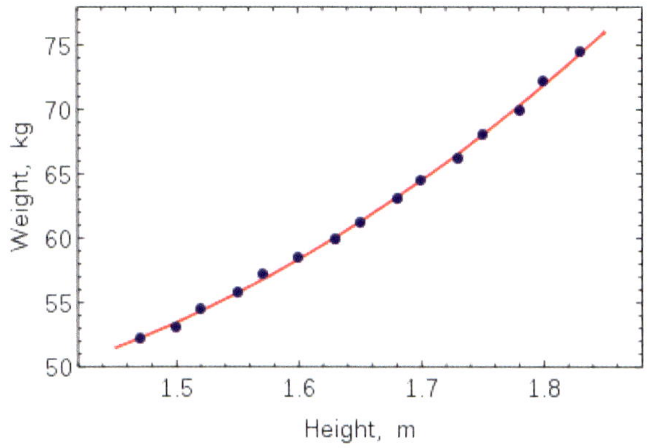

Fitted Regression

The output from most popular statistical packages will look similar to this:

Method	Least squares			
Dependent variable	WEIGHT			
Observations	15			
Parameter	**Value**	**Std error**	**t-statistic**	**p-value**
β_1	128.8128	16.3083	7.8986	0.0000
β_2	−143.1620	19.8332	−7.2183	0.0000
β_3	61.9603	6.0084	10.3122	0.0000
R^2	0.9989	S.E. of regression		0.2516
Adjusted R^2	0.9987	Model sum-of-sq.		692.61
Log-likelihood	1.0890	Residual sum-of-sq.		0.7595
Durbin–Watson stat.	2.1013	Total sum-of-sq.		693.37
Akaike criterion	0.2548	F-statistic		5471.2
Schwarz criterion	0.3964	p-value (F-stat)		0.0000

In this table:

- The *Value* column gives the least squares estimates of parameters β_j
- The *Std error* column shows standard errors of each coefficient estimate: $\hat{\sigma}_j = \left(\hat{\sigma}^2 \left[Q_{xx}^{-1}\right]_{jj}\right)^{\frac{1}{2}}$
- The *t-statistic* and *p-value* columns are testing whether any of the coefficients might be equal to zero. The *t*-statistic is calculated simply as $t = \hat{\beta}_j / \hat{\sigma}_j$. If the errors ε follow a normal distribution, *t* follows a Student-t distribution. Under weaker conditions, *t* is

asymptotically normal. Large values of t indicate that the null hypothesis can be rejected and that the corresponding coefficient is not zero. The second column, p-value, expresses the results of the hypothesis test as a significance level. Conventionally, p-values smaller than 0.05 are taken as evidence that the population coefficient is nonzero.

- *R-squared* is the coefficient of determination indicating goodness-of-fit of the regression. This statistic will be equal to one if fit is perfect, and to zero when regressors X have no explanatory power whatsoever. This is a biased estimate of the population *R-squared*, and will never decrease if additional regressors are added, even if they are irrelevant.
- *Adjusted R-squared* is a slightly modified version of R^2, designed to penalize for the excess number of regressors which do not add to the explanatory power of the regression. This statistic is always smaller than R^2, can decrease as new regressors are added, and even be negative for poorly fitting models:

$$\bar{R}^2 = 1 - \frac{n-1}{n-p}(1-R^2)$$

- *Log-likelihood* is calculated under the assumption that errors follow normal distribution. Even though the assumption is not very reasonable, this statistic may still find its use in conducting LR tests.
- *Durbin–Watson statistic* tests whether there is any evidence of serial correlation between the residuals. As a rule of thumb, the value smaller than 2 will be an evidence of positive correlation.
- *Akaike information criterion* and *Schwarz criterion* are both used for model selection. Generally when comparing two alternative models, smaller values of one of these criteria will indicate a better model.
- *Standard error of regression* is an estimate of σ, standard error of the error term.
- *Total sum of squares, model sum of squared,* and *residual sum of squares* tell us how much of the initial variation in the sample were explained by the regression.
- *F-statistic* tries to test the hypothesis that all coefficients (except the intercept) are equal to zero. This statistic has $F(p-1, n-p)$ distribution under the null hypothesis and normality assumption, and its *p-value* indicates probability that the hypothesis is indeed true. Note that when errors are not normal this statistic becomes invalid, and other tests such as Wald test or LR test should be used.

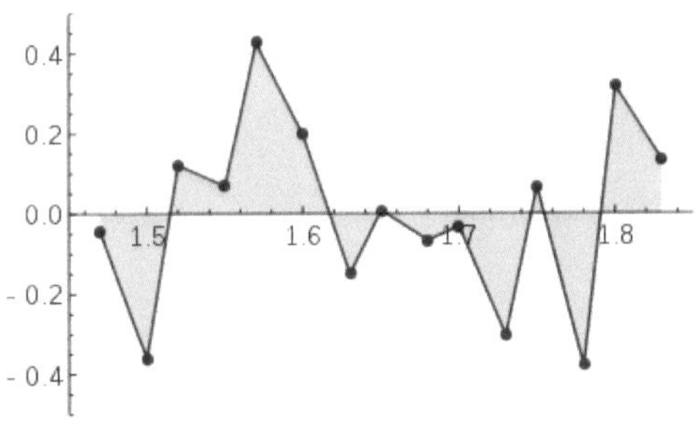

Residuals Plot

Ordinary least squares analysis often includes the use of diagnostic plots designed to detect departures of the data from the assumed form of the model. These are some of the common diagnostic plots:

- Residuals against the explanatory variables in the model. A non-linear relation between these variables suggests that the linearity of the conditional mean function may not hold. Different levels of variability in the residuals for different levels of the explanatory variables suggests possible heteroscedasticity.
- Residuals against explanatory variables not in the model. Any relation of the residuals to these variables would suggest considering these variables for inclusion in the model.
- Residuals against the fitted values, $\hat{y}$.
- Residuals against the preceding residual. This plot may identify serial correlations in the residuals.

An important consideration when carrying out statistical inference using regression models is how the data were sampled. In this example, the data are averages rather than measurements on individual women. The fit of the model is very good, but this does not imply that the weight of an individual woman can be predicted with high accuracy based only on her height.

Sensitivity to Rounding

This example also demonstrates that coefficients determined by these calculations are sensitive to how the data is prepared. The heights were originally given rounded to the nearest inch and have been converted and rounded to the nearest centimeter. Since the conversion factor is one inch to 2.54 cm this is *not* an exact conversion. The original inches can be recovered by Round(x/0.0254) and then re-converted to metric without rounding. If this is done the results become:

	Const	Height	Height^2
Converted to metric with rounding.	128.8128	−143.162	61.96033
Converted to metric without rounding.	119.0205	−131.5076	58.5046

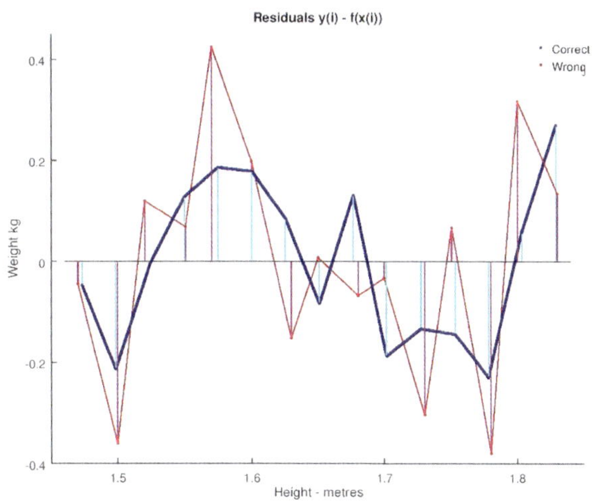

Residuals to a quadratic fit for correctly and incorrectly converted data.

Using either of these equations to predict the weight of a 5' 6" (1.6764m) woman gives similar values: 62.94 kg with rounding vs. 62.98 kg without rounding. Thus a seemingly small variation in the data has a real effect on the coefficients but a small effect on the results of the equation.

While this may look innocuous in the middle of the data range it could become significant at the extremes or in the case where the fitted model is used to project outside the data range (extrapolation).

This highlights a common error: this example is an abuse of OLS which inherently requires that the errors in the independent variable (in this case height) are zero or at least negligible. The initial rounding to nearest inch plus any actual measurement errors constitute a finite and non-negligible error. As a result, the fitted parameters are not the best estimates they are presumed to be. Though not totally spurious the error in the estimation will depend upon relative size of the x and y errors.

Another Example with Less Real Data

Problem Statement

We can use the least square mechanism to figure out the equation of a two body orbit in polar base co-ordinates. The equation typically used is $r(\theta)=\frac{p}{1-e\cos(\theta)}$ where $r(\theta)$ is the radius of how far the object is from one of the bodies. In the equation the parameters p and e are used to determine the path of the orbit. We have measured the following data.

θ (in degrees)	43	45	52	93	108	116
$r(\theta)$	4.7126	4.5542	4.0419	2.2187	1.8910	1.7599

We need to find the least-squares approximation of e and p for the given data.

Solution

First we need to represent e and p in a linear form. So we are going to rewrite the equation $r(\theta)$ as $\frac{1}{r(\theta)}=\frac{1}{p}-\frac{e}{p}\cos(\theta)$. Now we can use this form to represent our observational data as:

$A^T A\begin{pmatrix} x \\ y \end{pmatrix}=A^T b$ where x is $\frac{1}{p}$ and y is $\frac{e}{p}$ and A is constructed by the first column being the coefficient of $\frac{1}{p}$ and the second column being the coefficient of $\frac{e}{p}$ and b is the values for the respective $\frac{1}{r(\theta)}$ so

$$A=\begin{bmatrix} 1 & -0.731354 \\ 1 & -0.707107 \\ 1 & -0.615661 \\ 1 & 0.052336 \\ 1 & 0.309017 \\ 1 & 0.438371 \end{bmatrix} \text{ and } b=\begin{bmatrix} 0.21220 \\ 0.21958 \\ 0.24741 \\ 0.45071 \\ 0.52883 \\ 0.56820 \end{bmatrix}.$$

On solving we get $\begin{pmatrix} x \\ y \end{pmatrix} = \begin{pmatrix} 0.43478 \\ 0.30435 \end{pmatrix}$

so $p = \frac{1}{x} = 2.3000$ *and* $e = p \cdot y = 0.70001$

Probit Model

Probit regression, also called a probit model, is used to model dichotomous or binary outcome variables. In the probit model, the inverse standard normal distribution of the probability is modeled as a linear combination of the predictors.

```
+/- R Code
library(aod)
library(ggplot2)
```

Examples

- Example 1. Suppose that we are interested in the factors that influence whether a political candidate wins an election. The outcome (response) variable is binary (0/1); win or lose. The predictor variables of interest are the amount of money spent on the campaign, the amount of time spent campaigning negatively and whether or not the candidate is an incumbent.
- Example 2. A researcher is interested in how variables, such as GRE (Graduate Record Exam scores), GPA (grade point average) and prestige of the undergraduate institution, effect admission into graduate school. The response variable, admit/don't admit, is a binary variable.

Theoretical Background

Probit regression is a special type of the Generalized Linear Models (GLM; will be explained later). Here, the bivariate outcome Y has a Bernoulli distribution with parameter p (success probability $p \in (0,1)$). Recall that EY=p The *probit link function*

$$probit(EY) = \Phi^{-1}(p) = \Phi^{-1}(p[Y=1])$$

is used to transform the expectation of this 0/1 dependent variable. Then, the probit of the mean is modelled as a linear combination of the covariates (regressors) X, i.e., we have a *linear predictor*

$$probit(EY) = \mathbf{X}\beta,$$

where β is a vector of unknown parameters. The *maximum likelihood* based approach is used for the parameter estimation, where a version of the *IRLS algorithm* is applied .

Predicted probability can be obtained by the *inverse probit* (i.e., standard normal cdf) transformation

$$\hat{P}[Y_i = 1] = \Phi(\mathbf{X}_{i\cdot}\hat{\beta}).$$

References

- Harrell, Frank E. (2015). Regression Modeling Strategies. Springer Series in Statistics (2nd ed.). New York; Springer. doi:10.1007/978-3-319-19425-7. ISBN 978-3-319-19424-0.
- Regression-analysis: businessjargons.com, Retrieved 12 May 2018
- Grundel, R.; Pavlovic, N. B. (2007). "Response of bird species densities to habitat structure and fire history along a Midwestern open–forest Gradient". The Condor. 109 (4): 734–749. doi:10.1650/0010-5422(2007)109 [734:ROBSDT]2.0.CO;2.
- Comprehensive-guide-regression: analyticsvidhya.com, Retrieved 28 June 2018
- Ver Hoef, JAY M.; Boveng, Peter L. (2007-01-01). "Quasi-Poisson vs. Negative Binomial Regression: How should we model overdispersed count data?". Ecology. 88 (11): 2766–2772. doi:10.1890/07-0043.1. Retrieved 2016-09-01.
- What-is-linear-regression: statisticssolutions.com, Retrieved 09 July 2018
- McCullagh, Peter (1980). "Regression models for ordinal data". Journal of the Royal Statistical Society. Series B (Methodological). 42 (2): 109–142.
- Multinomial-logistic-regression: statisticshowto.datasciencecentral.com, Retrieved 27 June 2018
- Gelman, Andrew; Hill, Jennifer (2007). Data Analysis Using Regression and Multilevel/Hierarchical Models. New York: Cambridge University Press. pp. 79–108. ISBN 978-0-521-68689-1
- Semiparametric: statisticshowto.datasciencecentral.com, Retrieved 17 March 2018
- DeBano, S. J.; Hamm, P. B.; Jensen, A.; Rondon, S. I.; Landolt, P. J. (2010). "Spatial and temporal dynamics of potato tuberworm (Lepidoptera: Gelechiidae) in the Columbia Basin of the Pacific Northwest". Environmental Entomology. 39 (1): 1–14. doi:10.1603/EN08270.

Permissions

All chapters in this book are published with permission under the Creative Commons Attribution Share Alike License or equivalent. Every chapter published in this book has been scrutinized by our experts. Their significance has been extensively debated. The topics covered herein carry significant information for a comprehensive understanding. They may even be implemented as practical applications or may be referred to as a beginning point for further studies.

We would like to thank the editorial team for lending their expertise to make the book truly unique. They have played a crucial role in the development of this book. Without their invaluable contributions this book wouldn't have been possible. They have made vital efforts to compile up to date information on the varied aspects of this subject to make this book a valuable addition to the collection of many professionals and students.

This book was conceptualized with the vision of imparting up-to-date and integrated information in this field. To ensure the same, a matchless editorial board was set up. Every individual on the board went through rigorous rounds of assessment to prove their worth. After which they invested a large part of their time researching and compiling the most relevant data for our readers.

The editorial board has been involved in producing this book since its inception. They have spent rigorous hours researching and exploring the diverse topics which have resulted in the successful publishing of this book. They have passed on their knowledge of decades through this book. To expedite this challenging task, the publisher supported the team at every step. A small team of assistant editors was also appointed to further simplify the editing procedure and attain best results for the readers.

Apart from the editorial board, the designing team has also invested a significant amount of their time in understanding the subject and creating the most relevant covers. They scrutinized every image to scout for the most suitable representation of the subject and create an appropriate cover for the book.

The publishing team has been an ardent support to the editorial, designing and production team. Their endless efforts to recruit the best for this project, has resulted in the accomplishment of this book. They are a veteran in the field of academics and their pool of knowledge is as vast as their experience in printing. Their expertise and guidance has proved useful at every step. Their uncompromising quality standards have made this book an exceptional effort. Their encouragement from time to time has been an inspiration for everyone.

The publisher and the editorial board hope that this book will prove to be a valuable piece of knowledge for students, practitioners and scholars across the globe.

Index

O

P

Q

R

S

U

W

Z